Freshwater Ecosystems in Protected Areas

Freshwater ecosystems have the greatest species diversity per unit area and many endangered species. This book shows that, rather than being a marginal part of terrestrial protected area management, freshwater conservation is central to sustaining biodiversity. It focuses on better practices for conserving inland aquatic ecosystems in protected areas, including rivers, wetlands, peatlands, other freshwater and brackish ecosystems, and estuaries.

The authors define inland aquatic ecosystems, showing just how diverse and widespread they are. They examine the principles and processes that are essential for the conservation of freshwater ecosystems and aquatic species. Major categories of threats to freshwater ecosystems and the flow-on implications for protected area design are described. Practical case studies are used to illustrate principles and practices applied around the world. Specific management needs of the main types of freshwater ecosystems are considered, as well as the management of freshwaters in the broader landscape, showing how natural resource governance processes can be harnessed to better manage freshwater biodiversity. The book offers commentary on how to adapt freshwater conservation practices to climate change and ends with an insightful synthesis.

C. Max Finlayson is Director of the Institute for Land, Water and Society and Professor for Ecology and Biodiversity at Charles Sturt University, Australia, and the Ramsar Chair for the Wise Use of Wetlands at IHE Delft, The Netherlands. He is a visiting Professor at the Institute for Wetland Research in the China Academy of Forestry.

Angela H. Arthington is an Emeritus Professor in the Australian Rivers Institute at Griffith University, Australia.

Jamie Pittock is an Associate Professor in the Fenner School of Environment and Society at The Australian National University, Australia.

Earthscan Studies in Water Resource Management

Water Policy, Imagination and Innovation
Interdisciplinary Approaches
Edited by Robyn Bartel, Louise Noble, Jacqueline Williams and Stephen Harris

Rivers and Society
Landscapes, Governance and Livelihoods
Edited by Malcolm Cooper, Abhik Chakraborty and Shamik Chakraborty

Transboundary Water Governance and International Actors in South Asia
The Ganges-Brahmaputra-Meghna Basin
Paula Hanasz

The Grand Ethiopian Renaissance Dam and the Nile Basin
Implications for Transboundary Water Cooperation
Edited by Zeray Yihdego, Alistair Rieu-Clarke and Ana Cascao

Freshwater Ecosystems in Protected Areas
Conservation and Management
Edited by C. Max Finlayson, Angela H. Arthington and Jamie Pittock

Participation for Effective Environmental Governance
Evidence from European Water Framework Directive Implementation
Edited by Elisa Kochskämper, Edward Challies, Nicolas W. Jager and Jens Newig

China's International Transboundary Rivers
China's International Transboundary Rivers
By Lei Xie and Jia Shaofeng

For more information and to view forthcoming titles in this series, please visit the Routledge website: http://www.routledge.com/books/series/ECWRM/

Freshwater Ecosystems in Protected Areas

Conservation and Management

Edited by C. Max Finlayson, Angela H. Arthington and Jamie Pittock

Routledge
Taylor & Francis Group

LONDON AND NEW YORK

earthscan
from Routledge

First published 2018
by Routledge
2 Park Square, Milton Park, Abingdon, Oxon OX14 4RN

and by Routledge
711 Third Avenue, New York, NY 10017

Routledge is an imprint of the Taylor & Francis Group, an informa business

British Library Cataloguing-in-Publication Data
A catalogue record for this book is available from the British Library

Library of Congress Cataloging-in-Publication Data
A catalog record for this book has been requested.

ISBN: 978-0-415-78700-0 (hbk)
ISBN: 978-0-415-78714-7 (pbk)
ISBN: 978-1-315-22638-5 (ebk)

Typeset in Goudy
By Swales & Willis Ltd, Exeter, Devon, UK

Contents

Editor biographies

C. Max Finlayson is Director of the Institute for Land, Water and Society and Professor for Ecology and Biodiversity at Charles Sturt University, Australia, and the Ramsar Chair for the Wise Use of Wetlands at IHE Delft, The Netherlands. He is a visiting Professor at the Institute for Wetland Research in the China Academy of Forestry. His work covers the ecology and inventory of wetlands, including the impacts of invasive species, water pollution and climate change, and policy and management responses. He has formerly worked with the Australian Government, the International Water Management Institute and Wetlands International and has been a technical advisor to the Ramsar Convention on Wetlands since the early 1990s. He has been involved in global assessments on climate change, biodiversity and ecosystem services, water management, and the status of wetlands. In 2017 he was made a Fellow of the Society of Wetland Scientists.

Angela H. Arthington is an Emeritus Professor in the Australian Rivers Institute, Griffith University, Australia. She is a research ecologist focusing on river and fish conservation, especially through the science and management of environmental flows. Her research findings underpin several globally adopted environmental flow frameworks (DRIFT, ELOHA). Angela has edited three Special Issues on biodiversity conservation and environmental flows, produced over 220 papers and book chapters, and numerous research and consultancy reports for Australian and international agencies. Environmental flows research culminated in "Environmental Flows: Saving Rivers in the Third Millennium" (2012, University of California Press). Angela received the honorary "Making a Difference Award" (2015) from the US Instream Flow Council.

Jamie Pittock is an Associate Professor in the Fenner School of Environment and Society at The Australian National University, Australia. His work focuses on policies for conservation of freshwater biodiversity. In particular, he works on the positive synergies and conflicts among policies for conservation of biodiversity, responses to climate change and supply of energy, food and water. Prior to returning to academia in 2007, Jamie worked for a range of Australian and international environmental organisations. He was Director of WWF International's global freshwater program from 2001–2007.

Contributor biographies

Robin Abell is with The Nature Conservancy's Global Water Program. Her work focuses on developing and advancing tools and approaches for freshwater biodiversity conservation. She was the lead for Freshwater Ecoregions of the World, has explored how to incorporate freshwater biodiversity considerations into commodity certification programs, has developed a framework for freshwater protected areas, and most recently led an effort to quantify the co-benefits of source water protection globally. Previously, she worked in World Wildlife Fund US's Conservation Science Program.

Ivan Arismendi is an Assistant Professor in the Department of Fisheries and Wildlife at Oregon State University, USA. He is a quantitative aquatic ecologist interested in the role of natural variability and human-related disturbances on freshwaters across multiple spatial and temporal scales. His research focuses on freshwater–terrestrial links, invasive species, and the consequences of climate change on freshwaters. He is also interested in issues related to diversity and inclusion in science.

Lee Baumgartner is an Associate Research Professor specialising in fisheries and river management issues and is based within the Institute for Land, Water and Society at Charles Sturt University, Australia. Much of his research is applied and has fed back into adaptive management strategies which have resulted in state, national and global policy development. Recently, he has been involved in research activities in the lower Mekong Basin; specifically understanding mechanisms to help fisheries recover from human disturbance and quantifying the value of fish in a food security context.

Diego Juffe Bignoli is a Senior Programme Officer in the Protected Areas programme at the UN Environment World Conservation Monitoring Centre (UNEP-WCMC). He holds a bachelor degree in agronomy and a master's degree in environmental management. Diego has worked for five years in the agribusiness sector and for over eight in conservation of nature. Previously, he worked for the International Union for Conservation of Nature on The IUCN Red List of Threatened Species and on Key Biodiversity Areas and biodiversity policy. Today, his work focuses on the World Database on Protected Areas.

Esther Blom is an ecologist from the Utrecht University. At IUCN Netherlands she worked for the small grants for wetlands fund. She studies the high altitude wetlands of Nepal and the Tibetan Plateau for The Mountain Institute. Furthermore, she headed the freshwater team of WWF Netherlands for a decade with a specific focus on Chinese and Dutch river systems. Currently, she is the vice director of a Dutch NGO called ARK that is focused on nature development.

Jorge Cabrera is a forest engineer. He is a senior consultant in forestry economics, ecological economics, markets and forestry management. He was the head of the regional office of the Forestry Institute, Chile (INFOR) from 1979–2014 and leader of several projects related to ecosystem services payments.

Nick C. Davidson is an Adjunct Professor at Charles Sturt University, Australia and is a consultant on wetland conservation and wise use, and on migratory waterbirds. He was the Deputy Secretary General of the Ramsar Convention on Wetlands from 2000 to 2014, with responsibility for delivery of scientific, technical and policy guidance, and communications. Previously he worked for the UK's national government conservation agencies on coastal wetland inventory, assessment, information systems and communications, and as International Science Coordinator for the global NGO Wetlands International.

Nigel Dudley is a consultant ecologist, adjunct fellow at the University of Queensland and chair of the natural solutions theme of the IUCN World Commission on Protected Areas, focusing on the ecosystem services from protected areas. His work focuses particularly on ecosystem services, broad-scale approaches to conservation, and the planning and management of protected areas; it includes planning and implementing field and research projects, production of technical reports and research papers and international policy. He has worked in over 60 countries around the world.

Peter A. Gell is a professorial research fellow at Federation University Australia (Ballarat). His research focuses on wetland paleoecology where he examines records of ecological condition spanning centuries to provide temporal context for understanding the nature, causes and trajectories of change. He is a member of the scientific steering committee of Future Earth's core project Past Global Changes (PAGES) and leads the Aquatic Transitions working group that examines the impact of humans on the world's wetlands and the dynamics of the ecological response to disturbance.

Marianne Kettunen is Principal Policy Analyst at the Institute for European Environmental Policy and deputy chair of the natural solutions theme of the IUCN World Commission on Protected Areas. She has worked previously at research institutes and universities in Finland and on a development project in Peru. She was an integral member of The Economics of Ecosystems and Biodiversity initiative (TEEB). Her work includes editorship of a manual on assessing protected area benefits.

Christian Little is a forest engineer, leader of the national Forest and Water research program at Forest Institute (INFOR), Chile. His research work focuses on water quality and quantity as an ecosystem service from forested watersheds. He is particularly interested in the interaction of water and society, ecological restoration and forest policy.

Rebecca Flitcroft is a research fish biologist with the United States Forest Service at the Pacific Northwest Research Station, Oregon, USA. Her work focuses on broad-scale patterns of ecological systems, and interactions with natural and anthropogenic disturbance regimes. Rebecca is particularly interested in life history diversity of fishes and adaptive resilence to changing climate and habitat conditions.

Jamin P. Forbes is a member of the Institute for Land, Water and Society at Charles Sturt University, Australia, and a former Fisheries Scientist with the New South Wales Department of Primary Industries (Fisheries), Australia. Dr Forbes' research centres on freshwater recreational fisheries in Australia and using scientific evidence to underpin strategies for management of these fisheries.

Ray Froend is a Professor in Applied Ecology in the Centre for Ecosystem Management, School of Science at Edith Cowan University, Australia. His research focuses on the ecology of plant–water interactions, ecophysiology of phreatophytic plants, water requirements of groundwater dependent ecosystems and ecohydrology. Ray has advised NRM agencies in Australia and internationally on groundwater dependent ecosystems and his research continues to be the basis for the identifying and protecting the groundwater requirements of dependent ecosystems.

Virgilio Hermoso is a Ramón y Cajal research fellow at the Forest Science Centre of Catalonia, Spain, and an Adjunct Research Fellow at the Australian Rivers Institute, Griffith University, Australia. His work focuses on decision-making for conservation and restoration in freshwater ecosystems, including: systematic approaches to identify priority areas for conservation of freshwater biodiversity, monitoring networks for early detection of invasive species, monitoring ecological status, and prioritisation of management plans for endangered species. He collaborates with NGOs like WWF and WCS on conservation projects around the world.

Ritesh Kumar is the Conservation Programme Manager of Wetlands International South Asia, India. Over the last two decades, in collaboration with government and non-government agencies, he has worked on integrated management of wetlands in South Asia, with special emphasis on recognition of the contribution that these ecosystems make to development. He has served on the Scientific and Technical Review Panel of the Ramsar Convention since 2005, and is a coordinating lead author of the Asia Pacific Regional Assessment of the Intergovernmental Panel on Biodiversity and Ecosystem Services (IPBES).

Simon Linke is a senior research fellow of the Australian Rivers Institute, Griffith University, and a member of the IUCN WCPA Freshwater Task Force. He is one of the founders of the discipline of conservation planning in river systems. He co-edited (with Eren Turak) the first special issue on freshwater conservation planning for *Freshwater Biology*. Simon has been involved in large-scale conservation planning exercises for the Congo River and in Bhutan. He has developed tools to facilitate uptake of conservation science, for example a system to optimise environmental water allocations in Australia as well as integration of freshwater planning into conservation software.

Rob J. McInnes is Managing Director at RM Wetlands & Environment Ltd, Oxfordshire, UK. Through regular collaboration with governmental and non-governmental agencies, and the private sector, he specialises in working towards the wise use of wetlands and recognition of the multiple benefits they provide. His work ranges from practical management and restoration interventions on the ground through to working on national wetland policy development. He has served on the Ramsar Convention's Scientific and Technical Review Panel since 2005. Rob was formerly Head of Wetland Conservation at the Wildfowl & Wetlands Trust and President of the European Chapter of the Society of Wetland Scientists.

Randy Milton leads the Ecosystems and Habitats Program with Nova Scotia's Department of Natural Resources (Canada), is an Adjunct Professor with Acadia University (Canada) and an Adjunct Research Associate at Charles Sturt University (Australia). He works on a broad range of biodiversity conservation issues at the nexus of species and ecosystem protection, sustainable use, and socio-economic development. He has been a technical advisor to the Ramsar Convention on Wetlands (2000–2015), a contributing author to the Millennium Ecosystem Assessment, and served on the International Plan Committee (2005–2014) for the North American Waterfowl Management Plan.

Jeanne L. Nel is a senior scientist at the Vrije Universiteit Amsterdam in The Netherlands, and a Research Associate of Nelson Mandela Metropolitan University in South Africa. Her research focusses on freshwater ecosystems: their condition, conservation, links to social development and governance, and the implications of global change for these relationships. She embeds this research in practical application by working with scientists, policy makers and practitioners to co-produce knowledge that informs water resource management, conservation and development planning, disaster risk reduction and natural capital accounting.

Walter Rast is an Emeritus Professor and Director of International Watershed Studies, Meadows Center for Water and the Environment, Texas State University. He currently chairs the Scientific Committee of the International Lake Environment Committee (ILEC). He previously was Senior Environmental Advisor to the US/Canada International Joint Commission on border environmental issues. He was Deputy Director of the Water Branch, UNEP (Nairobi, Kenya) during 1993–2000. Upon returning to the USA, he was Professor and Director, Aquatic Resources Program, Texas State University. His current research focuses on: integrated water system management; sustainable water-related ecosystem goods and services; and environmental governance, including assessment and management of lakes, reservoirs, wetlands and other lentic water systems.

Dirk Roux is a Specialist Scientist at South African National Parks and Adjunct Professor of Nelson Mandela Metropolitan University in South Africa. His research interests include adaptive management, ecosystem services, and institutional learning processes, mostly in relation to the conservation of freshwater ecosystems. Before joining SANParks he was Principal Researcher at a research council (CSIR, 1995–2007) and Director of the Water Research Node at Monash South Africa (2008–2010).

Luiz G. M. Silva is an Adjunct Professor for Environmental Management and Sustainable Development at Federal University of São João del-Rei (UFSJ), Brazil, and a Visiting Academic at the Institute for Land, Water and Society, Charles Sturt University, Australia. He has been working on freshwater fish ecology in tropical systems for over 15 years, focusing on the investigation of impacts due to hydropower development and flow regulation. His research has been focused on developing and applying sustainable water infrastructure to balance water usage and freshwater fish conservation globally.

Jason Thiem is a fisheries scientist with the NSW Department of Primary Industries and an adjunct research associate at the Institute for Land, Water and Society, Charles Sturt University, Australia. Over the past 10 years he has undertaken fisheries research in Australia, Canada, USA and The Bahamas. His research encompasses a range of species and focusses on applied fisheries issues including habitat restoration, riverine connectivity, fishways, stocking and post-capture impairment.

Michele Thieme is Director of Freshwater Science for WWF-US. She currently leads an effort to provide an updated global assessment of where free-flowing rivers remain and also provides scientific support to the development of river Basin Report Cards. She was a key contributor to the development of the Freshwater Ecoregions of the World, edited a book on the freshwater ecoregions of Africa and Madagascar, and has authored numerous freshwater conservation science publications. She has a MSc in fisheries biology and a BSc in biology.

Eren Turak is a principal scientist at the Office of Environment and Heritage in New South Wales, Australia. His main areas of work are biodiversity monitoring, ecological modelling, freshwater conservation planning and bio-diversity observation networks. He is also interim co-chair of the Freshwater Biodiversity Observation Network (FWBON), a global network of experts and practitioners involved in monitoring and assessing the biodiversity of inland waters.

Jane Turpie has a BSc (Hons) in Zoology and a PhD (1994) on estuarine birds and behavioural ecology. As well as estuarine ecology and ornithology, her work includes valuation of ecosystem services, analysing community depend-ence and pressures on natural resources, modelling ecological–economic systems, incentive and financing systems for conservation, integrated con-servation and development planning, scenario analysis for integrated river basin management, and impact and adaptation studies relating to climate change and biodiversity. Jane developed the Estuary Health Index and also the socio-economics component of South Africa's classification process for aquatic ecosystems. Jane is director of Anchor Environmental Consultants, a Research Fellow of the Environmental Economics Policy Research Unit at the School of Economics, University of Cape Town and Honorary Research Associate of the South African Institute for Aquatic Biodiversity.

Lara van Niekerk is a senior scientist at the Council for Scientific and Industrial Research (CSIR) in South Africa. Lara contributed to development of the ecological flow requirement methods and policies for the effective manage-ment of South Africa's estuaries. She has been involved in over 50 estuarine freshwater flow requirement studies. Lara is the architect of the SA National Estuarine Management Protocol and guidelines. She led the team that assessed the ecosystem condition of all South Africa's estuaries as part of the SA National Biodiversity Assessment in 2011 and is in the process of refining this for 2018.

Daphne Willems is a river ecologist by profession and passion. She worked for Rijkswaterstaat (the Dutch public river manager) and ten years as a self-employed river consultant, on projects combining river safety, nature development and people. Since 2014 she has been working as an international freshwater expert for WWF-Netherlands.

Preface

This book is born of love of the freshwater ecosystems that are central to the lives of people, and a myriad of animals and plants that live in these habitats. Sadly, the debates on how best to conserve the planet's biodiversity have not provided effective conservation of our rivers, lakes and other wetlands. Fortunately, we can learn from our colleagues who have been advocating for the conservation of terrestrial, and more recently, marine ecosystems; they have made great strides in developing more effective approaches, setting targets, and achieving change through the establishment of extensive protected areas. However, the conservation of freshwater ecosystems has not benefitted from similar systematic planning and management efforts.

This book brings together a wealth of knowledge on how to better conserve inland aquatic ecosystems in protected areas, including rivers, other fresh water ecosystems, brackish waters and coastal estuaries. We hope that the knowledge provided by a talented team of expert authors from around the world will inspire and inform non-government and government officials, protected area managers, researchers and citizens to strive even harder to conserve these wonderful ecosystems that underpin so much of our biodiversity and our societies.

Most nature reserves are designated as 'terrestrial' or 'marine' and the obvious question for most managers is 'why should I worry about the (usually) small portion of my protected area that involves freshwater habitat, especially when water is difficult to manage?' On the contrary, we argue that freshwater and estuarine habitats are particularly significant for conserving biodiversity and that protected area managers need to apply the freshwater-specific principles and conservation tools outlined in these chapters. Freshwater ecosystems have the greatest species diversity per unit area, a larger portion of freshwater and estuarine species are threatened, and the ecosystem services of these biomes are used unsustainably to a greater extent than any other biomes. Many terrestrial species depend on freshwater and estuarine ecosystems. Rather than forming a marginal aspect of protected area management, freshwater conservation is central to sustaining global biodiversity and societal well-being.

We start by defining inland aquatic ecosystems, showing just how diverse they actually are, and that they occupy a larger area than commonly appreciated.

We then propose ecological principles and processes that are essential for the conservation of freshwater ecosystems and aquatic species, recognising fundamental differences from terrestrial ecosystems. The specific management needs of the main types of freshwater ecosystems are explored. Threats to freshwater ecosystems are introduced, such as climate change, and the flow-on implications for protected area design, management and interaction with people in the wider landscape are examined. Case studies from around the world are used to illustrate principles and practices.

Our wonderful array of freshwater ecosystems, flora and fauna need our help to thrive in the 21st century. We hope that you are further encouraged by the practical knowledge contained in these chapters to make an even greater contribution to conservation of freshwater ecosystems.

Professor C. Max Finlayson
Institute for Land, Water and Society, Charles Sturt University, Albury, Australia & IHE Delft, Institute for Water Education, Delft, The Netherlands

Emeritus Professor Angela H. Arthington
Australian Rivers Institute, Griffith University, Brisbane, Queensland, Australia

Associate Professor Jamie Pittock
Fenner School of Environment and Society, The Australian National University, Acton, Australia
June 2017

Figures

Tables

Boxes

Chapter 1

An introduction to issues for managing freshwater ecosystems in protected areas

C.M. Finlayson, A.H. Arthington and J. Pittock

Key messages

- Effective management of freshwater ecosystems is critical as they have the greatest species diversity per unit area, a larger portion of threatened species, and use of their ecosystem services is generally unsustainable compared with other ecosystems.

- Rather than being treated as a marginal part of Protected Area (PA) management, freshwater conservation is central to sustaining the global biodiversity contained within the world's PAs. Conserving freshwater PAs based on ecological principles will also require specific conservation and management tools, including systematic conservation planning, catchment management plans, appropriate water management practices such as the use of environmental flows, and natural resource governance processes to ensure effective co-ordinated management at landscape scale, taking into account on-site and off-site influences and distant connectivities.

- Decisions about the management of freshwater PAs in relation to national policy settings as well as the settings provided by the Ramsar Convention and the Aichi Biodiversity Targets will be particularly relevant to achieving the Sustainable Development Goals for 2030. These goals are likely to drive the conservation agenda in the coming decades. In the context of this new global framework effective freshwater and estuarine biodiversity conservation via PAs and surrounding landscape-scales of management could be enhanced by integrating an international framework and guidance for managing freshwater PAs under rising human populations, increasing demands for food, water, energy and space, and changing climatic regimes.

Managing freshwater ecosystems in protected areas

The loss and degradation of freshwater ecosystems, also referred to in a broad sense as freshwater wetlands, has been documented at site, regional and national levels in many parts of the world, with the most recent analyses showing ongoing decline of species populations and loss of ecosystems (Davidson, 2014; Dixon et al., 2016; Gardner et al., 2015). These losses have occurred despite the Ramsar Convention on Wetlands and the Convention on Biological Diversity being signed or ratified by 169 and 196 national governments respectively. The effort afforded to the conservation and management of freshwater ecosystems through these governmental commitments has not stemmed biodiversity loss and ecosystem degradation. In particular there have been many questions asked about the effectiveness of PA networks for freshwater ecosystems and species (Pittock et al., 2015). This situation prompted the compilation of the papers that are included in this book.

The Ramsar Convention, as the most prominent international mechanism for promoting the conservation and management of freshwater ecosystems, has made some far-reaching and forward-looking decisions for ensuring the wise use of wetlands, and in particular has provided guidance for managing wetlands of international importance (known as Ramsar sites) (Davidson, 2016). However, these achievements are at odds with the declining status of wetlands globally. This situation has created a paradox whereby the effort to develop and agree on international policy, along with a substantial and expanding information base, has not halted and reversed the global degradation and loss of wetlands. While global policy emphases have included those that have promoted the designation and management of PAs, much of the global effort has not targeted freshwater ecosystems specifically, but focused primarily on terrestrial conservation with freshwater ecosystems forming a part of these (Hermoso et al., 2016). This situation has been recognised by IUCN (2014) with a call for a specific focus on the coverage and management of freshwater ecosystems in their own right rather than as components of terrestrial ecosystems.

With this background we explore the practices for conserving freshwater ecosystems in PAs, including rivers and other freshwater ecosystems, brackish waters and estuaries. As most nature reserves are designated as 'terrestrial' or 'marine' an obvious question for many conservation planners or managers is 'why should we be concerned about the (usually) small portion of any PA that involves freshwater habitat, especially when aquatic systems are difficult to manage?' In response we extend the argument made in Pittock et al. (2015) that freshwater and estuarine habitats are significant for conserving global biodiversity, both resident and migratory, and that PA managers need to apply freshwater-specific conservation principles and tools to ensure that protected habitats are effective in conserving aquatic biodiversity and ecosystem services. This is particularly important given that freshwater ecosystems have the greatest species diversity per unit area, a larger portion of freshwater and estuarine species are threatened, and the ecosystem services of these biomes are used unsustainably to a greater extent than any other biomes (Millennium Ecosystem Assessment, 2005). As many terrestrial species depend to various degrees on freshwater ecosystems, rather than freshwater conservation being a marginal part of PA management, their conservation is central to sustaining global biodiversity.

Information for managing freshwater protected areas

The following text provides an overview of the issues raised in the papers included in this book on the conservation and management of freshwater ecosystems in PAs.

Freshwater ecosystem types and extents

Milton and Finlayson (Chapter 2, this volume) start by defining freshwater ecosystems, showing how diverse they are (Figure 1.1) and that they occupy more area than many estimates indicate. This is an important issue as recent data analyses have shown that freshwater ecosystems are more prevalent than

Figure 1.1 Freshwater ecosystems support a wide diversity of plants and animals
(photographs © C.M. Finlayson).

Figure 1.1 (continued)

previously thought and occur on all continents, although the data on wetland areas is still fraught with uncertainties over definitions and inaccuracies. Milton and Finlayson show how recent analyses using remote sensing approaches and GIS-based modelling have greatly extended the area and accuracy of areal estimates, although differences with definitions and understanding of what constitutes a freshwater ecosystem still remain. Combined estimates including those for rivers, lakes and reservoirs (≥1 ha) and rice fields suggest permanent, seasonal, and intermittent freshwater ecosystems globally cover $12,543.5–14,443.5 \times 10^3\,km^2$. They further illustrate that the spatial extent of freshwater wetlands listed under the Ramsar Convention maximally represents 13.6 to 15.6 per cent of the total freshwater wetlands globally. While further analyses of representativeness and effectiveness are needed these data suggest that globally there may be a sound basis for meeting the Aichi Biodiversity Target 11 whereby at least 17 per cent of inland water areas are conserved through effectively and equitably managed, ecologically representative and well-connected systems of PAs (Armenteras and Finlayson, 2012).

Freshwater ecological principles

There have been many calls for scientists to more effectively communicate with decision makers at all levels of governmental and society to make certain that the values of freshwater ecosystems are recognised and effective cases are made for their conservation. Arthington et al. (Chapter 3, this volume) propose five high level ecological principles as a basis for conserving the values of freshwater ecosystems, and also, possibly even more importantly, as a basis for informed communication including the development of more sophisticated messages and motivations for conservation. In brief, the principles relate to: i) the intimate links between freshwater ecosystems and the surrounding catchment (basin or watershed); ii) the importance of the water regime (standing, flowing or sub-surface); iii) the hydrological, biogeochemical and ecological connectivity within and between freshwater (and estuarine) ecosystems; iv) protecting evolved patterns and 'hot spots' of native aquatic biodiversity, endemism and endangered species; and v) maintenance of ecological resilience in the aquatic ecosystems of the present and the future. Freshwater species have long histories of exposure and adaptation to variable environmental conditions and extremes (e.g. drought and floods), conferring resistance and resilience at the individual, community and ecosystems levels. Maintaining catchment integrity, natural flow and standing water regimes, the spatial and temporal dimensions of connectivity, and native biodiversity 'hotspots' will help to maintain the biodiversity and ecological resilience of aquatic systems in PAs, and support societal adaptations to the shifting environmental and climatic regimes of the future (Dudgeon et al., 2006). The principles we have proposed speak to the relationships between people and water and the biodiversity components that support our livelihoods and cultures by providing settings for human developments, health and well-being.

Defining and enhancing freshwater protected areas

PAs are 'clearly defined geographical spaces, recognised, dedicated, and managed, through legal or other effective means (Figure 1.2), to achieve the long term conservation of nature with associated ecosystem services and cultural values' (Dudley, 2008). Hermoso et al. (Chapter 4, this volume) discuss the effectiveness of PAs and common problems that can compromise effectiveness, such as insufficient consideration of the special ecological requirements of freshwater ecosystems when designing and declaring PAs (see Chapter 3, this volume), limited resources devoted to freshwater conservation management, and poor understanding and capacity to address complex management problems that extend beyond the limits of the PA (see Chapter 12, this volume). This chapter reviews new mechanisms that encourage public and private funding to achieve improved outcomes for biodiversity conservation in freshwater systems. Examples include payments for ecosystem services, water reserves, biodiversity offsets and system-wide planning to limit the impacts of water infrastructure on aquatic ecosystems. Hermoso et al. conclude that despite the increasing and innovative efforts to implement conservation in freshwater systems over the last decades, there remains an urgent need for improved assessment of the effectiveness of freshwater PAs through tailored monitoring programs and appropriate management plans.

Figure 1.2 Freshwater Protected Areas comprise a variety of clearly defined geographical areas that are recognised and managed through legal or other effective means to ensure the conservation and wise use of freshwater ecosystems, species and ecosystem services animals (photographs © C.M. Finlayson).

Figure 1.2 (continued)

What is different about protected areas?

Dudley et al. (Chapter 5, this volume) ask 'what is different about freshwater protected areas?' In answering they contrast the focus of many terrestrial and marine PAs on the areas within their boundaries to the unique features that influence the roles and management of freshwater PAs. Although freshwater PAs have often been overlooked, or inappropriately subsumed into terrestrial and marine categories, the unique mechanisms for protecting them have been recognised in the IUCN's guidelines on PA categories, and for governance and management (Dudley, 2008; Pittock et al., 2015). This is important as freshwaters provide a broad range of essential ecosystem services, linked particularly to food and water provisioning and increasing security against extreme weather events, making them a key link between people and nature (Millennium Ecosystem Assessment, 2005). The importance of freshwater conservation is also evident in both the UN Sustainable Development Goals and the Convention on Biological Diversity's Aichi targets. Freshwater PAs are critical for reducing and reversing declines in freshwater ecosystems and in generating ecosystem services for people. Conservation of freshwaters requires catchment-wide management of land use, pollution prevention and hydrology that often requires harmonisation action across political boundaries. Freshwater PA managers therefore need to work with a wide range of stakeholders to enhance conservation management, emphasising that well-managed freshwater PAs can deliver a huge range of services to people.

Managing threats to freshwater systems with protected areas

Pittock et al. (Chapter 6, this volume) contend that freshwater and estuarine ecosystems are among the most threatened in the world, are under-represented in PA policy and have the highest portion of species threatened with extinction. They argue that freshwater biodiversity is particularly threatened because its conservation depends on: i) maintaining hydrological processes; ii) retaining longitudinal connectivity of water flows without barriers along rivers; iii) conserving lateral connectivity between a water body and its floodplain; iv) sustaining adequate groundwater–surface water interactions; managing exogenous threats that are propagated across catchments; and v) integrating governance by multiple management authorities. As many freshwater PAs are cultural landscapes the impacts of human uses need to be managed, notably those from agriculture, aquaculture and fishing. Within PAs there are particular challenges in minimising the impacts of water infrastructure and visitor facilities, controlling invasive species incursions, and preventing pollution. Despite public concern about floods, droughts and fires, these are natural processes that many freshwater ecosystems will normally tolerate. However, climate change is exacerbating these processes leading to changes in ecological character (see Chapter 13, this volume). Further, depending on the circumstances, freshwater ecosystems may attenuate or increase the impacts of natural disasters on people.

Conserving freshwater species in protected areas

Turak and Pittock (Chapter 7, this volume) describe how freshwater biodiversity accounts for almost 10 per cent of all recognised species on earth including over half of all fish species (28,900 species), 6,000 species of mammals, birds, reptiles and amphibians, and 2,600 species of macrophytes. Freshwaters also have a high diversity of invertebrates. These species may have different habitat and connectivity requirements, and vulnerability to invasive species and other threats. Consequently it is a challenge to know which freshwater species are present or may be re-introduced as part of recovery programs, and what actions are needed to maintain viable populations within PAs. Conserving freshwater species is one of the many management objectives within most PAs. Identifying the overlaps of actions needed to conserve freshwater species with those of other management objectives allows PA managers to prioritise actions that have multiple benefits and engender great support. Monitoring the status of freshwater biodiversity is essential to understanding the efficacy of conservation interventions, but there are hundreds of possible variables. Applying the Essential Biodiversity Variables (EBV) framework (Turak et al., 2016) allows identification of a set of variables that can accurately measure the success of such actions. Freshwater species are potentially affected by any action within both surface water and groundwater catchments, requiring use of conceptual models and mechanistic understanding to determine cause and effect relationships between management actions, freshwater biodiversity and catchment processes. Understanding these cause and effect relationships at the landscape scale helps to prioritise actions to better manage threats to species that originate from outside of PAs.

Managing specific freshwater ecosystems

In this chapter Arthington et al. (Chapter 8, this volume) outline general principles and a range of practices for effective management of specific freshwater ecosystems types, including rivers and estuaries, lakes and reservoirs, peatlands and groundwater-dependent ecosystems. A number of relatively straightforward changes to the way PAs are designed, expanded and managed can help to further improve their conservation benefits for freshwater ecosystems. Namely they are to: avoid using a river as the boundary of a PA; encourage expansion of existing PAs to incorporate natural large-scale catchment processes into PAs where possible; ensure that the water regimes of rivers and estuaries, lakes, peatlands and groundwater-dependent ecosystems are well managed within PAs and their catchments, enabling them to recover from the impact of activities upstream as water moves through the PA; avoid development of visitor infrastructure on priority freshwater ecosystems in PAs; encourage expansion of existing PAs to incorporate biodiversity hotspots, functional processes and connectivity pathways at relevant scales; and promote new PAs for the last remaining free-flowing rivers, lakes and other high priority freshwater ecosystems. The International Lake

Environment Committee has identified six major pillars for lake governance: i) policies, ii) institutions, iii) stakeholder participation iv) 'hard' and 'soft' technology, v) knowledge and information, and vi) finances (ILEC, International Lake Environment Committee, 2005). The complementary concepts of Integrated Lake Basin Management (ILBM) and Integrated Water Resources Management (IWRM) together address vital management and governance issues for PAs involving lakes, reservoirs, rivers, estuaries and linked groundwater dependent ecosystems.

Freshwater protected area corridors

Pittock et al. (Chapter 9, this volume) argue that riparian and floodplain corridors form key habitat for animals in the terrestrial landscape, and in most parts of the world they support more species of plants and animals than any other landscape unit. Consequently, maintenance and restoration of riparian corridors is a priority to conserve both freshwater and terrestrial ecosystems. In particular, there are considerable benefits to be gained from restoring riparian forests. These forests play key roles in providing organic matter that drives elements of the aquatic food chain, forming physical habitat, filtering out pollutants and providing shade for maintaining appropriate water temperatures. At a minimum restoration should form riparian corridors wide enough to enable full development of the vegetation canopy to maximize shade across the relevant water body and form an adequate mesic (moist, humid) micro-climate. For even better results for biodiversity conservation, the full width of the regularly inundated riparian land and floodplain should be restored.

A number of systems of river corridor PAs are reported by Pittock et al. from many jurisdictions around the world, applying criteria such as biological importance, maintaining free-flowing ecological processes and supporting cultural values. They note that dams and levee banks are being removed from rivers and their floodplains to restore ecosystem functions and services in many developed and developing countries. Restoration of functional riparian and floodplain systems may aid flood management and enhance other climate change adaptation measures to the benefit of both people and biodiversity conservation.

Planning ecologically: the importance of management at catchment scales

Flitcroft et al. (Chapter 10, this volume) consider the importance of integrating freshwater PAs with landscape management including an emphasis on addressing scientific questions (such as multi-scale ecosystem functions, processes and interactions), and applied challenges (such as management of land uses in diverse multi-user landscapes). In this respect rivers should not be treated as disconnected reaches where species and populations are analysed without a catchment or network context. Instead, the use of broader conceptual frameworks that link ecological communities to underlying geophysical systems is encouraged, such as: the river continuum concept; hierarchical organization of instream habitat; the natural flow regime; process domains; the network dynamics hypothesis; and the riverscape concept. Catchment-management plans are presented as a means of integrating the diverse land uses and owners who, in combination, may directly or indirectly influence the quality of a shared river system. As there is no clear template for catchment management that works under all circumstances a number of examples of successful projects, and tools, are presented to illustrate the diverse approaches used to solve complex water management issues at catchment scale.

Planning for the protection and management of freshwater ecosystems inside and outside protected areas

Continuing the theme of water planning, Nel and Roux (Chapter 11, this volume) stress the need for PA authorities to engage in water planning processes both within their PA boundary and beyond in associated catchments. Three water management approaches – Integrated Water Resource Management (IWRM), ecosystem-based management and adaptive management – provide useful guidance for water planning, whether PA authorities are planning inside or outside PAs (Schoeman et al., 2014). When negotiating trade-offs between different water development options, authorities should insist on scientifically-credible freshwater ecosystem assessments, such as environmental flow assessment, systematic conservation planning and ecosystem service assessment. By engaging in water planning processes within and beyond their area of responsibility, PA authorities can act as powerful stakeholders in the negotiation for freshwater ecosystem protection within regional water planning processes.

Managing freshwater protected areas in the global landscape

Finlayson et al. (Chapter 12, this volume) develop the need for management of freshwater PAs to extend across regional and continental landscapes to ensure the many benefits obtained from diverse freshwater wetlands are not further reduced. This landscape approach includes the designation of wetlands for inclusion in the Ramsar 'List of Wetlands of International Importance', which is the keystone of the Ramsar Convention (Davidson, 2016) and has been a success, with 1,687 inland freshwater wetlands covering an area of 195.5 million km^2 being listed. However, the responses to further recommendations to prepare and implement an appropriate management plan for protecting listed wetlands and to prepare a national wetland inventory illustrate less than adequate outcomes (Finlayson, 2012). Other legislative frameworks are also described, including the European Union's Water Framework Directive, which offers other approaches for managing wetlands across landscapes, although significant challenges remain to fully deliver an appropriate quality and quantity of freshwater to maintain the character PAs, including the ecosystem services that are provided. Similarly there has been a long history of the establishment of bilateral and multilateral agreements aimed specifically at conserving migratory waterbird flyways. Most flyway initiatives include the establishment of networks of wetland sites for migratory waterbirds with some being supported by legally binding arrangements, and many being supported by the activities of non-governmental organisations. A case is also made for a move away from a static view of the ecological condition of a wetland to accommodate greater variability while also ensuring the wetland retains its ecological functions, being aware that this is likely to require sustainable management approaches that extend across multiple landscapes and temporal scales.

Climate change and the management of freshwater protected areas

Continuing the theme of static versus dynamic freshwater ecosystems, Finlayson and Pittock (Chapter 13, this volume) point out that whilst PAs have played an important role in freshwater conservation efforts it may not be possible to manage them with the same rulebook for coming decades. In particular, strategies for promoting more climate-resilient approaches are needed with less focus on maintaining past reference states that are typically impossible to recover. The development of novel ecosystems, for example, may require new thinking and a range of new approaches to water management to cope with increasingly uncertain futures. Notwithstanding such uncertainties, many climate-change adaptation interventions are available to better conserve freshwater biodiversity, in particular the identification and management of refugia and the use of environmental flows to counter climate change impacts on aquatic ecosystems (Bond et al., 2008). At the same time it is recognized that all adaptation options involve risks and costs as well as benefits, and a suite of different but complementary interventions is likely to result in better practice by spreading the risk while seeking to maximize the benefits. These approaches could be facilitated by the development of an international framework and guidance for managing freshwater PAs under shifting climatic regimes – efforts to do this through the Ramsar Convention have not been fully successful (Finlayson et al., 2017).

Conclusions

In the final chapter of this book, Finlayson et al. (Chapter 14, this volume) provide a synthesis of the principles and processes that are essential for the conservation of freshwater ecosystems and aquatic species, especially as these are often fundamentally different from those for terrestrial ecosystems. This synthesis elaborates and extends the key issues raised by Pittock et al. (2015) about the importance of freshwater ecosystems and the need for concerted scientifically grounded management actions, including those for ensuring the effectiveness of the PA network. Effectiveness is governed by five ecological principles that relate to: i) the intimate links between freshwater ecosystems and the surrounding catchment (basin or watershed); ii) the importance of the water regime (standing, flowing or sub-surface); iii) the hydrological, biogeochemical and ecological connectivity within and between freshwater, terrestrial (and estuarine) ecosystems; iv) protecting evolved patterns and 'hot spots' of native aquatic biodiversity, endemism and endangered species; and v) maintenance of ecological resilience in the aquatic ecosystems of the present and the future (Chapter 3, this volume). Pittock et al. (2015) also emphasize the benefits of seeking to conserve freshwater biodiversity along the entire lengths of rivers given the trade-offs and interventions in both freshwater and terrestrial systems that unwittingly degrade freshwater habitats.

As the advent of climate change will exacerbate competition between people and ecosystems for fresh water in many parts of the world the complexities of determining trade-offs for water use will need to be included in planning both within PAs and across adjacent and distant landscapes. Water use trade-offs include addressing conflicts and positive synergies between different climate change mitigation and adaptation measures that affect the water needed for any particular freshwater PA, as well as being aware that rivers provide landscape corridors with variable gradients, flows of water and nutrients, and movement of species between reaches and, in many instances, between PAs (Chapter 9, this volume).

In addition to the Aichi Biodiversity Targets as described in the 2010–2020 Strategic Plan for Biodiversity (Armenteras and Finlayson, 2012) the world's governments have agreed on the UN Sustainable Development Goals (SDGs) which are likely to drive the conservation agenda in the coming decades (http://www.un.org/sustainabledevelopment/, accessed 19 March 2017). This global agenda specifically considers fresh water and biodiversity via Goal 6 covering access to clean water and sanitation and Goal 15 covering sustainable land management and halting the loss of biodiversity. In this respect decisions made about freshwater PAs in relation to the policy settings provided by the Ramsar Convention (Finlayson et al., 2017) and the Aichi Biodiversity Targets (Juffe-Bignoli et al., 2016) will be particularly relevant to achieving the SDGs. Failure to more effectively select and manage freshwater PAs could lead to further loss and degradation of freshwater ecosystems and the decline of species populations. Adverse outcomes for freshwater biodiversity will also have adverse outcomes for the many people who benefit from the many ecosystem services provided by these ecosystems. In the context of this new global framework for sustainability, effective freshwater and estuarine biodiversity conservation via PAs and surrounding landscape-scales of management could be enhanced by integrating an international framework and guidance for managing freshwater PAs under rising human populations, increasing demands for food, water, energy and space, and changing climatic regimes.

References

Armenteras, D. and Finlayson, C.M. (2012) 'Biodiversity'. In *UNEP, Keeping Track of Our Changing Environment: From Rio to Rio+20 (1992–2012)*. United Nations Environment Programme, Nairobi.

Bond, N., Lake, P. and Arthington, A. (2008) 'The impacts of drought on freshwater ecosystems: an Australian perspective', *Hydrobiologia*, 600, pp3–16.

Davidson, N.C. (2014) 'How much wetland has the world lost? Long-term and recent trends in global wetland area', *Marine and Freshwater Research*, vol 65, pp934–41.

Davidson, N.C. (2016) 'The Ramsar Convention on Wetlands'. In Finlayson, C.M., Everard, M., Irvine, K., McInnes, R.J., Middleton, B.A., van Dam, A.A. and Davidson, N.C. (eds), *The Wetland Book I: Structure and Function, Management and Methods*. Springer Publishers, Dordrecht. DOI 10.1007/978-94-007-6172-8_113-1.

Dixon, M.J.R., Loh, J., Davidson, N.C., Beltrame, C., Freeman, R. and Walpole, M. (2016) 'Tracking global change in ecosystem area: The Wetland Extent Trends index'. *Biological Conservation*, vol 193, pp27–35.

Dudgeon, D., Arthington, A.H., Gessner, M.O., Kawabata, Z.I., Knowler, D.J., Lévêque, C., Naiman, R.J., Prieur-Richard, A.H., Soto, D., Stiassny, M.L. and Sullivan, C.A. (2006) 'Freshwater biodiversity: importance, threats, status and conservation challenges'. *Biological Reviews*, vol 81, pp163–182.

Dudley, N. (Ed.) (2008) *Guidelines for Applying Protected Area Management Categories*. International Union for the Conservation of Nature, Gland, Switzerland.

Finlayson, C.M. (2012) 'Forty years of wetland conservation and wise use'. *Aquatic Conservation: Marine and Freshwater Ecosystems*, vol 22, pp139–143.

Finlayson, C.M., Capon, S.J., Rissik, D., Pittock, J., Fisk, G., Davidson, N.C., Bodmin K.A., Papas, P., Robertson, H.A., Schallenberg, M., Saintilan, N., Edyvane, K. and Bino, G. (2017) 'Policy considerations for managing wetlands under a changing climate', *Marine and Freshwater Research* (in press), published early online: https://doi.org/10.1071/MF16244.

Gardner, R.C., Barchiesi, S., Beltrame, C., Finlayson, C.M., Galewski, T., Harrison, I., Paganini, M., Perennou, C., Pritchard, D.E., Rosenqvist, A. and Walpole, M. (2015) 'State of the World's Wetlands and their Services to People: A compilation of recent analyses'. Ramsar Convention Secretariat, Ramsar Scientific and Technical Briefing Note No. 7, Gland, Switzerland.

Hermoso, V., Abell, R., Linke, S. and Boon, P. (2016) 'The role of protected areas for freshwater biodiversity conservation: challenges and opportunities in a rapidly changing world'. *Aquatic Conservation: Marine and Freshwater Ecosystems*, vol 26 (Suppl. 1), pp3–11.

ILEC (International Lake Environment Committee) (2005) 'Managing Lakes and Their Basins forSustainable Use. A Report for Lake Basin and Stakeholders'. International Lake Environment Committee, Kusatsu, Japan. http://www.ilec.or.jp/en/pubs/p2/lbmi.

IUCN (2014). 'The promise of Sydney: innovative approaches for change. A strategy of innovative approaches and recommendation to reach conservation goals in the next decade'. Available at: worldparkscongress.org/downloads/approaches/Stream1.pdf.

Juffe-Bignoli, D., Harrison, I., Butchart, S.H.M., Flitcroft, R., Hermoso, V., Jonas, H., Lukasiewicz, A., Thieme, M., Turak, E., Bingham, H., Dalton, J., Darwall, W., Deguignet, M., Dudley, N., Gardner, R., Higgins, J., Kumar, R., Linke, S., Milton, G.R., Pittock, J., Smith, K.G. and van Soesbergen, A. (2016) 'Achieving Aichi

Biodiversity Target 11 to improve the performance of protected areas and conserve freshwater biodiversity'. *Aquatic Conservation: Marine and Freshwater Ecosystems*, vol 26 (Suppl. 1), pp133–151.

Millennium Ecosystem Assessment (2005) 'Ecosystems and human well-being: wetlands and water synthesis'. World Resources Institute, Washington, DC.

Pittock, J., Finlayson, M., Arthington, A.H., Roux, D., Matthews, J.H., Biggs, H., Harrison, I., Blom, E., Flitcroft, R., Froend, R., Hermoso, V., Junk, W., Kumar, R., Linke, S., Nel, J., Nunes da Cunha, C., Pattnaik, A., Pollard, S., Rast, W., Thieme, M., Turak, E., Turpie, J., van Niekerk, L., Willems, D. and Viers, J. (2015) 'Managing freshwater, river, wetland and estuarine protected areas'. In Worboys, G.L., Lockwood, M., Kothari, A., Feary, S. and Pulsford, I. (eds), *Protected Area Governance and Management*. ANU Press, Canberra.

Schoeman, J., Allan, C. and Finlayson, C.M. (2014) 'A new paradigm for water? A comparative review of integrated, adaptive and ecosystem-based water management in the Anthropocene', *International Journal of Water Resources Development*, vol 30, pp377–390.

Turak. E., Harrison, I., Dudgeon, D., Abell, R., Bush, A., Darwall, W., Finlayson, C.M., Ferrier, S., Freyhof, J., Hermoso, V., Juffe-Bignoli, D., Linke, S., Nel, J., Patricio, H.C., Pittock, J., Raghavan, R., Revenga, C., Simaika, J.P. and De Wever, A. (2016) 'Essential Biodiversity Variables for measuring change in global freshwater biodiversity'. *Biological Conservation*, http://dx.doi.org.ezproxy.csu.edu.au/10.1016/j.biocon.2016.09.005.

Chapter 2

Freshwater ecosystem types and extents

G.R. Milton and C.M. Finlayson

Key messages

- Freshwater ecosystems, also referred to as freshwater wetlands, comprise many different types, geomorphic forms and sizes and have been defined and classified in different ways, although not all efforts are systematic or readily comparable. In a general sense wetland definition and classification has been based on particular features of the wetlands, including the landform and geomorphic setting, physical dimensions, vegetation, water quality and water regime, soils, and as habitat for specific fauna.

- Despite the effort to describe and map wetlands globally, an accurate and comprehensive inventory and global map does not exist for freshwater ecosystems in part due to limitations in observing smaller sized classes. Global estimates for lakes and reservoirs are variable depending upon the downscaling approach used for those ≤ 10 ha in area. Combined estimates including those for rivers, lakes and reservoirs (≥ 1 ha) and rice fields suggest permanent, seasonal, and intermittent freshwater ecosystems globally cover $12{,}543.5{-}14{,}443.5 \times 10^3 km^2$.

- Wetlands are found on all continents and global mapping initiatives have documented the latitudinal and longitudinal distribution of area as well as displayed a general distribution of wetland types. The latitudinal distribution of wetland area largely parallels the distribution of lakes with a preponderance in boreal and arctic latitudes corresponding to the extent of the last glacial maximum in northern North America and Scandinavia including parts of northern Russia; the abundance is much lower in southern latitudes where the continental area is lower and 50 per cent of lake area is located at elevations below 500 m above sea level.

- There is an increasing amount of information about individual wetlands or types of wetlands, including comparative studies of large sites, large reservoirs, or for wetland expanses, or for peatlands globally. Freshwater peatlands are widely distributed and possibly represent more than a third of global wetlands and are especially wide spread over mid–high latitudes of North America and Eurasia and areas of excess moisture, e.g., tropical regions where high evapotranspiration is counterbalanced by extremely high rainfall.

- A recent compilation of articles provides an overview of major wetland types and descriptions of individual wetlands distributed among nine regions. These show the complexity and variability within and among wetlands even when 'similarly' classified, and demonstrate that wetland heterogeneity is an outcome inter alia of the ecological components, topographical setting, hydrological regime, and processes interacting temporally and spatially within landscapes.

Introduction to freshwater ecosystems

Freshwater ecosystems, hereafter referred to as freshwater wetlands, comprise many different types, geomorphic forms and sizes and have been defined and classified in different ways (Mitsch 1994; Finlayson and van der Valk 1995a; Finlayson et al., 2016), although not all efforts are systematic or readily comparable (Gerbeaux et al., 2017). In a general sense wetland definition and classification has been based on particular features of the wetlands, including the landform and geomorphic setting, physical dimensions, vegetation, water quality and water regime, soils, and as habitat for specific fauna. This mix of features has resulted in many different emphases when defining what is meant by a freshwater wetland (Figure 2.1) and created confusion when clear definitions have not been used (Finlayson and van der Valk, 1995b; Finlayson et al., 1999, 2016; Gerbeaux et al., 2017).

One of the broadest definitions of wetlands is used by the Ramsar Convention on Wetlands, whereby: '. . . wetlands are areas of marsh, fen, peatland or water, whether natural of artificial, permanent or temporary, with water that is static or flowing, fresh, brackish or salt, including areas of marine water the depth of which at low tide does not exceed six metres'. The Convention also developed a typology to accompany this definition and to provide a listing of wetland types being designated as Wetlands of International Importance (commonly known as Ramsar sites). The typology contains a listing of a wide range of fresh, brackish or saline aquatic ecosystems, including freshwater rivers, lakes, swamps, marshes and caves (Ramsar Convention Secretariat, 2010).

Figure 2.1 Variety of freshwater wetlands: a) freshwater floodplain, Mexico; b) Lakenheath, United Kingdom; c) Gippsland lakes, Australia; and d) Yodha Wewa, Sri Lanka (photographs © C.M. Finlayson).

Figure 2.1 (continued)

Wetland extent

Differences in definitions and classification have presented some practical difficulty in determining the distribution and extent of freshwater wetlands despite the large effort directed at wetland inventory globally (Mitsch, 1994; Finlayson and van der Valk, 1995a; Finlayson et al., 1999, 2016). When comparing estimates it is important therefore to ascertain exactly what is included. It is also necessary to check whether or not such estimates are current and account for the rapid rates of wetland loss globally. This is important given Davidson (2014) reported that some 64–71per cent of wetlands have been lost globally since 1900 AD; and Dixon et al. (2016) reported natural wetlands have declined globally by about 30 per cent between 1970 and 2008.

A series of continental-scale wetland directories was produced in the 1980s and 1990s for the Neotropics (Scott and Carbonell, 1986), Asia (Scott, 1989), Oceania (Scott, 1993) and Australia (Usbank and James, 1993) using a similar format. Different formats were used for two directories of African wetlands (Burgis and Symoens, 1987; Hughes and Hughes, 1992). The data from the Neotropics, Asian and Oceanian directories were later made available on the internet by the International Water Management Institute (Rebelo et al., 2009).

Estimates by many authors do provide a valuable historical perspective on the global extent of wetlands, although the emphases and the level of detail vary considerably, with some possibly being misleading given inadequacies in approaches and documentation (Finlayson et al., 2016). The Global Review of Wetland Resources and Priorities for Wetland Inventory (Finlayson et al., 1999) estimated that the global area of wetlands (including coastal and marine) was more than 12.80 million km^2. Although this value is much larger than other estimates and the 8.2–10.1 million km^2 preferred by Mitsch and Gosselink (2015), it did include lakes and reservoirs that were estimated by Lehner and Döll (2004) to cover an additional 2.7 million km^2. A recent synthesis of papers on the world's wetlands and their future under climate change estimated a global area of 11.52 million km^2 while acknowledging the extent of wetland was not known for several major regions, including South America, Africa and Russia (Junk et al., 2013). Despite the effort to describe and map wetlands globally, an accurate and comprehensive inventory and global map does not exist for freshwater ecosystems in part due to limitations in observing smaller sized classes.

The global estimates for lakes and reservoirs (Table 2.1) are variable depending upon the downscaling approach used for those ≤10 ha in area. The estimate by Raymond et al. (2013) is at the lower boundary of global estimates for lakes and reservoirs (3,000 × 10^{3}km^2). Lewis's (2011) estimate of 3,100–4,200 × 10^{3}km^2 for lakes ≥0.1 ha accounts for the uncertainty in the downscaling approaches for systems ≤10 ha. This however does not include the area of reservoirs (excludes regulated lakes) and small farm impoundments (≥0.01 ha) that may approximate an additional 305.7 × 10^{3}km^2 (Lehner et al., 2011). The upper estimates (≥0.1 ha) of Lewis (2011) and Lehner et al. (2011) correspond with Verpoorter et al.'s (2014) satellite remote sensing based estimate of $5,000 \times 10^3 km^2$ (≥ 0.2 ha) for freshwater bodies (lakes and reservoirs). However, Messager et al. (2016) suggest Verpoorter et al.'s (2014) estimate for lakes and reservoirs ≥ 1 ha ($4,760 \times 10^3 km^2$) may be an overestimate due to '. . . misclassifications or differences in defining lakes versus rivers and wetlands'. Downscaling the HydroLAKES database, Messager et al. (2016) estimate natural lakes ≥ 1 ha cover $3,232 \times 10^3 km^2$. Several authors have provided areal estimates for rivers with Raymond et al. (2013) being the most exhaustive. These estimates together with Lehner and Döll's (2004) estimate for other freshwater wetlands, reservoirs ≥ 1 ha excluding regulated natural

Table 2.1 Recent estimates of global wetland area (adapted from Fluet-Chouinard et al., 2015).

Source	Size	Lakes (10³ km²)	Reservoirs (10³ km²)	Rivers (10³ km²)	Freshwater wetlands[h] (10³ km²)
Lehner and Döll 2004	≥10 ha	2,428[a]	251	360	6,764–8,664
	≥1 ha	3,169[ab]			
Lehner et al., 2011	≥1 ha		295.3[bd]		
	≥0.1 ha		301.8[bd]		
	≥0.01 ha		305.7[bd]		
Downing et al., 2006	≥1 ha	3,507[bc]	258.6[b]		
	≥0.1 ha	4,200[bc]	76.8[e]		
	Small[e]				
Downing et al., 2012				485–662	
Lewis, 2011	≥0.1 ha	3,100–4,200[bc]			
McDonald et al., 2012	≥0.1 ha	3,786[bf]			
Raymond et al., 2013	≥0.1 ha	2,740[bc]	260[b]		624
Verpoorter et al., 2014	≥1 ha	4,760[g]			
	≥0.2 ha	5,000[g]			
Messager et al., 2016	≥10 ha	2,677[ci]	250		
	≥1 ha	3,232[bci]			

Notes:
a   includes reservoirs that could not be identified as such in the database.
b   based on extrapolation method to small lakes or reservoirs, all other estimates are from data.
c   excludes man-made reservoirs.
d   excludes regulated natural lakes.
e   small farm pond impoundments based on extrapolation method, minimum and maximum areas not described.
f   includes reservoirs.
g   based on remote sensing and includes all lakes and reservoirs but not rivers.
h   does not include lakes, reservoirs, rivers or coastal/brackish systems.
i   includes regulated natural lakes and small unreported reservoirs.

lakes (295.3 × 10^3km²; Lehner et al. 2011), and a current estimate for rice fields otherwise not included (~ 1,628.2 × 10^3km², IRRI 2016), suggest permanent, seasonal and intermittent freshwater ecosystems globally cover 12,543.5–14,443.5 × 10^3km². This estimate does not include the hundreds of freshwater lakes of Antarctic's coastal oases that range in surface area to 20 km² (Laybourn-Parry and Wadham, 2014) or its 400+ subglacial lakes varying in length from less than one km to Lake Vostok's approximate 250 km and 14,000 km² area (Siegert et al., 2005, 2016) of which Lake Vostok is the largest.

Every estimated area is an approximation as the data sources used in the analyses do not account for the temporal and spatial variability due to climate and hydrology. A first approximation of this is the downscaling of the Global Inundation Extent from Multi-Satellites (GIEMS) to produce an inundation probability map corresponding to mean annual seasonal low (MA_{Min} – 6,538 × 10^3km²), high (MA_{Max} – 12,089 × 10^3km²) and long term maximum (LT_{Max} – 17,255 × 10^3km²) areas (Fluet-Chouinard et al., 2015). Data sets used to develop the downscaled probability map have a variety of uncertainties, errors and inconsistencies, but Fluet-Chouinard et al. (2015) consider them to be an improvement over existing maps. However, the MA_{Max}–LT_{Max} includes the range of the higher resolution global freshwater estimate above and thus suggests the lower resolution of the GIEMS may underestimate the global area of permanent and seasonal freshwater wetlands.

Wetland distribution

Wetlands are found on all continents and global mapping initiatives have documented the latitudinal and longitudinal distribution of area as well as displayed a general distribution of wetland types (Figure 2.2). Lehner and Döll (2004) note the latitudinal distribution of wetland area largely parallels the distribution of lakes with a preponderance in boreal and arctic latitudes (50°–70° N) corresponding to the extent of the last glacial maximum in northern North America and Scandinavia including parts of northern Russia (Messager et al., 2016); the abundance is much lower in southern latitudes where the continental area is lower and 50 per cent of lake area is located at elevations below 500 m above sea level (Verpoorter et al., 2014). Fluet-Chouinard et al. (2015) compare the latitudinal and longitudinal distribution of inundated area from their downscaling of GIEMS with the GLWD (Lehner and Döll 2004) and Global Land Cover 2000 (GLC2000 – Bartholomé and Belward, 2005). Substantial departure occurs from the generally expected pattern over parts of Asia due to GIEMS seasonal documentation of inundated rice cultivation (Fluet-Chouinard et al., 2015). Although peatlands are widely distributed (Schurmann and Joosten, 2008) and possibly represent more than a third of global wetlands (Joosten, 2010), they are especially wide spread over mid–high latitudes of North America and Eurasia and areas of excess moisture, e.g., tropical regions where high evapotranspiration is counterbalanced by extremely high rainfall (Joosten, 2004).

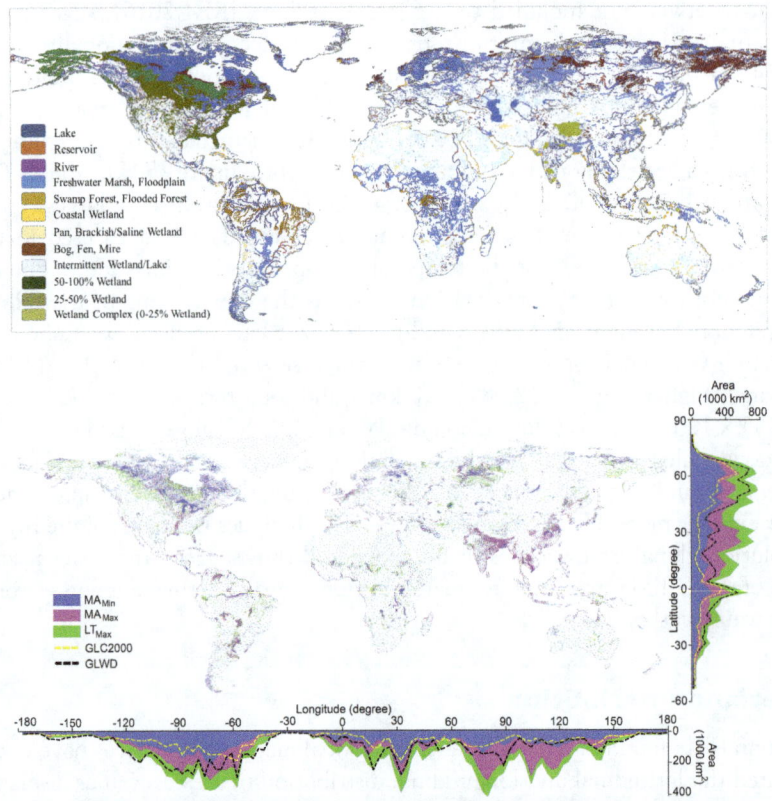

Figure 2.2 (a) Generalized distribution of wetland types (Figure 5 from Lehner and Döll, 2004, with permission); (b) latitudinal and longitudinal distributions from downscaled GIEMS for the three states of inundation compared with GLWD and GLC2000 wetland areas. The three states are average minimum (MA_{Min}) and maximum (MA_{Max}) inundation area and long-term maximum inundation area (LT_{Max}) (Figure 5b from Fluet-Chouinard et al., 2015, with permission).

Wetland diversity

Whilst usage of the term 'wetland' has increased in recent decades there is a diversity of terms that have been used to describe freshwater ecosystems (Mitsch and Gosselink, 2015; Gerbeaux et al., 2017), including those in the Ramsar Convention's definition and classification of wetland types (Ramsar Convention Secretariat, 2010). Even common terms used by the scientific community can signify different wetland types due to regional or continental norms of usage. Some common English names for freshwater wetlands (Table 2.2) convey this diversity.

Table 2.2 Some common names for freshwater wetlands in English (taken largely from information contained in Finlayson and van der Valk, 1995b and Mitsch and Gosselink, 2015).

Billabong	Australian term that is loosely used to describe lagoons in cut off meanders and remnant pools in stream channels.
Bog	A peat-accumulating wetland that has no significant inflows or outflows and supports acidophilic mosses, particularly sphagnum. Includes domed and blanket bogs.
Bottomland	Lowlands, usually forested, along streams and rivers, usually on alluvial floodplains that are periodically flooded.
Carr	Wetland dominated by woody vegetation.
Delta	Fan-shaped accumulation of alluvial sediments usually at the mouth of a river. Inland deltas also occur.
Fen	A peat-accumulating wetland that receives some drainage from surrounding mineral soil and usually supports marsh vegetation.
Lark	Permanently or temporarily flooded depression containing sparse peat forming vegetation in peatlands.
Lagg	Marginal stream or swamp fringing a domed bog.
Lake	Large body of standing water surrounded by land.
Lagoon	Term frequently used to denote deep-water enclosed or partially opened aquatic system, especially in coastal regions.
Loch/Lough	Irish and Scottish term for a lake or sea inlet. A lochan is a small loch.
Marsh	A frequently or continually inundated wetland with emergent herbaceous vegetation in a mineral soil substrate.
Mire	Synonymous with any actively peat-accumulating wetland.
Moor	Synonymous with any peatland. A 'highmoor' is a raised bog, while a 'lowmoor' is a peatland in a basin that is not elevated above its perimeter.
Muskeg	North-American term for large expanses of peatlands or bogs.
Peatland	A generic term for a wetland with a naturally accumulated layer of partly decayed plant matter greater than 30 cm deep under permanent water saturation.
Playa	Flat-bottomed depression in arid or semi-arid regions with distinct wet-dry seasons.
Polje	Flat-floored enclosed depressions with karstic drainage which may be dry or occur as seasonally flooded wetland or permanent lakes.
Pond	Small pool of standing water.
Pothole	Shallow marsh-like ponds, particularly in the USA and Canadian prairie provinces.
Reed swamp	Marsh dominated by *Phragmites* (common reed); term used particularly in Europe.
Riparian system	Ecosystems with a high water table because of proximity to an aquatic system, usually along a stream or river.

(continued)

Table 2.2 (continued)

Slough	A swamp or shallow lake system in the northern and midwestern United States. A slowly flowing shallow swamp or marsh in southeastern United States.
Swale	Elongated depression between dunes or coastal ridges roughly parallel to the coast.
Swamp	Wetlands dominated by trees or shrubs. In Europe forested fens and areas dominated by reeds (*Phragmites*) are also called swamps.
Tarn	Small lake in the mountains, often with no significant tributaries.
Tidal freshwater marsh	Marshes along rivers and estuaries close enough to the coastline to experience significant tides by non-saline water.
Turlough	Shallow depressions in a Carboniferous karst landscape subject to periodic flooding mainly from groundwater.
Vernal pool	Shallow temporary pools with seasonal flooding.
Vlei	Southern-African term for a shallow, seasonal or intermittently flooded lake.
Wet meadow	Waterlogged grassland but without standing water for most of the year.
Wet prairie	Similar to a marsh but with water levels usually intermediate between a marsh and a wet meadow.

There is an increasing amount of information about individual wetlands or types of wetlands, including comparative studies of large sites, such as that provided by Junk et al. (2006) for a number of internationally important wetlands, large reservoirs (Lehner et al., 2011) or for wetland expanses, such as that provided by Lukacs and Finlayson (2010) for northern Australia, or by Joosten (2010) for peatlands globally. Efforts to collate information on individual wetland types to compile a global estimate is limited by differences in definitions and classifications (Finlayson et al., 2016), between freshwater and saline lakes, for example, and whether forested wetlands have been included, as well as by the limitations of various methods, including whether the extent of inundation reflects the distribution of wetlands, or whether flooding under tree canopies has been determined. The extent of a number of wetland types sourced from individual reviews and based on different approaches for determining the areal estimates is provided in Table 2.3, although in some instances these data are not mutually exclusive.

While not providing a systematic analysis of freshwater wetlands, Finlayson et al.'s (2017a) compilation of articles provides an overview of major wetland types and descriptions of individual wetlands distributed among nine regions. Nearly 140 of the articles address freshwater ecosystems, e.g., describing the features of river basins and their wetlands, landscapes in which wetlands are a dominant feature, and globally and regionally significant wetland sites. In their summary of the literature, Milton et al. (2016) note '. . . the complexity and variability within and among wetlands even when "similarly" classified'. Furthermore, 'wetland heterogeneity is

Table 2.3 Estimates of area of wetland types globally (adapted from Finlayson et al., 2016 using information from multiple sources).

Wetland type	Area km²	Reference
Peatlands	4,000,000	Multiple sources compiled by Joosten 2010.
Tropical freshwater swamps (mineral soil)	1,460,000	Multiple sources summarised by Giessen 2016.
Tropical peat swamps	441,025	Page et al., 2011.
Seasonal rice fields	1,628,200	IRRI, 2016.
Freshwater lakes	2.7–5,000,000	Multiple sources, refer to Table 1.
Reservoirs > 10 ha	284,700	Lehner et al., 2011.
Salt/brackish marsh	67,580	From Schuyt and Brander, 2004 and reported by Tiner 2013.

an outcome *inter alia* of the ecological components, topographical setting, hydrological regime, and processes interacting temporally and spatially within landscapes directly or indirectly impacted by natural and anthropogenic disturbances'.

Freshwater protected areas

A specific analysis of the extent and effectiveness of freshwater PAs has not been undertaken, but has been increasingly discussed given the call under the Strategic Plan for Biodiversity (2011–2020) and Aichi Biodiversity Target 11 whereby at least 17 per cent of inland water areas are conserved through effectively and equitably managed, ecologically representative, and well-connected systems of protected areas (Juffe-Bignoli et al., 2016). This in part is due to protected area networks historically being established for terrestrial conservation with freshwater ecosystems as a part of these. This situation has been recognised by IUCN (2014) with a call for a specific focus on the coverage and management of freshwater ecosystems in their own right rather than as components of terrestrial ecosystems.

Assessing inland wetlands Juffe-Bignoli et al. (2014) report 19.6 per cent of inland freshwater area (7,940 × 10³km²: excludes coastal and brackish/saline) included in the Global Lakes and Wetlands Database (GLWD: Lehner and Döll, 2004) is within PAs; and suggests the 17 per cent area Aichi 11 target has been achieved for inland [fresh]water area (Juffe-Bignoli et al., 2016) but not for all wetland types (Juffe-Bignoli et al., 2014). The GLWD area applied in their analyses is however 55–63 per cent of the estimated permanent, seasonal and intermittent freshwater area provided above and thus likely overestimates the Aichi 11 target achievement.

Finlayson et al. (this volume) provide a summary of the number of freshwater wetlands included in the Ramsar List of Wetlands of International Importance – in October 2016 there were 1,687 (93 per cent of the total number listed) sites that included inland freshwater wetlands, covering a total area of 1.96 million km². The largest number of sites were in Europe (827 sites, 52 per cent of the total),

Figure 2.3 Freshwater wetlands listed as internationally important comprise 13.6 to 15.6 per cent of the total freshwater wetlands globally, and include: a) Koeledo National Park, India; b) Macquarie Marshes Nature Reserve, Australia; c) Okavango delta, Botswana; and d) Pantanal Matogrossense, Brazil (photographs © C.M. Finlayson).

although these comprised only 12 per cent (0.24 million km²) of the total area globally. The largest area was in Africa (0.90 million km², 46 per cent of the total area) from a much smaller number of sites (295 sites, 17 per cent of the total). The number and area of sites reflects the geopolitical differences between regions, with many smaller wetland sites remaining in most European countries. A smaller proportion of the Ramsar sites are exclusively inland freshwater and cover an area of 1.0 million km² (Pittock et al., 2015). Based on the area of freshwater wetlands presented above these sites (if predominantly freshwater) maximally represent 13.6 to 15.6 per cent of the total freshwater wetlands globally (Figure 2.3).

This value is not far below the 17 per cent area being sought under the Aichi Biodiversity Target, although the global wetland areas have a variety of uncertainties, errors and inconsistencies, and may underestimate the global area of permanent and seasonal freshwater wetlands.

Beyond showing the uneven distribution of Ramsar sites globally the representativeness and connectedness of these areas, as also required under the Aichi Target, have not been assessed. The information provided in the Ramsar sites database does not include specific areas for the freshwater wetland types found in each listed site. While it is possible to determine which sites have management plans it is not possible to determine the effectiveness of the management at each site using the data available from the Convention. However, as only 57 per cent of the European sites and 28 per cent of the African sites have management plans (Finlayson et al., 2017a) there is no certainty whether the network of sites is being effectively managed even if individual sites with a plan were managed effectively.

Although the Ramsar Convention has a database containing information on internationally important wetlands it cannot be used to readily determine the area covered only by wetland or to collate an overview of area by wetland type as a contribution to assessing the contribution to achieving the Aichi Biodiversity 11 Target for freshwater extent and representivity. An accurate and comprehensive inventory and global map is furthermore required to assess how well inland freshwater ecosystems are covered by protected areas. These seem remarkable gaps left by the parties that have signed the Ramsar Convention for the resource they have undertaken to maintain.

References

Bartholomé, E. and Belward, A.S. (2005) 'GLC2000: A new approach to global land cover mapping from earth observation data', *International Journal of Remote Sensing*, vol 26, pp1959–1977.

Burgis, M.J. and Symoens, J.J. (eds) (1987) *African wetlands and shallow water bodies*, ORSTOM, Paris.

Davidson, N.C. (2014) 'How much wetland has the world lost? Long-term and recent trends in global wetland area', *Marine and Freshwater Research*, vol 65, pp934–941.

Dixon, M.J.R., Loh, J., Davidson, N.C., Beltrame, C., Freeman, R. and Walpole, M. (2016) 'Tracking global change in ecosystem area: The Wetland Extent Trends index', *Biological Conservation*, vol 193, pp27–35.

Downing, J.A., Cole, J.J., Duarte, C.M., Middelburg, J.J., Melack, J.M., Prairie, Y.T., Kortelainen, P., Striegl, R.G., McDowell, W.H. and Tranvik, L.J. (2012) 'Global abundance and size distribution of streams and rivers', *Inland Waters*, vol 2, pp229–236.

Downing, J.A., Prairie, Y.T., Cole, J.J., Duarte, C.M., Tranvik, L.J., Striegel, R.G., McDowell, W.H., Kortelainen, P., Caraco, N.F., Melack, J.M. and Middelburg, J.J. (2006) 'The global abundance and size distribution of lakes, ponds, and impoundments', *Limnology and Oceanography*, vol 51, pp2388–2397.

Finlayson, C.M., Davidson, N.C., Spiers, A.G. and Stevenson, N.J. (1999) 'Global wetland inventory – status and priorities', *Marine and Freshwater Research*, vol 50, pp717–727.

Finlayson, C.M., Milton, G.R. and Prentice, R.C. (2016) 'Wetland Types and Distribution', in Finlayson, C.M., Milton, G.R., Prentice, R.C. and Davidson, N.C. (eds) *The Wetland Book II: Distribution, Description and Conservation*, Springer, Dordrecht (in press), available online: doi 10.1007/978-94-007-6173-5_186-1.

Finlayson, C.M., Milton, G.R., Prentice, R.C. and Davidson, N.C. (eds) (2017a) *The Wetland Book II: Distribution, Description and Conservation*, Springer, Dordrecht (in press).

Finlayson, C.M. and van der Valk, A.G. (eds) (1995a) 'Classification and Inventory of the World's Wetlands', *Advances in Vegetation Science 16*, Kluwer Academic Publishers, Dordrecht.

Finlayson, C.M. and van der Valk, A.G. (1995b) 'Wetlands classification and inventory: A summary', *Vegetatio*, vol 118, pp185–192.

Fluet-Chouinard, E., Lehner, B., Rebelo, L.-M., Papa, F. and Hamilton, S.K. (2015) 'Development of a global inundation map at high spatial resolution from topographic downscaling of coarse-scale remote sensing data', *Remote Sensing of Environment*, vol 158, pp348–361.

Gerbeaux, P., Finlayson, C.M. and van Dam, A.A. (2017). 'Wetland classification, overview', in Finlayson, C.M., Everard, M., Irvine, K., McInnes, R.J., Middleton, B.A., van Dam, A.A. and Davidson, N.C. (eds) *The Wetland Book I: Structure and Function, Management and Methods*, Springer, Dordrecht (in press), available online: doi 10.1007/978-94-007-6172-8_329-1.

Giessen, W. (2016) 'Tropical Freshwater Swamps (Mineral Soils)', in Finlayson, C.M., Milton, G.R., Prentice, R.C. and Davidson, N.C. (eds) *The Wetland Book II: Distribution, Description and Conservation*, Springer, Dordrecht.

Hughes, R.H. and Hughes, S. (1992) *A Directory of African Wetlands*, UNEP, Nairobi (Kenya), Joint publication with IUCN, Gland (Switzerland)/WCMC, Cambridge (UK).

International Rice Research Institute (IRRI) World Rice Statistic Online Query Facility. http://ricestat.irri.org:8080/wrsv3/entrypoint.htm [Accessed: 10 December 2016].

IUCN (2014) *The promise of Sydney: innovative approaches for change. A strategy of innovative approaches and recommendation to reach conservation goals in the next decade*. Available at: worldparkscongress.org/downloads/approaches/Stream1.pdf.

Joosten, H. (2004) *The IMCG Global Peatland Database*, www.imcg.net/gpd/gpd.htm [Accessed: 7 June 2016].

Joosten, H. (2010) *The Global Peatland CO$_2$ Picture: Peatland status and drainage related emissions in all countries of the world*, Wetlands International, Ede (Netherlands).

Juffe-Bignoli, D., Burgess, N.D., Bingham, H., Belle, E.M., de Lima, M.G., Deguignet, M., Bertzky, B., Milam, A.N., Martinez-Lopez, J., Lewis, E., Eassom, A., Wicander, S., Geldmann, J., van Soesbergen, A., Arnell, A.P., O'Connor, B., Park, S., Shi, Y.N., Danks, F.S., MacSharry, B. and Kingston, N. (2014) *Protected Planet Report 2014*, UNEP-WCMC, Cambridge, UK.

Juffe-Bignoli, D., Harrison, I., Butchart, S.H.M., Flitcroft, R., Hermoso, V., Jonas, H., Lukasiewicz, A., Thieme, M., Turak, E., Bingham, H., Dalton, J., Darwall, W., Deguignet, M., Dudley, N., Gardner, R., Higgins, J., Kumar, R., Linke, S., Milton, G.R., Pittock, J., Smith, K.G. and van Soesbergen, A. (2016) 'Achieving Aichi Biodiversity Target 11 to improve the performance of protected areas and conserve freshwater biodiversity', *Aquatic Conservation: Marine and Freshwater Ecosystems*, vol 26 (Suppl. 1), pp133–151.

Junk, W.J., Brown, M., Campbell, I.C., Finlayson, C.M., Gopal, B., Ramberg, L. and Warner, B.G. (2006) 'Comparative biodiversity of large wetlands: a synthesis', *Aquatic Sciences*, vol 68, pp400–414.

Junk, W.J., Shuqing, A., Finlayson, C.M., Gopal, B., Kvĕt, J., Mitchell, S.A., Mitsch, W.J. and Robarts, R.D. (2013) 'Current state of knowledge regarding the world's wetlands and their future under global climate change: a synthesis', *Aquatic Sciences*, vol 75, pp151–167.

Laybourn-Parry, J. and Wadham, J.L. (2014) *Antarctic Lakes*, Oxford University Press, New York.

Lehner, B. and Döll, P. (2004) 'Development and validation of a global database of lakes, reservoirs and wetlands', *Journal of Hydrology*, vol 296, pp1–22.

Lehner, B., Liermann, C.R., Revenga, C., Vörösmarty, C., Fekete, B., Crouzet, P., Döll, P., Endejan, M., Magome, J., Nilsson, C., Robertson, J.D., Rödel, R., Sindorf, N. and Wisser, D. (2011) 'High-resolution mapping of the world's reservoirs and dams for sustainable river-flow management', *Frontiers in Ecology and the Environment*, vol 9, pp494–502.

Lewis Jr., W.M. (2011) 'Global primary production of lates: 19th Baldi Memorial Lecture', *Inland Waters*, vol 1, pp1–28.

Lukacs, G.P. and Finlayson, C.M. (2010) 'An evaluation of ecological information on Australia's northern tropical rivers and wetlands', *Wetlands Ecology and Management*, vol 18, pp597–625.

McDonald, C.P., Rover, J.A., Stets, E.G. and Striegl, R.G. (2012) 'The regional abundance and size distribution of lakes and reservoirs in the United States and implications for estimates of global lake extent', *Limnology and Oceanography*, vol 57, pp597–606.

Messager, M.L., Lehner, B., Grill, G., Nedava, I. and Schemitt, O. (2016) 'How much water resides in lakes? Estimating the abundance and age of global lake volume', *Nature Communications*, vol 7, pp13603, DOI: 10.1038/nscomms13608 | www.nature.com/naturecommunications.

Milton, G.R, Prentice, R.C. and Finlayson, C.M. (2016) 'Wetlands of the world', in Finlayson, C.M., Milton, G.R., Prentice, R.C. and Davidson, N.C. (eds) *The Wetland Book II: Distribution, Description and Conservation*, Springer, Dordrecht.

Mitsch, W.J. (ed) (1994) *Global Wetlands: Old World and New*. Elsevier, Amsterdam, The Netherlands.

Mitsch, W.J. and Gosselink, J.G. (2015) *Wetlands 5th Edition*, John Wiley & Sons, Inc., Hoboken, New Jersey.

Page, S.E., Rieley, J.O. and Banks, C.J. (2011) 'Global and regional importance of the tropical peatland carbon pool', *Global Change Biology*, vol 17, pp798–818.

Pittock, J., Finlayson, M., Arthington, A.H., Roux, D., Matthews, J.H., Biggs, H., Harrison, I., Blom, E., Flitcroft, R., Froend, R. and Hermoso, V. (2015) 'Managing freshwater, river, wetland and estuarine protected areas', in Worboys, G.L., Lockwood, M., Kothari, A., Ferry, S. and Pulsford, I. (eds) *Protected Area Governance and Management*, ANU Press, Canberra, Australia, pp569–608.

Ramsar Convention Secretariat (2010) 'Designating Ramsar Sites: Strategic framework and guidelines for the future development of the List of Wetlands of International Importance', *Ramsar Handbooks for the Wise Use of Wetlands*, 4th edition, vol. 17, Ramsar Convention Secretariat, Gland, Switzerland.

Raymond, P.A., Hartmann, J., Lauerwald, R., Sobek, S., McDonald, C., Hoover, M., Butman, D., Striegl, R., Mayorga, E., Humborg, C., Kortelainen, P., Dürr, H.,

Meybeck, M., Cias, P. and Guth, P. (2013) 'Global carbon dioxide emissions from inland waters', *Nature*, vol 503(7476), pp355–359.

Rebelo, L.-M., Finlayson, C.M. and Nagabhatla, N. (2009) 'Remote sensing and GIS for wetland inventory, mapping and change analysis', *Journal of Environmental Management*, vol 90, pp2144–2153, doi:10.1016/j.jenvman.2007.06.027.

Schumann, M. and Joosten, H. (2008) *Global Peatland Restoration Manual*, Institute of Botany and Landscape Ecology, Greifswald University, Greifswald, (DE), www.imcg. net/media/download_gallery/books/gprm_01.pdf [Accessed: 10 June 2016].

Schuyt, K. and Brander, L. (2004) *The Economic Values of the World's Wetlands*, World Wildlife Fund, Gland, Switzerland, joint publication with Institute for Environmental Studies, Vrije Universiteit, Amsterdam (The Netherlands).

Scott, D.A. (1989) *A Directory of Asian Wetlands*, IUCN, Gland, Switzerland and Cambridge, UK.

Scott, D.A. (ed.) (1993) *A Directory of Wetlands in Oceania*, IUCN, Gland and AWB, Kuala Lumpur.

Scott, D.A. and Carbonell, M. (1986) *A Directory of Neotropical Wetlands*, IUCN, Gland, Switzerland.

Siegert, M., Carter, S., Tobacco, I., Popov, S. and Blankenship, D.D. (2005) 'A revised inventory of Antarctic subglacial lakes', *Antarctic Science*, vol 17 (3), pp453–460.

Siegert, M.J., Ross, R. and Le Brocq, A.M. (2016) 'Recent advances in understanding Antarctic subglacial lakes and hydrology', *Philosophical Transactions Royal Society A*, 374(2059), pp20140306 (doi: 10.1098/rsta.2014.0306).

Tiner, R.W. (2013) *Tidal Wetlands Primer: An Introduction to Their Ecology, Natural History, Status, and Conservation*, University of Massachusetts Press, Amherst, MA.

Usbank, S. and James, R. (1993) *A Directory of Important Wetlands in Australia*, Australian Nature Conservation Agency, Canberra (AU), p687.

Verpoorter, C., Kutser, T. and Tranvik, L. (2014) 'Automated mapping of water bodies using Landsat multispectral data', *Limnology and Oceanography Methods*, vol 10, pp1037–1050.

Chapter 3

Freshwater ecological principles

A.H. Arthington, C.M. Finlayson and J. Pittock

Key messages

- Five high level ecological principles common to all freshwater ecosystems are presented. Although they have different ramifications for each ecosystem type these principles are fundamental to the design and management of all freshwater Protected Areas (PAs) and the conservation of aquatic biodiversity.
- A universal guiding principle for PA management is that the entire catchment with its land, water, biogeochemical resources and processes is the ideal unit to be protected and managed. Where full protection is not possible, the catchment needs to be managed in a sustainable way that minimises threats and impacts from non-reserved lands. Protected area management that fails to recognise and address the threats and pressures arising in the catchment risks loss of environmental quality, species diversity and ecological resilience.
- The flow of water is one of five dynamic environmental regimes that regulate much of the structure and functioning of every running water ecosystem and many aspects of lentic and groundwater systems. The naturally dynamic flow regime plays a critical role in sustaining native biodiversity and ecosystem integrity in streams and rivers. Likewise the characteristics and variability of lentic (lakes and estuaries) and subsurface water regimes are critical to their dynamics, management and conservation.
- The spatial and temporal connectivity patterns and processes of aquatic ecosystems in their natural state are important elements for consideration in PA design and management. Connectivity in rivers is defined in three spatial dimensions: longitudinal (upstream–downstream), lateral (interactions between channel and riparian/floodplain systems), and vertical (connections between the surface and groundwater systems), with

temporal dynamics influencing all spatial dimensions of connectivity. Thus minimising the impacts of dams and levee banks as barriers, and changes to water flows, is crucial for conservation. The hydrological connectivity between lakes, streams, estuaries and subsurface environments also requires special attention in PA design and management, and forms a central pillar of Integrated Lake Basin Management (ILBM).

- A primary goal of biodiversity conservation is to delineate PAs that conserve species-rich habitats and vital resources, important species radiations and the greatest number of threatened endemic species. Significant interbasin differences in biodiversity and levels of endemism mean a lack of 'substitutability' among freshwater habitat units, adding to the complexity of freshwater biodiversity conservation. The tools of systematic conservation planning lend themselves to identification of the most beneficial options for biodiversity protection.

- Freshwater species have long histories of exposure and adaptation to variable environmental conditions and extremes (e.g., drought and flood cycles), conferring resistance and resilience at the individual, community and ecosystems levels. Maintaining catchment integrity, natural flow and standing water regimes, the spatial and temporal dimensions of connectivity, and native biodiversity hotspots will help to maintain the ecological resilience of aquatic systems in protected areas, and support societal adaptations to shifting environmental and climatic regimes.

Characteristics of freshwater ecosystems

Freshwater ecosystems, also often referred to as freshwater wetlands, encompass many different types, geomorphic forms and sizes and have been defined and classified in different ways, although not all efforts are systematic or readily comparable (see overviews in Milton and Finlayson, Chapter 2, this volume; Gerbeaux et al., 2017). They develop where water that is fresh is the primary factor controlling plant and animal life and the wider environment, where the water table is at or near the land surface, or where water covers the land. While there is uncertainty about the areal extent of surface freshwater wetlands, global estimates suggest that they contain only a small portion of the world's water and surface area, with Gleick's (1996) estimates being about 0.01 per cent and 0.8 per cent, respectively.

Wetlands arise from the atmospheric and precipitation component of the global water cycle that percolates into soils and superficial groundwater, or runs off the land into rivers and other aquatic systems. Flowing or 'lotic' surface waters – streams and rivers – are prominent directional features of most landscapes, even deserts. They exert a significant influence on landscape form and function through erosion, transport and deposition of materials from the mountains to valleys, to inland wetlands and lakes, and to estuaries and the oceans. Lakes, as well as ponds, swamps and estuaries, are 'lentic' water systems that pool or store freshwater. These ecosystems are often linked hydrologically to upstream rivers or tributaries flowing into them, to downstream water systems into which they discharge, and often to sub-surface groundwater. Peatlands are areas of land with a naturally accumulated layer of peat formed under a wide range of vegetation types including lowland or upland fens, reed beds, wet woodland, bogs, and mangroves. Rivers and springs, lakes, wetlands, peatlands and estuaries may be partly or fully dependent on the surface expression of groundwater, for example as river baseflow and the water in window lakes and desert springs. The hydrologic linkages between these lentic and lotic ecosystems regulate physical, biogeochemical and ecological connectivity, and have important implications for the design and management of PAs.

Five high level ecological principles relevant to all freshwater ecosystems are proposed and presented here. In brief, they relate to i) the intimate links between freshwater ecosystems and the surrounding catchment (basin or watershed); ii) the importance of the water regime (standing, flowing or sub-surface); iii) the hydrological, biogeochemical and ecological connectivity within and between freshwater (and estuarine) ecosystems; iv) protecting evolved patterns and 'hot spots' of native aquatic biodiversity and endemism; and v) the importance of these principles to the maintenance of ecological resilience in the aquatic ecosystems of the present and the future. Although these principles have different ramifications for each ecosystem type, they are fundamental to the design and management of all freshwater PAs and the conservation of aquatic biodiversity.

Ecological principles for freshwater ecosystems

Catchment characteristics and processes

Freshwater ecosystems are deeply influenced by the characteristics of the catchment (basin or watershed) in which they are situated and from which most receive their water. 'In every respect, the valley rules the stream' (Hynes, 1975). Their low position in landscape valleys means that rivers, lakes, groundwater dependent ecosystems and estuaries are the recipients of water, sediments, nutrients and organic matter generated from their catchments, as well as wastes and pollutants carried in runoff (Baron et al., 2002). Climate, geology, geomorphology and vegetation significantly influence these natural delivery processes. However, the natural environmental regimes of catchments are vulnerable to disturbance arising from human activities, such as deforestation and logging, livestock grazing, cropping, salinization, industrial developments, transport infrastructure and urbanization (Figure 3.1; Allan, 2004; Pittock et al., Chapters 6 and 9, this volume). Furthermore, the disturbances wrought by catchment landuse, regulated river flows and degraded water quality often propagate downstream into estuaries, deltas and coastal habitats, affecting sediment and nutrient dynamics, estuarine and near shore salinity, habitat diversity, biodiversity and fishery stocks (Adams, 2013).

Figure 3.1 Land use practices within catchments have an important influence on the management of freshwater ecosystems (photograph © C.M. Finlayson).

A universal guiding principle for PA management is that the entire catchment with its land, water, biogeochemical resources and processes is the ideal unit to be protected and managed (Finlayson et al., Chapter 12, this volume). Management plans targeting the catchment scale can offer opportunities for protected-area managers to favourably influence stakeholders and neighbouring land use, as well as promoting resilient ecosystems and ecosystem services (Flitcroft et al., Chapter 10, this volume). Although catchment scale management is widely advocated and clearly beneficial, gaining management control over relatively large areas of land in order to protect relatively small waterbodies, or the entire length of connected waterways (including standing waters, groundwater and estuaries), is complex and challenging in practice (Dudgeon et al., 2006). Alternative principles for freshwater protected area design and catchment management have been proposed, including identification of multiple-use zones, use of vegetated buffer strips, attention to ecological flow requirements, eradication of alien species (Saunders et al., 2002), the concepts of freshwater focal area, critical management zone and catchment management zone (Abell et al., 2007), and maintaining intact tributaries to achieve river conservation targets since tributaries are generally less regulated than main rivers (Nel et al., 2007).

Although difficult, catchment scale management can be achieved, such as for Kakadu National Park in northern Australia, and even for large downstream protected areas such as coastal waters and reefs. In Australia, Reef Plan (2013) aims to reduce damage from land-based agricultural practices on the Great Barrier Reef by managing land use in upland catchments to reduce the input of sediment and nutrients into freshwater wetlands, rivers and the far distant waters of the reef ecosystem (Brodie, 2014; Pearson et al., 2013). Other examples of successful catchment-scale planning have been documented in the USA (Flitcroft et al., 2009; Margerum, 2012) and Europe (Warner et al., 2012).

An ideal solution is management that integrates protection of catchment and freshwater environments (including estuaries) and connects PAs in these eco-logical realms, generating benefits over and above those achieved by individual terrestrial and freshwater PAs (Beger et al., 2010). This approach would have the advantage of protecting important parts of the catchment, the riparian zone and the floodplain as well as the biodiversity and habitats within river chan-nels and the processes that link these systems (Dudgeon et al., 2006; Pusey and Arthington, 2003). Further downstream, estuaries should be managed as part of comprehensive plans for catchments, rivers and coastal habitats, rather than as isolated systems (van Niekerk and Turpie, 2012). As an example, removal of levees in the Salmon River estuary, Oregon, expanded rearing habitats for juve-nile salmonids and restored pathways connecting freshwater to saltmarsh habitat, enhancing population resilience at the catchment scale (Flitcroft et al., 2016).

Protected area management that fails to recognise and address the threats and pressures arising in the catchment risks loss of environmental quality, biodiversity and ecological resilience, as has been highlighted with freshwater PAs along the Murray-Darling River in south-eastern Australia (Pittock and Finlayson, 2011) and in Kruger National Park, South Africa, where upper tributaries of the park's rivers flow through developed catchments (Roux et al., 2008).

Water regimes and eco-hydrological principles

The flow of water is one of five dynamic environmental regimes that regulate much of the structure and functioning of every stream and river ecosystem and many aspects of lakes, estuaries and groundwater systems – water, sedi-ments, chemicals and nutrients, organic matter, light and temperature (Baron et al., 2002). A river's flow regime defines the rates and pathways by which precipitation enters and circulates within river channels, wetlands, floodplains and connected groundwater systems, and the residence time of water in the surface and groundwater compartments of the fluvial hydrosphere. Ecologists recognise five ecologically relevant characteristics of natural stream and river flow regimes – the magnitude, frequency, timing, duration, rate of change of hydrological conditions, and the predictability of these flow patterns (Richter et al., 1996; Poff et al., 1997). Taken together, the five facets of a river's flow regime provide a summary of the hydrological character of the system, as well as details of individual flow events and temporal patterns of profound ecological significance: for example, seasonal patterns of discharge (e.g., summer or winter floods, timing of dry spells), whether the flow is ephemeral (flowing briefly only after storm events), intermittent (flowing after storm events and during wet seasons when fed by groundwater) or perennial (with year round flows), the seasonal (or intermittent) timing, extent and duration of floods and dry periods, and the overall variability and predictability of flows over days, months and years (Arthington, 2012).

Flow has been called the 'master variable' or the 'maestro . . . that orchestrates pattern and process in rivers' (Walker et al., 1995). The naturally dynamic flow regime plays a critical role in sustaining native biodiversity and ecosystem integrity in streams and rivers, and this 'natural flows paradigm' underpins universal principles of river conservation, restoration and environmental flow management (Poff et al., 1997; Bunn and Arthington, 2002). These issues have been widely explored for the Amazon River and its tributaries where natural pulse flows are seen as critical for sustaining the high biodiversity and productivity of the floodplains (Junk, 2000). Likewise the characteristics and variability of lakes and subsurface water regimes are critical to their dynamics, management and conservation (Arthington et al., Chapter 8, this volume).

Bunn and Arthington (2002) proposed four guiding eco-hydrological principles driving freshwater biodiversity in unaltered rivers and their floodplains, and illustrated how altered flow regimes can affect aquatic biodiversity.

1   Flow is a major determinant of physical habitat in streams, which in turn is a major determinant of biotic composition. Modified flow regimes alter habitat at varying spatial scales and influence the distribution and abundance of species and the composition and diversity of aquatic communities. Bunn and Arthington (2002) record numerous examples of biotic responses to altered flow regimes in relation to flow induced changes in habitat. Dams managed to generate hydroelectricity can be particularly damaging by severely altering downstream flows to the extent of dewatering entire river reaches and aquatic habitats, or by imposing daily water level fluctuations that strand aquatic species or disrupt fish spawning and larval emergence (Renöfält et al., 2010).

2   Aquatic species have evolved life history strategies primarily in direct response to the natural flow or water regime. Flow pattern and water temperature have a major influence on shaping the life history strategies of aquatic species and alteration of flow regimes can lead to recruitment failure, such as for salmon in North American streams (Lenhardt et al., 2006) and sturgeon in Europe and Russia (Safford and Norman, 2011). Cold-water releases from the deep hypolimnion of impoundments can lead to reductions in spring–summer temperatures downstream and cause significant population declines or local extinctions of native fishes, such as those of the Colorado River, USA (Olden and Naiman, 2010). The management of environmental flow releases from stratified impoundments must give consideration to downstream water temperatures in tandem with flow patterns (Rolls et al., 2013).

3   Maintenance of natural patterns of longitudinal and lateral connectivity is essential to the vitality of populations of many riverine species. Loss of longitudinal and lateral connectivity through construction of barriers (Figure 3.2) can lead to isolation of populations, failed recruitment and local extinction of fish and other aquatic biota. Many threatened diadromous species, such as salmonids, eels (*Anguilla* spp.), striped bass, shads (*Alosa* spp.), river sharks (*Glyphis* spp.), largetooth sawfish (*Pristis pristis*), and sturgeons (*Acipenser* spp.), are prevented from migrating by the longitudinal fragmentation of rivers by dams (Arthington et al., 2016). Large hydroelectric dams proposed or planned for the Andean tributaries of the Amazon River would cause the first major break in connectivity between protected Andean headwaters and the lowland Amazon (Finer and Jenkins, 2012). In the lateral dimension, the regular flood pulses of tropical rivers create hydrological connections with backwater habitats and the floodplain, generating food-rich habitats used for rearing of young fish and adult growth. Numerous studies highlight how loss of these connections can cause declines in fish diversity along rivers and reduced floodplain fisheries production (Junk, 2000; Welcomme et al., 2006).

4   The invasion and success of alien and translocated species in rivers is facilitated by the alteration of flow regimes and IBTs. Flow regulation and the creation of large human-made lakes can affect the establishment, spread and persistence of introduced species. Examples include the establishment of large mats of the floating fern, *Salvinia molesta*, in human-made lakes, such as Lake Kariba in southern Africa (Mitchell and Tur, 1975). Introduced species of *Gambusia* (Poeciliidae) have established in many regulated rivers and wetlands (often in association with alien vegetation) where their competitive predatory behaviour can seriously depress native fish and invertebrate populations (Pyke, 2008). Inter-basin transfers of water (IBTs) and artificial connectivity can act as a major mechanism for the transfer of alien and pest species between catchments. The movement of species through the Volga-Baltic Waterway connecting the Caspian basin with the Baltic region, is one example that has enabled translocation of many aquatic species, as well as the Main-Danube Canal that has allowed many Ponto-Caspian invertebrate species to reach the Rhine basin and from there to disperse further (Finlayson and D'Cruz, 2005). Lake trout (*Salvelinus namaycush* declined in the upper Great Lakes after the establishment of the parasitic sea lamprey (*Petromyzon marinus*) introduced originally via the Welland Canal between the Welland and Niagara rivers. The zebra mussel (*Dreissena polymorpha*), a prohibited invasive species native to Eastern Europe and Western Russia, was transported to the Great Lakes in the ballast water of ships: their filter-feeding behaviour causes 'a withering of the pelagic food web and a flourishing of the littoral food web' (Strayer, 2010). By forming large colonies attached to native American unionid clams this tiny mussel has caused their near extinction in Lake St. Clair and the western basin of Lake Erie (Schloesser and Nalepa, 1994).

Figure 3.2 Construction of in-stream barriers such as weirs and dams has had major impacts on the movement of aquatic species (photograph © C.M. Finlayson).

The four eco-hydrological principles outlined above with examples of their implications for river management have provided the foundations for holistic environmental flow assessment methods designed to achieve river conservation or restoration through provision of water allocations and other restoration techniques (Arthington, 2012).

Connectivity of freshwater ecosystems

Connectivity in rivers is defined in three spatial dimensions: longitudinal (upstream–downstream), lateral (interactions between channel and riparian/floodplain systems), and vertical (connections between the surface and groundwater systems), with temporal dynamics influencing all spatial dimensions of connectivity (Ward and Stanford, 1989). Population viability of many aquatic species depends on their ability to move freely through the stream network to access habitat, food resources, mates and spawning grounds (Bunn and Arthington, 2002). Aquatic species from shrimps to fish and river dolphins use different habitats at different life history stages, and longitudinal migrations may be an obligatory component of life histories, particularly when migration is associated with breeding (Welcomme et al., 2006). Dudgeon (2010) provides a summary of the impact of dams, along with pollution, on the migration of sturgeon and paddlefish in the Yangtze River with a pessimistic view for the future of many species in this large river. For example, dam construction, reduced access to large tributary lakes, boat collisions and overfishing have been implicated in the decline of the Yangtze River dolphin or baiji, *Lipotes vexillifer*, now declared extinct (Turvey et al., 2007). Diadromous fishes migrate across the freshwater–marine interface, thus connectivity pathways between these aquatic realms must be recognized and protected from fragmentation by floodplain levees, degraded habitat and upstream dams (Arthington et al., 2016; Flitcroft et al., 2016).

In the lateral dimension, the regular flood pulses of tropical rivers create hydrological connections with backwater habitats and the floodplain, generating food-rich habitats used for rearing of young fish and adult growth (Junk, 2000). Lateral exchanges with the river continue when flood levels fall and water, nutrients, sediments and aquatic species are returned to the channel and its aquatic habitats (Welcomme et al., 2006). Closer to rivers, the riparian zone performs a suite of vital physical and ecological functions, including food and habitat for many species, the regulation of the light environment of flowing and standing water, water temperature, nutrient transfer and sedimentary processes such as bank erosion and deposition (Naiman et al., 2005; Davies, 2010; Capon et al., 2013). Riparian and littoral vegetation stands and root systems contribute physical structure to stream banks and beds, constrain bank erosion and shape channels and wetland aquatic habitat (Figure 3.3). Logs and other plant fragments contribute to habitat complexity and

Figure 3.3 Riparian zones provide an important buffer between many freshwater systems and surrounding land uses, yet have been degraded or totally removed in many instances (photograph © C.M. Finlayson).

influence water flows and channel formation. In forested headwater catchments, riparian inputs (leaves, flowers, fruits) provide energy to aquatic food webs through biological processing and decay (Vannote et al., 1980). Large woody debris (LWD) derived from riparian trees and shrubs provides submerged logs and leaf packs where invertebrates and fish find refuge from thermal extremes, protection from predators and safe spawning sites (Pusey and Arthington, 2003). Fish and aquatic invertebrates often decline after riparian clearing and removal of LWD and increase locally in response to LWD additions (Lyon et al., 2009). Salmon enhancement programs over much of the western United States and Canada are replacing wood by direct reintroduction or by leaving large trees growing alongside streams for future supply of timber to waterways.

Many terrestrial and semi-aquatic animals and birds feed and breed on flood-plains and in the riparian zone, and depend on access to water during dry periods (Dudgeon et al., 2006). In recognition of the seasonal habitat use and activities of terrestrial, riparian and amphibiotic fauna (frogs, water dragons and snakes, platypus, otters and many water birds) their water needs are often included in holistic environmental flow assessments (e.g., King et al., 2003).

Surface–groundwater connectivity is the invisible dimension linking aquatic ecosystems. Lakes, rivers, riparian zones, alluvial floodplains and estuaries have various dependencies on groundwater and on water derived from the unsaturated zone and surface runoff from land (Arthington et al., Chapter 8, this volume). The longitudinal, lateral and vertical dimensions of groundwater systems associated with rivers and other GDEs are critical to their functionality. Connectivity with groundwater influences water chemistry, thermal properties, biological composition of subsurface (hyporheic) and surface water communities, and ecological processes therein (Boulton and Hancock, 2006). Hyporheic and stygofauna communities are often surprisingly diverse. Human activities contributing to water-table decline include groundwater abstraction for irrigation, domestic water use and mining, whereas groundwater recharge can be altered by land use (deforestation, afforestation, cropping, urbanisation). Dam construction, water impoundment, river flow regulation, channelisation and the construction of drainage ditches can also interfere with groundwater tables and impact GDEs. Eamus and Froend (2006) stress that 'there is no level of groundwater extraction that will not, in the long run, result in declines of natural discharges, with consequent environmental impacts'.

The spatial and temporal connectivity patterns and processes of aquatic ecosystems in their natural state are vital elements for consideration in PA design and management, as discussed above in relation to hydrological connectivity. Large connected areas are especially important for biodiversity conservation because they allow gene flow and opportunities for adaptation to disturbances (e.g., by migration to more favourable habitats). Paradoxically, artificial connectivity, for example through inter-basin transfers (IBTs) of water, can be detrimental to biodiversity conservation. Inter-basin transfers can alter the physical and chemical features of the giving and receiving waterbody, and enable species and propagules (seeds, insect eggs, fish larvae) to become established beyond their natural distributional boundaries

(Ghassemi and White, 2007). Water transfers within a river system (intra-basin transfers) can also disrupt natural patterns and processes along the river continuum from headwaters to the middle and lower reaches of rivers. The consequences of 'unwitting experiments in genetics' (Davies and Day, 1998), when species previously separated in different catchments or tributaries are brought together, are perhaps the most difficult to predict and manage. Biodiversity decline and the loss of future evolutionary potential are likely outcomes.

Recent developments in systematic conservation planning for rivers include methods to incorporate longitudinal, lateral (river to floodplain), vertical (surface–groundwater) and temporal connectivity as well as accounting for threatening processes that may compromise biodiversity protection (Hermoso et al., 2012; Linke et al., 2012), including climate change (Pittock et al., 2008). The hydrological connectivity between lentic and lotic environments also requires special attention in PA design and management, and forms a central pillar of the Integrated Lake Basin Management (ILBM) Platform Process (Arthington et al., Chapter 8, this volume).

The principle of connectivity can be broadened further to consider networks of rivers, wetlands and lakes that are never connected hydrologically but form landscape mosaics of aquatic habitat used by mobile water dependent taxa such as insects, turtles and waterbirds (Hermoso et al., 2012) with the latter in particular providing an example where the links can occur within (Bellio et al., 2016) and between continental landscapes (e.g., Boere and Stroud, 2006). Protecting these forms of connectivity, such as migratory bird flyways and the waterbodies needed for resting and feeding along migration routes, is an important dimension of freshwater biodiversity conservation (Finlayson et al., Chapter 12, this volume).

Aquatic biodiversity, endemism and conservation

Freshwater ecosystems support extraordinary biological diversity given that they contain a small amount only of the world's water and cover a very small area of the Earth's surface. At least 15,150 Actinopterygian fishes live in fresh water, and they constitute over 50 per cent of global fish diversity (Carrete Vega and Wiens, 2012) and 25 per cent of global vertebrate diversity (Dudgeon et al., 2006). Freshwater biodiversity in surface ecosystems represents about 2.4 per cent of total global biodiversity, compared with 15 per cent for marine ecosystems, but taking into account the smaller area involved they contain a disproportionately higher number of species (Millennium Ecosystem Assessment, 2005), even relative to more recent analyses that estimate that marine ecosystems contain 25 per cent of all species (Mora et al., 2011). Estimates for global freshwater biodiversity are likely to increase as more species are described from poorly studied habitats such as groundwater and wetlands in remote regions (Finlayson et al., Chapter 12, this volume).

The global distribution of freshwater biodiversity and the biotic communities of freshwater ecosystems reflect a long evolutionary history of adaptations to dynamic heterogeneous environments and the biological processes of colonization, succession and extinction, competition and predation, played out at multiple spatial and temporal scales. The totality of species (biodiversity) in any river or lake at a particular location can be envisaged as a function of the number of species in the regional species 'pool' and the effects of various historical and contemporary processes that selectively remove species from the pool (Smith and Powell, 1971; Jackson and Harvey, 1989; Tonn, 1990). Climate, landscape features and biotic processes act like a series of 'filters' through which species in the regional pool must pass to have a good chance of persisting at a particular locality (Poff, 1997). Species can only pass through each filter if they possess certain ecological characteristics or 'traits' like thermal or velocity tolerances, or their habitat and reproductive needs are met (Poff and Allan, 1995; Poff et al., 2006). The filter model was originally applied to lake fish assemblages (Tonn, 1990). Lake faunas are shaped by past and recent climatic regimes, barriers to dispersal (e.g., catchment divides, saltwater barriers), geomorphology, lake size and habitat diversity, water sources, thermal and chemical regimes, optical characteristics and biotic processes. The isolation of lakes by catchment divides and salt-water barriers limits gene flow and promotes local adaptations, radiations and high levels of endemism. Lake Malawi in the western African Rift Valley has 'outstanding fish diversity including 400–800 cichlid species (99 per cent endemic), a radiation of 17 deep-water clariid catfishes (*Bathyclarias* spp.) and many other endemic species' (Dudgeon et al., 2006, based on data from Thieme et al., 2005).

A primary goal of biodiversity conservation is to delineate PAs that conserve species-rich habitats and vital resources, important species radiations and the greatest number of threatened endemic species. Significant interbasin differences in biodiversity and levels of endemism may mean a lack of 'substitutability' among freshwater habitat units, adding to the complexity of freshwater biodiversity conservation (Dudgeon et al., 2006), especially in regions with many isolated species-rich aquatic habitats and ancient lineages. Protection of a few such waterbodies may rarely protect all freshwater biodiversity within a region, or even a small proportion of endemic or endangered species.

The tools of systematic conservation planning lend themselves to identification of the most beneficial options for biodiversity protection, especially in that they can simultaneously address threatening processes that may compromise success (Linke et al., 2012) such as the effects of new dams and flow regulation. For example, Winemiller et al. (2016) call for strategic planning of new hydropower dams to prevent species extinctions and basin-wide declines in fisheries and other ecosystem services in mega-diverse tropical rivers of the Mekong, Congo and Amazon basins.

Ecological resilience

Freshwater ecosystems are exposed to numerous threats and pressures for which the general parameters of ecological outcomes from individual pressures (such as water pollution, flow regulation, habitat loss and alien species) are well established (Ormerod et al., 2010; Pittock et al., Chapter 6, this volume). Although there are many scientific challenges, understanding of the interactions among individual forms of pressure is increasing steadily; for example, altered flow patterns and lowered water temperature below dams can inhibit fish spawning (Rolls et al., 2013). However, new pressures are emerging, such as nanoparticles and different categories of chemicals (endocrines, etc), producing novel combinations of pressures and unforeseen interactions among pressures. Adding to the complexity of disentangling the ecological impacts of interactions is that shifting environmental regimes, such as those associated with climate change (water availability and temperature) and alien species invasions (Figure 3.4), create new contexts for the impacts of individual and interacting pressures, and are generally expected to exacerbate ecological impacts (Kopf et al., 2015; Capon et al., 2015).

Figure 3.4 Floating invasive plants such as *Salvinia molesta* have the potential to rapidly cover areas of open water and result in ecological changes including the physio-chemical characteristics as well as impeding water flows (photograph © C.M. Finlayson).

Freshwater species have long histories of exposure and adaptation to variable environmental conditions and extremes (e.g., drought and flood cycles), conferring resistance and resilience at the individual, community and ecosystems levels. Maintaining catchment integrity, natural surface flows and standing water regimes, and the spatial and temporal dimensions of connectivity, will help to enhance the ecological resilience of aquatic systems in protected areas. Nevertheless there are limits to ecosystem capacity to return to former more natural states, and this may be particularly so where the envelope of environmental conditions is shifting in response to human demands, novel pollutants, invasive alien species and/or climate change (Strayer, 2010; Acreman et al., 2014). How to envisage, plan, configure and manage PAs under climate change is discussed in greater detail by Finlayson and Pittock (Chapter 13, this volume).

Recommendations to apply ecological principles in PAs

While recognizing that conserving whole river systems in PAs has seldom been a practical management option (Nel et al., 2007), there are ways in which the representation of freshwater ecosystems within PAs could be improved. These include four steps proposed by Nel et al. (2007) and listed here: i) giving explicit consideration to representing freshwater ecosystems in protected areas; ii) understanding the relative contribution different land makes to freshwater conservation in consolidating land around existing protected areas; iii) avoiding the use of rivers to delineate boundaries of protected areas; and iv) using alternative design and management strategies, in combination with existing PAs, to protect rivers before they enter the PA. In effect, a mix of strategies will need to be considered, including those that ensure the protection of key biodiverse freshwater ecosystems and their catchments, programs to avoid or eliminate invasive alien species, and plans that focus on particular species or habitats while reconciling biodiversity conservation with human uses of water (Dudgeon et al., 2006).

There have been many calls for scientists to more effectively communicate with decision makers at all levels of governmental and society to make certain that the values of freshwater ecosystems are recognised and effective cases are made for their conservation. These messages were central to the outcomes of the Millennium Ecosystem Assessment (2005) and given the continuing decline and loss of freshwater ecosystems (Davidson, 2014; Dixon et al., 2016) they are as important now as they were then, possibly more so. The five high level principles proposed and developed here provide a basis for assessing the values of freshwater ecosystems, and also, possibly even more importantly, provide a basis for further communication, including the development of more sophisticated messages and motivations for conservation.

The principles we have proposed speak to the relationships between people and water and the biodiversity components that support our livelihoods and cultures by providing settings for human developments, health and well-being (*sensu* Horwitz and Finlayson, 2011). There are challenges in fostering these relationships but by following the proposed principles we have the basis for a way forward from the ineffective policies of the past. While focussed on ways to conserve freshwater ecosystems and their native species within and beyond PAs, the principles also provide guidance for ensuring beneficial relationships between people, as individuals and societies, and freshwater ecosystems in all landscapes.

In brief, the principles relate to the:

- intimate links between freshwater ecosystems and the surrounding catchment (basin or watershed);
- ecological roles of the water regime (standing, flowing or sub-surface);
- hydrological, biogeochemical and ecological connectivity within and between freshwater (and estuarine) ecosystems;
- protection of evolved patterns and 'hot spots' of native aquatic biodiversity and endemism; and
- maintenance of ecological resilience in the aquatic ecosystems of the present and the future.

Although these principles have different ramifications for different freshwater ecosystems, they are fundamental to the design and management of all freshwater PAs and the conservation of aquatic biodiversity, and vital to the maintenance and strengthening of the important relations that occur worldwide between people, fresh water and ecosystems.

References

Abell, R., David Allan, J.D. and Lehner, B. (2007) 'Unlocking the potential of protected areas for freshwaters', *Biological Conservation*, vol 134, pp48–63.

Acreman, M., Arthington, A.H., Colloff, M.J., Couch, C., Crossman, N., Dyer, F., Overton, I., Pollino, C.A., Stewardson, M. and Young, W. (2014) 'Environmental flows for natural, hybrid and novel riverine ecosystems in a changing world', *Frontiers in Ecology and Environment*, vol 12, pp466–473.

Adams, J.B. (2013) 'A review of methods and frameworks used to determine the environmental water requirements of estuaries', *Hydrological Sciences Journal*, vol 59, pp451–465.

Allan, J.D. (2004) 'Landscape and riverscapes: The influence of land use on river ecosystems', *Annual Reviews of Ecology, Evolution and Systematics*, vol 35, pp257–284.

Arthington, A.H. (2012) *Environmental Flows. Saving Rivers in the Third Millennium*, University of California Press, Berkeley.

Arthington, A.H., Dulvy N.K., Gladstone W., Winfield I.J. (2016) 'Fish conservation in freshwater and marine realms: status, threats and management', *Aquatic Conservation: Marine and Freshwater Ecosystems*, vol 26, pp838–857.

Baron, J.S., Poff, N.L., Angermeier, P.L., Dahm, C.N., Gleick, P.H., Hairston, N.G., Jackson, R.B., Johnston, C.A., Richter, B.D. and Steinman, A.D. (2002) 'Meeting ecological and societal needs for freshwater', *Ecological Applications*, vol 12, pp1247–1260.

Beger, M., Grantham, H.S., Pressey, R.L., Wilson, K.A., Peterson, E.L., Dorfman, D., Mumbye, P.J., Lourival, R., Brumbaugh, D.R. and Possingham, H.P. (2010) 'Conservation planning for connectivity across marine, freshwater, and terrestrial realms', *Biological Conservation*, vol 143, pp565–575.

Bellio, M., Minton, C. and Veltheim, I. (2016) 'Challenges faced by shorebird species using the inland wetlands of the East Asian-Australasian flyway: the little curlew example', *Marine and Freshwater Research*, dx.doi.org/10.1071/MF15240.

Boere, G.C. and Stroud, D.A. (2006) 'The flyway concept: what it is and what it isn't'. In Boere, G.C., Galbraith, C.A. and Stroud, D.A. (eds), *Waterbirds around the World*, pp40–47. The Stationery Office, Edinburgh, UK.

Boulton, A.J. and Hancock, P.J. (2006) 'Rivers as groundwater-dependent ecosystems: a review of degrees of dependency, riverine processes and management implications', *Australian Journal of Botany*, vol 54, pp133–144.

Brodie, J. (2014) 'Dredging the Great Barrier Reef: Use and misuse of science', *Estuarine, Coastal and Shelf Science*, vol 142, pp1–3.

Bunn, S.E. and Arthington, A.H. (2002) 'Basic principles and ecological consequences of altered flow regimes for aquatic biodiversity', *Environmental Management*, vol 30, pp492–507.

Capon, S., Chambers, L., MacNally, R., Naiman, R., Davies, P., Marshall, N., Pittock, J., Reid, M., Capon, T., Douglas, M., Catford, J., Baldwin, D., Stewardson, M., Roberts, J., Parsons, M. and Williams, S. (2013) 'Riparian ecosystems in the 21st century: hotspots for climate change adaptation?' *Ecosystems* vol 16(3), pp359–381.

Capon, S.J., Lynch, J.J., Bond, N., Chessman, B.C., Davis, J. Davison, N., Finlayson, C.M., Gell, P.A., Hohnberg, D., Humphrey, C., Kingsford, R.T., Nielsen, D., Thomson, J.R., Ward, K. and MacNally, R. (2015) 'Regime shifts, thresholds and multiple stable states in freshwater ecosystems; a critical appraisal of the evidence', *Science of the Total Environment*, vol 534, pp122–130.

Carrete Vega, G. and Wiens, J.J. (2012) 'Why are there so few fish in the sea?' *Proceedings of the Royal Society B*, vol 279, pp2323–2329.

Davidson, N.C. (2014) 'How much wetland has the world lost? Long-term and recent trends in global wetland area', *Marine and Freshwater Research*, vol 65, pp934–941.

Davies, P.M. (2010) 'Climate change implications for river restoration in global biodiversity hotspots', *Restoration Ecology*, vol 18(3), pp261–268.

Davies, B. and Day, J. (1998) *Vanishing Waters*, University of Cape Town Press, Cape Town, South Africa.

Dixon, M.J.R., Loh, J., Davidson, N.C., Beltrame, C., Freeman, R. and Walpole, M. (2016) 'Tracking global change in ecosystem area: The Wetland Extent Trends index', *Biological Conservation*, vol 193, pp27–35.

Dudgeon, D., Arthington, A.H., Gessner, M.O, Kawabata, Z., Knowler, D., Lévêque, C., Naiman, R.J., Prieur-Richard, A.-H., Soto, D., Stiassny, M.L.J. & Sullivan C.A. (2006). Freshwater biodiversity: importance, threats, status, and conservation challenges. *Biological Reviews* 81 (2): 163-182.

Dudgeon, D. (2010) 'Requiem for a river: extinctions, climate change and the last of the Yangtze', *Aquatic Conservation: Marine and Freshwater Ecosystems*, vol 20, pp127–131.

Eamus, D. and Froend, R. (2006) 'Groundwater-dependent ecosystems: the where, what and why of GDEs', *Australian Journal of Botany*, vol 54, pp91–96.

Finer, M. and Jenkins, C.N. (2012) 'Proliferation of hydroelectric dams in the Andean Amazon and implications for Andes–Amazon connectivity', *Plos ONE*, vol 7, e35126.

Finlayson, C.M. and D'Cruz, R. (2005) 'Inland water systems'. In Hassan, R., Scholes, R., Ash, N. (eds), *Ecosystems and Human Well-being: Current State and Trends: Findings of the Condition and Trends Working Group*. Island Press, Washington, DC, pp. 551–583.

Flitcroft, R.L., Dedrick, D.C., Smith, C.L., Thieman, C.A. and Bolte, J.P. (2009) 'Social infrastructure to integrate science and practice: the experience of the Long Tom Watershed Council', *Ecology and Society*, vol 14, no 2, 36, www.ecologyandsociety.org/vol14/iss2/art36/.

Flitcroft, R.L., Bottom, D.L., Haberman, K.L., Bierly, K.F., Jones, K.K., Simenstad, C.A., Gray, A., Ellingson, K,S., Baumgartner, E., Cornwell, T.J. and Campbell, L.A. (2016) 'Expect the unexpected: place-based protections can lead to unforeseen benefits', *Aquatic Conservation: Marine and Freshwater Ecosystems*, vol 26, pp39–59.

Gerbeaux, P., Finlayson, C.M. and van Dam, A.A. (2017) 'Wetland classification, overview'. In Finlayson, C.M., Everard, M., Irvine, K., McInnes, R.J., Middleton, B.A., van Dam, A.A. and Davidson, N.C. (eds), *The Wetland Book I: Structure and Function, Management and Methods*, Springer, Dordrecht (in press).

Ghassemi, F. and White, I. (2007) Inter-Basin Water Transfer: Case Studies from Australia, United States, Canada, China and India, Cambridge University Press, Cambridge, UK.

Gleick, P.H. (1996). 'Water resources'. In Schneider, S.H. (ed.), *Encyclopedia of Climate and Weather*, Oxford University Press, New York, USA, pp817–823.

Hermoso, V., Kennard, M.J. and Linke, S. (2012) 'Integrating multidirectional connectivity requirements in systematic conservation planning for freshwater systems', *Diversity and Distributions*, vol 18, pp448–458.

Horwitz, P. and Finlayson, C.M. (2011) 'Wetlands as settings: ecosystem services and health impact assessment for wetland and water resource management', *BioScience*, vol 61, pp678–688.

Hynes, H.B.N. (1975) 'The stream and its valley', *Verhandlungen des Internationalen Verein Limnologie*, vol 19, pp1–15.

Jackson, D.A. and Harvey, H.H. (1989) 'Biogeographic associations in fish assemblages – local vs regional processes', *Ecology*, vol 70(5), pp1472–1484.

Junk, W.K. (2000) 'The central Amazon river floodplain: concepts for the sustainable use of its resources'. In Junk, W.J., Ohly, J.J., Piedade, M.T.F. and Soares, M.G.M. (eds), *The Central Amazon Floodplain: Actual use and options for a sustainable management*. Backhuys Publishers, Leiden, The Netherlands, pp75–94.

King, J., Brown, C. and Sabet, H. (2003) 'A scenario-based holistic approach to environmental flow assessments for rivers', *River Research and Applications*, vol 19, pp619–639.

Kopf, R.K., Finlayson, C.M., Humphries, P., Sims, N.C. and Hladyz, S. (2015) 'Anthropocene baselines: human-induced changes to global freshwater biodiversity restoration potential', *BioScience*, vol 65, pp798–811.

Lenhardt, M., Jaric, I., Kalauzi, A. and Cvijanovic, G. (2006) 'Assessment of extinction risk and reasons for decline in sturgeon', *Biodiversity and Conservation*, vol 15, pp1967–1976.

Linke, S., Kennard, M.J., Hermoso, V., Olden, J.D., Stein, J. and Pusey, B.J. (2012) 'Merging connectivity rules and large-scale condition assessment improves conservation adequacy in river systems', *Journal of Applied Ecology*, vol 49, pp1036–1045.

Lyon, J.P., Nicol, S.J., Lieschke, J.A. and Ramsey, D.S.L. (2009) 'Does wood type influence the colonisation of this habitat by macroinvertebrates in large lowland rivers?' *Marine and Freshwater Research*, vol 60, pp384–393.

Margerum, R. (2012) 'Integrated water resources management in the United States: The Rogue and Willamette River cases'. In Warner, J. F., Van Buuren, A. and Edelenbos J. (eds), *Making Space for the River: Governance Experiences with Multifunctional River Flood Management in the US and Europe*. IWA Publishing, London, UK.

Millennium Ecosystem Assessment (2005) *Ecosystems and Human Well-being: Wetands and Water Synthesis*. World Resources Institute, Washington, DC.

Mitchell, D.S. and Tur, N.M. (1975) 'The rate of growth of *Salvinia molesta* (*S. auriculata* Auct.) in laboratory and natural conditions', *Journal of Applied Ecology*, vol 12, pp213–225.

Mora, C., Tittensor, D.P., Adl, S., Simpson, A.G.B. and Worm, B. (2011) 'How many species are there on Earth and in the Ocean?' *PLoS Biol* 9(8), e1001127.

Naiman, R.J, Décamps, H., McClain, M.C. (2005) *Riparia*. Academic Press, San Diego, USA.

Nel, J.L., Roux, D.J., Maree, G., Kleynhans, C.J., Moolman, J., Reyers, B., Rouget, M. and Cowling, R.M. (2007) 'Rivers in peril inside and outside protected areas: a systematic approach to conservation assessment of river ecosystems', *Diversity and Distributions*, vol 13, pp341–352.

Olden, J.D. and Naiman, R.J. (2010) 'Incorporating thermal regimes into environmental flows assessments: modifying dam operations to restore freshwater ecosystem integrity', *Freshwater Biology*, vol 55, pp86–107.

Ormerod, S.J., Dobson, M., Hildrew, A.G. and Townsend, C.R. (2010) 'Multiple stressors in freshwater ecosystems', *Freshwater Biology*, vol 55, pp1–4.

Pearson R.P., Godfrey, P., Arthington, A.H., et al. (2013) 'Biophysical status of remnant lagoons on a tropical floodplain in the Great Barrier Reef catchment: a challenge for assessment and monitoring', *Marine and Freshwater Research*, vol 64(3), pp208–222.

Pittock J. and Finlayson C.M. (2011) 'Australia's Murray Darling Basin: freshwater ecosystem conservation options in an era of climate change', *Marine and Freshwater Research*, vol 62, pp232–243.

Pittock, J., Hansen, L.J. and Abell, R. (2008) 'Running dry: freshwater biodiversity, protected areas and climate change', *Tropical Conservancy Biodiversity*, vol 9, pp30–38.

Poff, N.L. (1997) 'Landscape filters and species traits: towards mechanistic understanding and prediction in stream ecology', *Journal of the North American Benthological Society*, vol 16, pp391–409.

Poff, N. L. and Allan, J.D. (1995) 'Functional organization of stream fish assemblages in relation to hydrological variability', *Ecology*, vol 76(2), pp606–627.

Poff, N.L., Allan, J.D., Bain, M.B., Karr, J.R., Prestegaard, K.L., Richter, B.D., Sparks, R.E. and Stromberg, J.C. (1997) 'The natural flow regime: a paradigm for river conservation and restoration', *Bioscience*, vol 47, pp769–784.

Poff, N.L., Olden, J.D., Vieira, N.K.M., Finn, D.S., Simmons, M.P. and Kondratieff, B.C. (2006) 'Functional trait niches of North American lotic insects: Traits-based ecological applications in light of phylogenetic relationships', *Journal of the North American Benthological Society*, vol 25(4), pp730–755.

Pusey, B.J. and Arthington, A.H. (2003) 'Importance of the riparian zone to the conservation and management of freshwater fish: a review', *Marine and Freshwater Research*, vol 54, pp1–16.

Pyke, G.H. (2008) 'Plague minnow or mosquitofish? A review of the biology and impacts of introduced Gambusia species', *Annual Review of Ecology, Evolution and Systematics*, vol 39, pp171–191.

Reef Plan (2013) www.reefplan.qld.gov.au/about.aspx.

Renofalt, B. M., Jansson, R. and Nilsson, C.: 2010. 'Effects of hydropower generation and opportunities for environmental flow management in Swedish riverine ecosystems', *Freshwater Biology*, 55, pp49–67.

Richter, B.D., Baumgartner, J.V., Powell, J. and Braun, D.P. (1996) 'A method for assessing hydrologic alteration within ecosystems', *Conservation Biology*, vol 10, pp1163–1174.

Rolls, R.J., Growns, I.O., Khan, R.A., Wilson, G.G., Ellison, T.L., Prior, A. and Waring, C.C. (2013) 'Fish recruitment in rivers with modified discharge depends on the interacting effects of flow and thermal regimes', *Freshwater Biology*, vol 58, pp1804–1819.

Roux, D.J., Nel, J.L., Ashton, P.J., Deacon, A.R., de Moor, F.C., Hardwick, D., Hill, L., Kleynhans, C.J., Maree, G.A., Moolman, J. and Scholes, R.J. (2008) 'Designing protected areas to conserve riverine biodiversity: Lessons from a hypothetical redesign of the Kruger National Park', *Biological Conservation*, vol 141, pp100–117.

Safford, T.G. and Norman, K.C. (2011) 'Water water everywhere, but not enough for salmon? Organizing integrated water and fisheries management in Puget Sound', *Journal of Environmental Management*, vol 92, pp838–847.

Saunders, D.L., Meeuwig, J.J. and Vincent, C.J. (2002) 'Freshwater protected areas: strategies for conservation', *Conservation Biology*, vol 16, pp30–41.

Schloesser, D.W. and Nalepa, T.F. (1994) 'Dramatic decline of unionid bivalves in offshore waters of western Lake Erie after infestation by the zebra mussel, Dreissena polymorpha', USGS Great Lakes Science Center.

Smith, C.L. and Powell, C.R. (1971) 'The summer fish communities of Brier Creek, Marshall County, Oklahoma', *American Museum Novitates*, vol 2458, pp1–30.

Strayer, D.L. (2010) 'Alien species in fresh waters: ecological effects, interactions with other stressors, and prospects for the future', *Freshwater Biology*, vol 55, pp152–174.

Thieme, M.L., Abell, R., Stiassny, M.L.J., Lehner, B., Skelton, P., Teugels, G., Dinerstein, E., Kamden Toham, A., Burgess, B. and Olson, D. (2005) *Freshwater Ecoregions of Africa and Madagascar. A Conservation Assessment.* Island Press, Washington, DC.

Tonn, W. M. (1990) 'Climate change and fish communities: a conceptual framework', *Transactions of the American Fisheries Society*, vol 119, pp337–352.

Turvey, S.T., Rovbert, P.L., Taylor, B.L., Barlow, J., Akamatsu, T., Barrett, L.A., Zhao, X., Reeves, R.R., Stewart, B.S., Wang, K., Wei, Z., Zhang, X., Pusser, L.T., Richlen, M., Brandon, J.R. and Wang D. (2007) 'First human-caused extinction of a cetacean species?', *Biology Letters*, vol 3, pp537–540.

van Niekerk, L. and Turpie, J.K. (eds) (2012) 'South African National Biodiversity Assessment 2011: Technical report. Volume 3: Estuary component', Council for Scientific and Industrial Research, Stellenbosch, South Africa.

Vannote, R.L., Minshall, G.W., Cummins, K.W., Sedell, J.R. and Cushing, C.E. (1980) 'The river continuum concept', *Canadian Journal of Fisheries and Aquatic Sciences*, vol 37, pp130–137.

Walker, K.F., Sheldon, F. and Puckridge, J.T. (1995) 'An ecological perspective on dryland river ecosystems', *Regulated Rivers: Research and Management*, vol 11, pp85–104.

Ward, J.V. and Stanford, J.A. (1989) 'Riverine ecosystems: the influence of man on catchment dynamics and fish ecology'. In D.P. Dodge (ed.), Proceedings of the International Large River Symposium. *Canadian Special Publication of Fisheries and Aquatic Sciences*, 106, pp56–64.

Warner, J.F., Van Buuren, A. and Edelenbos J. (eds) (2012) 'Making space for the river: Governance experiences with multifunctional river flood management in the US and Europe', IWA Publishing, London, UK.

Welcomme, R.L., Bene, C., Brown, C.A., Arthington, A., Dugan, P., King, J.M. and Sugunan, V. (2006) 'Predicting the water requirements of river fisheries'. In Verhoeven, J.T.A., Beltman, B., Bobbink, R. and Whigham, D.F. (eds), *Wetlands and Natural Resource Management*, Ecological Studies vol 190. Springer-Verlag, Berlin, Heidelberg, pp123–154.

Winemiller, K.O., McIntyre, P.B., Castello, L., Fluet-Chouinard, E., Giarrizzo, T., Nam, S., Baird, I.G., Darwall, W., Lujan, N.K., Harrison, I., et al. (2016) 'Balancing hydropower and biodiversity in the Amazon, Congo, and Mekong', *Science*, 351: pp128–129.

Chapter 4

Defining and enhancing freshwater protected areas

V. Hermoso, M. Thieme, R. Abell, S. Linke and E. Turak

Key messages

- To be effective for freshwater biodiversity, conservation efforts must consider the particularities of these systems, such as the key role of spatial–temporal connectivity for maintaining ecological processes (e.g., periodic migrations or dispersal from refuge areas, gene flow, or transport of energy and matter essential for the persistence of populations and species) and the effective propagation of threats along these systems.
- New mechanisms are arising to encourage/support public and private funding to achieve improved outcomes for biodiversity conservation in freshwater systems (e.g., payments for ecosystem services, water reserves, biodiversity offsets and system-wide planning to limit the impacts of water infrastructure on aquatic ecosystems).
- Despite the increasing and innovative efforts for implementing conservation in freshwater systems in recent decades, there remains an urgent need for improved assessment of the effectiveness of freshwater Protected Areas (PA) through tailored monitoring programs. It will also be necessary to redouble efforts to ensure that freshwater PAs are effectively implemented with appropriate management plans.

Place-based conservation in Protected Areas

PAs, as defined by the International Union for the Conservation of Nature (IUCN), are 'clearly defined geographical spaces, recognised, dedicated, and managed, through legal or other effective means, to achieve the long term conservation of nature with associated ecosystem services and cultural values' (IUCN, 2008). There is a wide variety of management categories defined by IUCN, ranging from strict protection, such as Nature Reserves or Wilderness Areas where human uses are strictly controlled, to other categories, such as places where human intervention is not only allowed, but should be fostered as a key element under sustainable development principles (Box 4.1). Freshwater PAs fall within the broad concept of PAs, although they do not represent a distinctive management category, and hereafter freshwater PA will be used to refer to PAs that were designed to protect freshwater features (biodiversity, ecological processes or freshwater resources like drinking water).

Box 4.1   IUCN's protected area management categories (source: Bower et al., 2014; UNEP-WCMC and IUCN, 2016).

Category Ia: Strict Nature Reserve. Human access is strictly controlled with the aim of conserving biodiversity and/or protecting landscape/ seascape features (e.g., The Wallaby Creek catchment and reference area within Kinglake National Park in Victoria, Australia).

Category Ib: Wilderness Area. Generally applied to areas with low to no anthropogenic impact. Human access is limited to preserve the natural condition (e.g., Craig Headwaters Protected Area, Canada).

Category II: National Park. Applied to large areas to preserve large-scale ecosystem processes and features. Human activity is permitted under regulatory guidance. May contain areas of strict protection resembling Categories 1a, 1b (e.g., Doñana National Park, Spain).

Category III: National Monument or Feature. Applied to specific locations that represent a natural or culturally significant feature or monument to preserve its state. Human traffic is regulated, but generally high volume (e.g., Fish River Canyon Conservation Area, Namibia).

Category IV: Habitat/Species Management Area. Applies to localized areas protected to promote restoration, conservation or maintenance of specific species or habitats. Human traffic is generally uncontrolled (e.g., Rubondo Island National Park, Tanzania).

Category V: Protected Landscape/Seascape. Applies to areas of land and sea with distinct scenic and cultural features where traditional land-use has played a role in maintaining system integrity. Maintenance of current human use is a goal of this category (e.g., Danube Delta Biosphere Reserve, Romania).

Category VI: Protected Area with Sustainable Use of Resources. Applies generally to large, natural areas with the aim of maintaining sustainable natural resource use and low-level industrial use. Human traffic is usually uncontrolled. "No-take" zones are recommended (e.g., Fraser Heritage River, Canada).

Since the first declaration of a National Park in the USA (i.e., Yellowstone National Park in 1872; although there are examples of former conservation efforts around springs and ponds), place-based conservation efforts via PAs have remained the cornerstone of worldwide efforts to halt and reverse the global loss of biodiversity (Gaston et al., 2008; Rands et al., 2010). The number and extent of PAs has continued to increase in the last century to a total of 202,467 terrestrial and inland water PAs recorded in the World Database on Protected Areas (WDPA), covering 19.8 million km^2 or 14.7 per cent of the ice-free terrestrial surface of the Earth (UNEP-WCMC and IUCN, 2016). This network of PAs covers 16.0 per cent of the length of rivers, either completely within PAs or forming their borders (Abell et al., 2016). However, the true extent of PAs remains unknown, as not all of the world's terrestrial and inland aquatic PAs are yet captured in the WDPA (UNEP-WCMC and IUCN, 2016). Moreover, the true conservation impact of these PAs remains unclear for several reasons: 1) monitoring of conservation outcomes of PAs is not common; 2) even when such monitoring happens it is unlikely to include comprehensive monitoring of freshwater species and ecosystems; and 3) a significant portion of these PAs has not been assigned to an IUCN management category so it is not possible to determine how much consideration is given to nature conservation (UNEP-WCMC and IUCN, 2016).

In 2010 the Parties to the Convention on Biological Diversity (CBD) adopted the Strategic Plan for Biodiversity 2010–2020 and its 20 Aichi Biodiversity Targets. This agreement continues to support place-based conservation efforts, by committing to further increase the extent of PAs worldwide to cover at least 17 per cent of all terrestrial and inland waters by 2020 (Target 11; Convention on Biological Diversity, 2010). Analyses of the effectiveness of PAs at protecting species and habitats show that, on average, they have been successful in reducing habitat loss, had positive effects on a broad range of species and have lowered the risk of extinction (UNEP-WCMC and IUCN, 2016). However, the effectiveness of existing PAs for freshwater systems and species is uncertain in most of cases (Thieme et al., 2012; Hermoso et al., 2015a; Abell et al., 2016), given that there has been little emphasis on declaring or managing PAs for the primary purpose of conserving freshwater ecosystems and biodiversity (Saunders et al., 2002; Abell et al., 2007), and that there is little information available to evaluate this effectiveness. Ramsar sites are an exception to this overall pattern as they are specifically declared to protect wetlands (not only freshwater). The list of Ramsar sites has been increasing since 1974 and as at October 2016 it included some 242 sites covering 2.15 million km^2, which represents 17 per cent of the estimated 12.1 million km^2 of global wetlands, although the area of global wetlands is itself an underestimate (Finlayson et al., Chapter 12, this volume). In order to ensure effectiveness of these sites for protecting biodiversity the Ramsar Convention has developed guidelines and a framework for managing Wetlands of International Importance and other wetland sites.

However, despite the existence of such guidance, as well as many national guidelines or regulations in support of management planning, some 49 per cent of 861 freshwater Ramsar sites did not have management plans in 2016 (Finlayson et al., Chapter 12, this volume). There is no certainty on whether or not the network of sites is being effectively managed even if individual sites with a plan were managed effectively.

In this chapter we describe the distinctive characteristics of freshwater ecosystems that must be considered when designing PAs in freshwater systems. We briefly describe the latest advances on how to integrate these ecological requirements. Finally, we review some alternative approaches to conventional PAs that are being used to implement conservation efforts in freshwater systems and that could help enhance the implementation and effectiveness of conservation in freshwater systems.

Features that define effective freshwater PAs

Protected Area Management Effectiveness (PAME) can be defined as 'the assessment of how well PAs are being managed, or to what extent management is protecting values and achieving the objectives expected for the PA' (Hockings et al., 2006). These assessments account for three main components: 1) design and planning issues; 2) appropriateness of management systems and processes; and 3) delivery of PA objectives.

There are examples of the positive effects that PAs can have on freshwater biodiversity (e.g., increased local abundance or size classes of some fish species; Cucherousset et al., 2007; Suski and Cooke, 2007; Penha et al., 2015). However, the benefits reported in these studies are normally restricted to a limited number of species. On the other hand, examples of the poor performance of PAs for freshwater biodiversity have also been documented (Mancini et al., 2005; Srinoparatwatana and Hyndes, 2011; Snyder et al., 2013). Common problems that can compromise the effectiveness of freshwater PAs are the insufficient consideration of the ecological requirements of freshwater species (e.g., Roux et al., 2008; Abell et al., 2016) when designing and declaring PAs, insufficient resources devoted to freshwater conservation management (Thieme et al., 2012), and poor understanding and capacity to address complex management problems that extend beyond the limits of the PA (Hermoso et al., 2015a). For example, freshwater PAs are commonly affected by disconnection from upstream or downstream areas imposed by dams and other infrastructure (e.g., Hermoso et al. 2015a; Thieme et al., 2016) and very often have no PAs in their upstream catchments. This occurs for 69.5 per cent of river length globally (Abell et al., 2016). Moreover, important information necessary to allow adequate decision-making for freshwater biodiversity within PAs, such as whether PAs contain fresh waters of ecological significance or if water quality is impaired, is usually limited (e.g., Thieme et al., 2012).

To be effective for freshwater conservation, PAs must consider some particularities of freshwater ecosystems which pose unique challenges to the implementation of effective conservation (Abell, 2002). Spatial–temporal connectivity plays a key role in maintaining important ecological processes, such as periodic migrations or dispersal from refuge areas, gene flow, or transport of energy and matter essential for the persistence of populations and species (Fagan, 2002). Connectivity in freshwater ecosystems is defined in four dimensions (Ward and Stanford, 1989): longitudinal (upstream–downstream), lateral (interactions between channel and riparian/floodplain systems), vertical (connections between the surface and groundwater systems), all of which are subject to temporal dynamics. All of these dimensions have been deeply modified through the construction of barriers such as dams, road crossings or culverts that affect the longitudinal movement of individuals (Pépino et al., 2012), reduce the flow of nutrients and sediments (Stanley and Doyle, 2003) and affect lateral connectivity by diminishing flood pulses (Arthington, 2012). To ensure conservation efforts in freshwater systems are effective these connectivity components need to be considered when designing and managing freshwater PAs (Arthington et al., Chapter 3, this volume).

Freshwater ecosystems are extremely dependent on the surrounding terrestrial landscape (e.g., as a source of material or vulnerability to threats originating in terrestrial systems, such as sediments or pollutants). Moreover, connectivity can facilitate the propagation of threats along river systems (Linke et al., 2011). For this reason, adequate management for conservation of freshwater biodiversity needs to go beyond the limits of the freshwater ecosystem and extend to the terrestrial realm, and sometimes even far distant areas (Finlayson et al., Chapter 12, this volume).

An additional key aspect to conservation in freshwater ecosystems is the need for water, often flowing but also standing water in lakes and connections to underground sources (Arthington et al., Chapter 8, this volume). Natural hydrological patterns define healthy freshwater ecosystems and sustain important features such as temperature, geomorphological and channel character, habitat diversity and processes such as transport of nutrients – keys for the maintenance of freshwater biodiversity (Arthington, 2012). However, as for connectivity, natural flows have been deeply transformed by human intervention: modification of the quantity, timing and variability of flows due to river regulation by dams, water abstraction and overexploitation of aquifers (Bunn and Arthington, 2002).

Incorporating ecological criteria when designing freshwater PAs

The last few years have witnessed a surge of novel ways to integrate these special needs of freshwater systems into well-established systematic conservation planning methods previously developed and widely applied in marine and terrestrial environments (Linke et al., 2011; Collier, 2011). Systematic conservation planning (Margules and Pressey, 2000) aims to inform decision-makers on how to achieve conservation goals (e.g., representing all species in a certain area within the set of priority PAs) in the most effective and efficient way. This is achieved by explicitly defining conservation objectives and integrating socio-economic (e.g., conservation costs and benefits) and ecological (e.g., connectivity) aspects when seeking optimal allocation of priority areas for conservation. Different approaches have been proposed to address the ecological requirements highlighted in the previous section, such as longitudinal (Linke et al., 2007; Moilanen et al., 2008; Hermoso et al., 2011), lateral (Thieme et al., 2007; Ausseil et al., 2007; Hermoso et al., 2012a), vertical (Nel et al., 2011) and temporal (Hermoso et al., 2012b) components of connectivity within systematic planning frameworks. The main aim of all these approaches is to identify spatially and/or temporally clustered priority areas that maintain connectivity in appropriate dimensions.

As a result of the unique spatial relationships of freshwater ecosystems, conservation recommendations delivered by these novel applications of systematic planning approaches usually extend over large areas, which makes their implementation difficult. For example, a common solution to address the propagation of upstream threats into freshwater PAs is to include whole (Linke et al. 2007; Thieme et al. 2007) or large portions (Hermoso et al. 2011) of the upstream catchment under the label of priority area for conservation (Figure 4.1a). However, little is recommended in terms of the actual management regime required in such areas. This constrains the value of systematic conservation planning because large areas cannot realistically be managed effectively under conventional conservation regimes (e.g., strict protection by designation as a National Park) due to potential conflicts with existing land uses. In an attempt to make conservation in freshwater ecosystems more practical, Abell et al. (2007) proposed a multi-zoning approach to help fulfil the spatial needs and ensure effective protection in a more flexible way (Figure 4.1b). This zoning is composed of: 1) 'freshwater focal zones', which are key areas for the protection of freshwater biodiversity, similar to PAs in terrestrial or marine realms; 2) 'critical management zones', as areas that need to be managed to maintain the ecological functionality of a focal area (e.g., connectivity to allow movement of individuals and gene exchange) and where uses that do not interfere with the purpose of this area are allowed; and, 3) 'catchment management zones', which link the entire upstream catchment to a critical management zone, where human uses are not constrained, but best practices (e.g., treat wastewater disposals, maintain riparian buffers in good condition or restrict the use of pesticides) are required (Figure 4.1b). This approach is increasingly accepted as an appropriate framework for freshwater conservation planning and design (e.g. Linke et al., 2011; Nel et al., 2011; Esselman et al., 2013) and is being implemented in common systematic conservation planning tools (e.g., Hermoso et al., 2015b). There have also been efforts devoted to the prioritisation of actual management actions required to enhance the persistence of biodiversity, although not necessarily within freshwater PAs. For example, Turak et al. (2011) and Cattarino et al. (2015, 2016) determined catchment scale priorities for restoration using systematic planning tools similar to those used in conservation planning. These priority maps were used by Catchment Management Authorities to prioritise their investment in protection or restoration actions (Turak et al., 2011). Actions tailored to achieve greatest possible improvement to freshwater biodiversity consider multiple factors such as current land use, potential future land use and location in relation to protected areas and major sources of disturbance (Turak et al., 2011, Cattarino et al., 2015, 2016). This scheme based on multiple management zones could help harmonise the implementation of conservation for biodiversity and the maintenance of other land uses (Hermoso et al., 2016a) and fits within other ongoing conservation efforts such as the declaration of heritage or wild rivers (Pittock et al., Chapter 9, this volume) that could play an important role as critical management zones or water reserves (see below) that could fit under critical management zones.

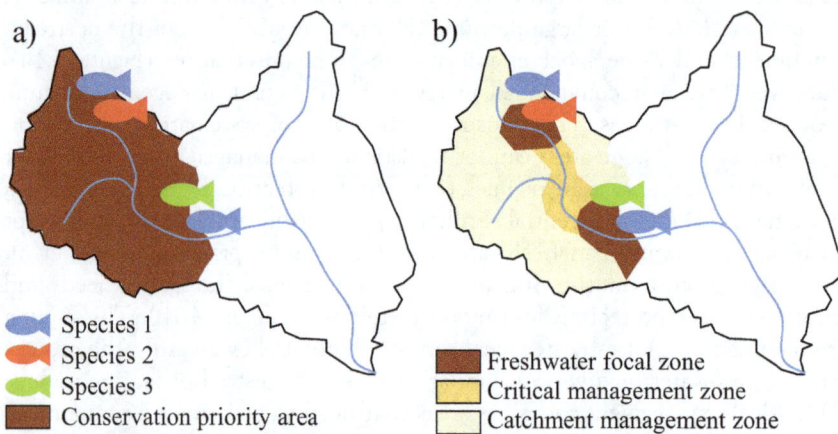

Figure 4.1 Schematic representation of (a) conventional conservation priority area and (b) catchment zoning proposed by Abell et al. (2007) designed to protect three species. The conventional conservation priority area in (a) aims to encompass at least part of the distribution of all three species, including upstream contributing catchments to ensure the representation of key ecological processes and avoid the negative effects of threats. The multi-zoning approach (b) tries to make the implementation of conservation more flexible by allocating zones under different management regimes, minimising the area under strict need of protection and addressing the ecological needs as in (a) (modified from Hermoso et al., 2015b).

Alternative approaches to implementing conservation efforts in freshwater systems

Different alternatives to conventional conservation approaches for implementing conservation objectives are quickly escalating in freshwater systems. Some of these alternatives arise from recognition of the strategic value of freshwater resources for human wellbeing and development, and also to make the most of the opportunity of new markets that arise from the payment for ecosystem services. These alternatives go beyond designing place-based conservation efforts for biodiversity values alone and incorporate key ecosystem services provided by freshwater systems, for example water provision through different initiatives such as Water Funds and Water Reserves (Abell et al., 2017). Some others focus on creating new conservation opportunities by channelling funds from compensatory projects (see biodiversity offsetting) or private sources (see Private PAs).

Water funds

Initiatives to improve or maintain the quality of drinking water, principally but not only for domestic use in municipalities, through land-based activities are expanding around the world and present an important opportunity for freshwater biodiversity conservation (Wickham and Flather, 2011; Bremer et al., 2016). Source water protection activities can include targeted land protection (with a focus on forests), revegetation (with a focus on reforestation, often in riparian zones), and agricultural best management practices such as the planting of cover crops, among others. Source water protection programs are designed largely to reduce sediment and nutrient pollution and consequently water treatment costs, with New York City's protection of its source watersheds among the best-known examples (Dudley and Stolton, 2003; McDonald and Shemie, 2014). Potential benefits for aquatic species include not only less polluted water but also more reliable base flows due to enhanced infiltration, plus the habitat values and ecological subsidies provided by riparian vegetation (Abell et al., 2017).

Many source water protection efforts fall under the umbrella of investments for watershed services programs, and one promising mechanism is the water fund. Water funds, in essence, pool funding from downstream beneficiaries and channel them to upstream land stewards in exchange for source water protection in the form of good land management (Bremer et al., 2016). Examples of existing water funds include those in Quito, Ecuador, the Rio Grande in the southwest US, and Nairobi, Kenya (Abell et al., 2017). Although the biodiversity benefits of water fund programs are in large part still to be measured, there is evidence that activities like riparian and upland forest protection common to source water protection efforts have potential for benefitting freshwater species and habitats if thoughtfully located, designed, and implemented (Broadmeadow and Nisbet, 2004; Hermoso et al., 2016b).

However, this use of funds from water supply agencies and protected area designations to protect upper catchment water sources has also been criticised, if poorly implemented, for trading off further degradation of lowland riverine ecosystems through changes to water flows for reservation of upper catchments (Pittock et al., 2015). Lowland rivers are often especially biodiverse (Tockner et al., 2008), are subject to intense human pressures and are thus more challenging to conserve. Often the upper catchments of rivers are less biodiverse and are located in forested and mountainous regions that are subject to the least human pressures. Consequently, good practice for water funds should include supporting environmental flows and funding of activities to conserve lowland aquatic ecosystems.

Water reserves

In order to safeguard important ecosystems, maintain invaluable ecological services, and adapt to water scarcity in a changing climate, several countries are formally recognizing and allocating a portion of the flow regime as a water reserve. For example, South Africa allocates or 'reserves' a portion of the water regime for basic human needs and the environment before water is allocated for other uses (Republic of South Africa, 1998). In Mexico, a public-private initiative of the National Water Commission is working to establish water reserve decrees in 189 river basins throughout the country. Under the National Water Law (Article 41), the Federal Executive can decree the total or partial reserve of the national waters with the purpose to guarantee flows for ecological protection, including the conservation or restoration of vital ecosystems (Barrios et al., 2015). Water reserves may also be established for urban public use and for generating energy for public consumption.

The process of establishing an environmental water reserve involves calculating the water requirements of critical ecosystems, including implementing a national standard for determining environmental flows (the water regime provided within a river to maintain ecosystems and their benefits, (Arthington, 2012)). An environmental water reserve is the volume of water that is excluded from the total amount to be allocated to various management purposes. It is the legal amount set aside for conservation or restoration of vital ecosystems, including maintenance of the functionality of the water cycle and its environmental services, and as a means of providing a reserve in a changing climate (Ramsar Convention, 2015).

The San Pedro Mezquital was the first Environmental Water Reserve designated by Mexico with around 80 per cent of its mean annual runoff allocated to ensure water and nutrients are supplied to the Marismas Nacionales (Barrios et al., 2015). In September 2014, the President of Mexico signed a decree for the 11 sub-basins that constitute the San Pedro Mezquital Basin (Mexico Secretariat of Environment and Natural Resources, 2014). The San Pedro Mezquital Basin is one of six pilots designed to test the effectiveness of implementing water reserves and associated flow regimes (Barrios et al., 2015).

The Water Reserve initiative was presented at the 12th Meeting of the Contracting Parties to the Convention on Wetlands in 2015 and adopted by Resolution XII.12 (Ramsar Convention, 2015) with a call to action to ensure and protect the water requirements of wetlands for the present and the future.

Biodiversity offsets and comprehensive planning

Several authors have suggested that system-wide planning is essential to limit the impacts of hydropower infrastructure on aquatic ecosystems (Opperman et al., 2015; Winemiller et al., 2016). There are a few examples where countries have either put such a system-wide approach in place or have taken steps towards offsetting impacts from specific projects through licensing and mitigation. As summarized in Moir et al. (2016), two such examples come from Norway and Costa Rica.

In Costa Rica, the national power company recently declared the Parismina River protected in perpetuity in order to offset the impacts of building an additional dam on the already dammed Reventazón River (Trujillo et al., 2012). The agreement guarantees that 'artificial modifications, including dams that would block migrations, will be prohibited and that the Parismina's natural flow pattern and integrity will be preserved or restored' (Inter-American Development Bank, 2015). The mitigation strategy also includes the development and implementation of a protected area along the 100.5-km length of the river with a protection zone and buffer zone (Chaves et al., 2015). While the Parismina River offers similar habitats for migratory fish, it is considerably smaller than the Reventazón in terms of width and flow. However, the Parismina is one of the first offsets implemented for a hydropower project, illustrating the potential value of this approach to direct mitigation toward the formal protection of a free-flowing river.

Norway has a complex legal framework regarding river protection, hydropower licensing and water management. Through multiple legislative actions in the 1970s and 1980s, Norway created a national Protection Plan for Watercourses which, by 1986, had designated nearly 200 rivers or stretches of river for protection, including removal from eligibility for future hydropower licenses (Huse, 1987). There is also a Master Plan for Water Resources (or Hydropower Development) to provide a national assessment of river resources and to target future hydropower projects to meet an energy goal with the lowest impacts on other resources. In part reflecting categories from the Master Plan, the Protection Plan for Watercourses has now grown to include 389 rivers or parts of rivers representing approximately 25 per cent of Norway's hydropower potential (Norwegian Water Resources and Energy Directorate, 2016). The Water Resources Act 2000 made these river protections statutory.

Non-government and private PAs

The final category that should not be overlooked is the value of non-government and private PAs – which are growing in importance globally as a response to perceived government inaction or to overcome bureaucratic hurdles. While non-government PAs are often viewed as potentially open to future disturbances such as mining leases or water allocations, private PAs can have significant advantages. Management regimes can be more flexible – for example, hunting could be allowed in a freshwater reserve as it might not impact fish populations. Also, non-government initiatives can provide faster action, they can mobilise additional resources from private donors and importantly can engage stakeholders suspicious of government agencies. This is crucial when implementing more systematic conservation interventions over large scales – such as continental connectivity corridors. Futhermore, Ramsar and World Heritage sites, as well as Biosphere reserves, are usually across a mix of tenures.

Some freshwater private PAs are nowadays acknowledged under the IUCN. A historical example is the Doñana National Park in Spain (Box 4.1), first purchased privately by WWF, and over 10 years led to the declaration of one of the world's most famous wetland reserves in a National Park. Examples for co-management with aboriginal traditional owners include the Fish River Station reserve in the Northern Territory, Australia, which is currently managed by a consortium of NGOs and government agencies and has been awarded IUCN Category II status (Fitzsimons and Looker, 2012). Also the Paruku Indigenous Protected area in Western Australia is co-managed by traditional owners and government, adhering to IUCN Category II principles. This form of co-management is not uncommon – sections of the Avalon Marshes in the United Kingdom are managed by two government agencies and five private trusts. Equally important as private protected areas are incentive-based mechanisms to reduce pressures on freshwater systems, increasing steadily. Some incentive-based mechanisms are supported by central governments, for example the restoration of the Yellow River Basin, China (Wohlfart et al., 2016).

Conclusions

In brief, the last two decades have witnessed increasing efforts globally to address the traditional limited consideration of freshwater systems in reserve design and identify freshwater PAs that account for the special needs of these ecosystems. Advances include novel methods to incorporate connectivity in reserve design, planning to designate multiple management zones, and tools to help prescribe optimal allocation of restoration efforts, among others. Moreover, new mechanisms are arising to encourage/support public and private funding to achieve improved outcomes for biodiversity conservation in freshwater systems. Both new methods for designing reserves and mechanisms for funding them should help create new conservation opportunities, enhance the implementation of conservation plans and improve the effectiveness of freshwater PAs. Despite these advances, it will be necessary to put more effort into monitoring programs to evaluate the effectiveness of conservation and restoration programs and help guide future decision-making.

Acknowledgement

We acknowledge funding support provided by the Spanish Government through a Ramón y Cajal contract (RYC-2013-13979) to VH.

References

Abell, R. (2002) 'Conservation biology for the biodiversity crisis: a freshwater follow up'. *Conservation Biology*, vol 16, pp1435–1437.

Abell, R., Allan, J.D. and Lehner, B. (2007) 'Unlocking the potential of protected areas for freshwaters'. *Biological Conservation*, vol 134, pp48–63.

Abell, R., Asquith, N., Boccaletti, G., Bremer, L., Chapin, E. et al. (2017) *Beyond the Source: The Environmental, Economic, and Community Benefits of Source Water Protection*. Arlington, VA: The Nature Conservancy.

Abell, R., Lehner, B., Thieme, M. and Linke, S. (2016) 'Looking beyond the fenceline: assessing protection gaps for the world's rivers'. *Conservation Letters*, DOI: 10.1111/conl.12312.

Arthington, A.H. (2012) *Environmental Flows: Saving Rivers in the Third Millennium*. Berkeley, CA: University of California Press.

Ausseil, A.G.E., Dymond, J.R. and Shepherd, J.D. (2007) 'Rapid mapping and prioritisation of wetland sites in the Manawatu–Wanganui Region, New Zealand'. *Environmental Management*, vol 39, pp316–325.

Barrios Ordóñez J.E., Salinas Rodríguez S.A., López Pérez M., Villón Bracamonte R.A., Rosales Ángeles F., et al. (2015) *National Water Reserve Program in Mexico: Experiences of ecological flow and allocation of water to environment*. Washington, DC, USA: Inter-American Development Bank.

Bower, S.D., Lennox, R.J. and Cooke, S.J. (2014) 'Is there a role for freshwater protected areas in the conservation of migratory fish?' *Inland Waters*, vol 5, pp1–6.

Bremer, L.L., Auerbach, D., Goldstein, J.H., Vogl, A.L., Shemie, D. et al. (2016) 'One size does not fit all: Natural infrastructure investments within the Latin American Water Funds Partnership'. *Ecosystem Services*, vol 17, pp217–236.

Broadmeadow, S. and Nisbet, T.R. (2004) 'The effects of riparian forest management on the freshwater environment: a literature review of best management practice'. *Hydrology and Earth System Sciences*, vol 8: pp286–305.

Bunn, S.E. and Arthington, A.H. (2002) 'Basic principles and ecological consequences of altered flow regimes for aquatic biodiversity'. *Environmental Management*, vol 30: pp492–507.

Cattarino, L., Hermoso, V., Bradford, L.W., Carwardine, J., Wilson, K.A., Kennard, M.J. and Linke, S. (2016) 'Accounting for continuous species' responses to management effort enhances cost-effectiveness of conservation decisions'. *Biological Conservation*, vol 197, pp116–123.

Cattarino, L., Hermoso, V., Carwardine, J., Kennard, M.J. and Linke, S. (2015) 'Multi-action planning for threat management: a novel approach for the spatial prioritization of conservation actions'. *PlosONE*, 10(5): e0128027. doi:10.1371/journal.pone.0128027.

Chaves, A., Fallas, J. and Campos, F. (2015) *Parismina River offset project*. http://previous. espconference.org/downloadattachment/83095/92052/21%20Offset-ParisminaICE-eng.pdf [accessed May 2016].

Convention on Biological Diversity (2010) Aichi Biodiversity Targets, www.cbd.int/sp/targets/.

Collier, K.J. (2011) 'The rapid rise of streams and rivers in conservation assessment'. *Aquatic Conservation: Marine and Freshwater Ecosystems*, vol 21, pp397–400.

Cucherousset, J., Paillisson, J.M., Carpentier, A., Thoby, V., Damien, J.P., Eybert, M.C., Feunteun, E. and Robinet, T. (2007) 'Freshwater protected areas: an effective measure to reconcile conservation and exploitation of the threatened European eels (Anguilla anguilla)?' *Ecology of Freshwater Fish*, vol 16: pp528–538.

Dudley, N. and Stolton, S. (2003) 'Running pure: The importance of forest protected areas to drinking water'. *World Bank/WWF Alliance for Forest Conservation and Sustainable Use*. Gland, Switzerland and Washington, DC, USA.

Esselman, P.C., Edgar, M., Breck, J., Hay-Chmielewski, E.M. and Wang, L. (2013) 'Riverine connectivity, upstream influences, and multi-taxa representation in a conservation area network for the fishes of Michigan, USA'. *Aquatic Conservation: Marine and Freshwater Ecosystems*, vol 23, pp7–22.

Fagan, W.F. (2002) 'Connectivity, fragmentation, and extinction risk in dendritic metapopulations'. *Ecology*, vol 83, pp3243–3249.

Fitzsimons, J. and Looker, M. (2012) 'Innovative approaches to land acquisition and conservation management: the case of Fish River Station, Northern Territory'. In P. Figgis, J. Fitzsimons and J. Irving) (eds), *Innovation for 21st Century Conservation*, pp. 78–85. Sydney: Australian Committee for IUCN.

Gaston, K.J., Jackson, S.E., Cantu-Salazar, L. and Cruz-Pinon, G. (2008) 'The ecological performance of protected areas'. *Annual Review of Ecology, Evolution, and Systematics*, vol 39, pp93–113.

Hermoso, V., Filipe, A.F., Segurado, P. and Beja, P. (2016a) 'Catchment zoning to unlock freshwater conservation opportunities in the Iberian Peninsula'. *Diversity and Distributions*, vol 22, pp960–969.

Hermoso, V., Abell, R., Linke, S. and Boon, S. (2016b) 'The role of protected areas for freshwater biodiversity conservation: challenges and opportunities in a rapidly changing world'. *Aquatic Conservation: Marine and Freshwater Ecosystems*, vol 26, pp3–11.

Hermoso, V., Filipe, A.F., Segurado, P. and Beja, P. (2015a) 'Effectiveness of a large reserve network in protecting freshwater biodiversity: a test for the Iberian Peninsula'. *Freshwater Biology*, vol 60: pp698–710.

Hermoso, V., Cattarino, L., Kennard, M.J. and Linke, S. (2015b) 'Catchment zoning for freshwater conservation: refining plans to enhance action on the ground'. *Journal of Applied Ecology*, vol 52, pp940–949.

Hermoso, V., Kennard, M.J. and Linke, S. (2012a) 'Integrating multi-directional connectivity requirements in systematic conservation planning to prioritise fish and waterbird habitat in freshwater systems'. *Diversity and Distributions*, vol 18, pp448–458.

Hermoso, V., Ward, D.P. and Kennard, M.J. (2012b) 'Using water residency time to enhance spatio-temporal connectivity for conservation planning in seasonally dynamic freshwater ecosystems'. *Journal of Applied Ecology*, vol 49, pp1028–1035.

Hermoso, V., Linke, S., Prenda, J. and Possingham, H.P. (2011) 'Addressing longitudinal connectivity in the systematic conservation planning of fresh waters'. *Freshwater Biology*, vol 56, pp57–70.

Hockings, M., Stolton, S., Leverington, F., Dudley, N. and Courrau, J. (2006) *Evaluating Effectiveness: A Framework for Assessing Management Effectiveness of Protected Areas*. 2nd edition. IUCN.

Huse, S. (1987) 'Norwegian river protection scheme: A remarkable achievement of environmental conservation'. *Ambio*, vol 16: pp304–308.

Inter-American Development Bank (2015) 'Project snapshot: Hydroelectric project takes unprecedented measures to protect habitat'. Available online at: www.iadb.org/en/topics/sustainability/project-snapshot-hydroelectric-project-takes-unprecedented-measures-to-protect-habitat,7998.html [accessed May 2016].

IUCN (2008) International Union for the Conservation of Nature. 'Defining protected areas: an international conference in Almeria, Spain'. Dudley, N. and Stolton, S. (eds). Gland, Switzerland, p220.

Linke, S., Pressey, R.L., Bailey, R.C. and Norris, R.H. (2007) 'Management options for river conservation planning: condition and conservation re-visited'. *Freshwater Biology*, vol 52, pp918–938.

Linke, S., Turak, E. and Nel, J. (2011) 'Freshwater conservation planning: the case for systematic approaches'. *Freshwater Biology*, vol 56, pp6–20.

Mancini, L., Formichetti, P., Anselmo, A., Tancioni, L., Marchini, S. and Sorace, A. (2005) 'Biological quality of running waters in protected areas: the influence of size and land use'. *Biodiversity and Conservation*, vol 14: pp351–364.

Margules, C.R. and Pressey, R.L. (2000) 'Systematic conservation planning'. *Nature*, vol 405, pp243–253.

McDonald, R.I. and Shemie, D. (2014). *Urban Water Blueprint: mapping conservation solutions to the global water challenge*. The Nature Conservancy, Washington, DC.

Mexico Secretariat of Environment and Natural Resources (2014) *Decree of Water Reserves in the San Pedro Mezquital River Basin*. Available online at: http://reservasdeagua.com.mx/documentos-tecnicos/.

Moilanen, A., Leathwick, J. and Elith, J. (2008) 'A method for spatial freshwater conservation prioritization'. *Freshwater Biology*, vol 53, pp577–592.

Moir, K., Thieme, M. and Opperman, J. (2016) *Securing A Future that Flows: Case Studies of Protection Mechanisms for Rivers*. World Wildlife Fund and The Nature Conservancy. Washington, DC.

Nel, J.L., Reyers, B., Roux, D.J., Impson, N.D. and Cowling, R.M. (2011) 'Designing a conservation area network that supports the representation and persistence of freshwater biodiversity'. *Freshwater Biology*, vol 56, pp106–124.

Norwegian Water Resources and Energy Directorate (2016) *Conservation Plan for Waterways*. Available online at: www.nve.no/vann-vassdrag-og-miljo/verneplan-for-vassdrag/ (accessed May 2016).

Opperman, J., Grill, G. and Hartmann, J. (2015) *The Power of Rivers: Finding Balance between Energy and Conservation in Hydropower Development*. The Nature Conservancy, Washington, DC, USA.

Penha, J., Médice Fernandes, I., Súarez, Y.R., Silveira, R.M.L., Florentino, A.C. and Mateus, L. (2015) 'Assessing the potential of a protected area for fish conservation in a Neotropical wetland'. *Biodiversity and Conservation*, vol 23, pp3185–3198.

Pépino, M., Rodríguez, A. and Magnan, P. (2012) 'Impacts of highway crossings on density of brook charr in streams'. *Journal of Applied Ecology*, vol 49, pp395–403.

Pittock, J., Finlayson, M., Arthington, A.H., Roux, D., Matthews, J.H., et al. (2015) 'Managing freshwater, river, wetland and estuarine protected areas'. In G.L. Worboys, M. Lockwood, A. Kothari, S. Feary and I. Pulsford (eds) *Protected Area Governance and Management*, pp. 569–608. ANU Press, Canberra.

Ramsar Convention (2015) Resolution XII.12. 'Call to action to ensure and protect the water requirements of wetlands for the present and the future'. Convention on Wetlands (Ramsar, Iran, 1971) Secretariat.

Rands, M.R., Adams, W.M., Bennun, L., Butchart, S.H., Clements, A., Coomes, D., et al. (2010) 'Biodiversity conservation: challenges beyond 2010'. *Science*, vol 329, pp1298–1303.

Republic of South Africa (1998) 'National Water Act No. 36 of 1998'. Cape Town, South Africa.

Roux, D.J., Nel, J.L., Ashton, P.J., Deacon, A.R., de Moor, F.C., Hardwick, D., Hill, L., Kleynhans, C.J., Maree, G.A., Moolman, J. and Scholes, R.J. (2008) 'Designing protected areas to conserve riverine biodiversity: lessons from a hypothetical redesign of the Kruger National Park'. *Biological Conservation*, vol 141: pp100–117.

Saunders, D.L., Meeuwig, J.J. and Vincent, A.C.J. (2002) 'Freshwater protected areas: strategies for conservation'. *Conservation Biology*, vol 16, pp30–41.

Snyder, M.N., Pringle, C.M. and Tiffer-Sotomayo, R. (2013) 'Landscape-scale disturbance and protected areas: long-term dynamics of populations of the shrimp, Macrobrachium olfersi in lowland Neotropical streams, Costa Rica'. *Journal of Tropical Ecology*, vol 29, pp81–85.

Srinoparatwatana, C. and Hyndes, G. (2011) 'Inconsistent benefits of a freshwater protected area for artisanal fisheries and biodiversity in a South-east Asian wetland'. *Marine and Freshwater Research*, vol 62: pp462–470.

Suski, C.D. and Cooke, S.J. (2007) 'Conservation of aquatic resources through the use of freshwater protected areas: opportunities and challenges'. *Biodiversity and Conservation*, vol 16, pp2015–2029.

Stanley, E.H. and Doyle, M.W. (2003) 'Trading off: the ecological effects of dam removal'. *Frontiers in Ecology and the Environment*, vol 1, pp15–22.

Thieme, M., Sindorf, N., Higgins, J., Abell, R., Takats, J., Naidoo, R. and Barnett, A. (2016) 'Freshwater conservation potential of protected areas in the Tennessee and Cumberland River Basins, USA'. *Aquatic Conservation: Marine and Freshwater Ecosystems*, vol 26, pp60–77.

Thieme, M.L., Rudulph, J., Higgins, J. and Takats, J.A. (2012) 'Protected areas and freshwater conservation: a survey of protected area managers in the Tennessee and Cumberland River Basins, USA'. *Journal of Environmental Management*, vol 109, pp189–199.

Thieme, M., Lehner, B., Abell, R., Hamilton, S.K., Kellndorfer, J., Powell, G. and Riveros, J.C. (2007) 'Freshwater conservation planning in data-poor areas: An example from a remote Amazonian basin (Madre de Dios River, Peru and Bolivia)'. *Biological Conservation*, vol 135, pp484–501.

Tockner, K., Bunn, S., Gordon, C., Naiman, R.J., Quinn, G. P. and Stanford, J.A. (2008) 'Flood plains: Critically threatened ecosystems', in N.V.C. Polunin (ed.), *Aquatic Ecosystems. Trends and Global Prospects*, pp. 45–61. London: Cambridge University Press.

Trujillo, C., Rodriguez, E., Franco Carassale, G.F., Boulet, E., Watkins, G., et al. (2012) *Reventazón Hydropower Project: Environmental and Social Management Report*. Costa Rica: Inter-American Development Bank.

Turak, E., Ferrier, S., Barrett, T., Mesley, E., Drielsma, M., Manion, G., Doyle, G., Stein, J. and Gordon, G. (2011) 'Planning for persistence of river biodiversity: exploring alternative futures using process-based models'. *Freshwater Biology*, vol 56, pp39–56.

UNEP-WCMC and IUCN (2016) *Protected Planet Report 2016*. UNEP-WCMC and IUCN: Cambridge UK and Gland, Switzerland.

Ward, J.V. and Stanford, J.A. (1989) 'The four-dimensional nature of lotic systems'. *Journal of the North American Benthological Society*, vol 8, pp2–8.

Wickham, J.D. and Flather, C.H. (2011) 'Integrating biodiversity and drinking water protection goals through geographic analysis', *Diversity and Distributions*, 19: pp1198–1207.

Winemiller, K.O., McIntyre, P.B., Castello, L., Fluet-Chouinard, E., Giarrizzo, T., Nam, S., Baird, I.G., Darwall, W., Lujan, N.K., Harrison, I., et al. (2016) 'Balancing hydropower and biodiversity in the Amazon, Congo, and Mekong'. *Science*, vol 351, pp128–129.

Wohlfart, C., Kuenzer, C., Chen, C. and Liu, G. (2016) 'Social–ecological challenges in the Yellow River basin (China): a review'. *Environmental Earth Sciences*, vol 75, pp1066.

Chapter 5

What is different about freshwater protected areas?

N. Dudley, D. Juffe Bignoli and M. Kettunen

Key messages

- Freshwaters contain a disproportionately high level of biodiversity, yet have experienced very high levels of loss over the last few decades: Protected Areas (PAs) play a critical role in reducing and reversing this decline.
- Despite these losses, freshwaters provide a unique range of essential ecosystem services for humanity, linked particularly to food and water provisioning and increasing security against extreme weather events.
- Freshwater PAs are particularly influenced beyond their boundaries by watershed-scale, regional or international influences, making landscape-scale approaches an essential component of management.
- This means that managing with other stakeholders, rather than in conflict with these stakeholders, is particularly important in freshwaters, and experience in these approaches is developing fast.

The multiple roles of freshwater protected areas

Although the earliest of the modern-style national parks were set up at the end of the nineteenth century, the rapid growth of national and international PA networks is a modern phenomenon (UNEP-WCMC, 2016); anyone old enough to be reading this book will have been alive during the establishment of over half the area currently under protection. Targets, such as the Sustainable Development Goals[1] (SDGs) and those set by the Convention on Biological Diversity[2] (CBD), include both explicit and implicit reference to many issues relating to freshwater protection and are driving further expansion of PAs, in response to increasing recognition of the importance of biodiversity conservation and ecosystem services. For example, SDG 6 (Ensure access to water and sanitation for all) recognises the need to "By 2020, protect and restore water-related ecosystems, including mountains, forests, wetlands, rivers, aquifers and lakes" to achieve such goal, while SDG 15 (Sustainably manage forests, combat desertification, halt and reverse land degradation, halt biodiversity loss) calls for "By 2020, ensure the conservation, restoration and sustainable use of terrestrial and inland freshwater ecosystems and their services, in particular forests, wetlands, mountains and drylands, in line with obligations under international agreements". Similarly, CBD Aichi Biodiversity Target 11 aims to conserve "By 2020, at least 17 per cent of terrestrial and inland water areas . . ." through effectively and equitably managed and well-connected systems of protected areas; conditions which have particular relevance for freshwater protected areas (Juffe-Bignoli et al., 2016; Arthington et al., Chapter 3, this volume; Finlayson et al., Chapter 12, this volume).

Freshwater PAs play a critical role in nature conservation, as evidenced by the success of sites such as the Everglades in the United States, Kakadu in Australia and the Pantanal in Brazil (Junk et al., 2006). However, during their brief history the roles of PAs have expanded to include a host of ecosystem services and related socio-economic benefits such as tourism, health benefits and spiritual and cultural values (Watson et al., 2014; Balmford et al., 2015). Today PAs are also increasingly expected to pay their own way, so that managers often have to report on their contributions to poverty alleviation or the local job market while proactively seeking funding for management measures from a range of different sources.

Therefore, although in most people's minds PAs have rather a specialised niche as places supporting nature conservation; in reality they play a multiplicity of roles. Some of these are illustrated for freshwater PAs in Table 5.1 below.

Table 5.1 Key functions provided by freshwater protected areas.

Role	Example
Biodiversity conservation	Freshwater makes up only 0.01% of the planet's water and 0.8% of the planet's surface, yet supports a relatively high proportion of species. Many of these are under threat (Collen et al., 2011); PAs provide critical habitats to stabilise species populations. For example, 21% of Important Bird Areas and Alliance for Zero Extinction sites that hold important freshwater sites are currently completely covered by PAs (Juffe-Bignoli et al., 2016; Butchart et al., 2015).
Supporting natural processes that maintain life on Earth	Refers to primary productivity, nutrient recycling, soil formation and the basic diversity of life-forms and ecological interactions. Large scale changes in freshwaters, including wholesale draining and destruction through alterations to flow and the impacts of pollution and over-harvesting are undermining these services in many places (Dudgeon et al., 2006; Gardner et al., 2015). For example, a meta-analysis of ecosystem services in wetlands in agricultural landscapes found a mean value of US$5,788/ha/ year for nutrient cycling (Brander et al., 2013).

(continued)

Table 5.1 (continued)

Role	Example
Maintenance of freshwater fish populations	PAs can provide safe spawning and nursery areas for fish, ensuring that sufficient populations survive to support both subsistence and commercial fisheries. Freshwater fish contribute almost 80% of animal protein for people in Cambodia (Hortle, 2007), much of it from Siem Reap UNESCO Man and Biosphere Reserve, where managers work with fishing communities to ensure a sustainable harvest (Cooperman et al., 2012).
Other food from freshwaters	Freshwater plants and invertebrates (MacAdam and Stockan, 2015) collected from the wild are also very important food sources for humans and are harvested from many PAs.
Groundwater recharge	Protecting groundwater catchments from industrial use or intensive agriculture can prevent contamination and ensure a sustainable supply (UNEP, 2014).
Provision of raw materials	Many less strict PAs permit the managed extraction of reeds, thatch, timber and other materials from wetland areas for local building, crafts, etc. For example, harvesting of the reed *Phragmites australis* is regionally important in many parts of the Mediterranean (Rhazi et al., 2012).
Water purification	Freshwater ecosystems can act as natural water filters, either mechanically through reed beds and similar vegetation or through biological actions by water plants; for example, in Florida's cypress swamps, 98% of all nitrogen and 97% of phosphorous entering from waste water were removed before this water entered the groundwater reservoirs (Ramsar, 2008).
Medicinal products	Freshwater plant species are used as medicines, for example at least 90% of freshwater plants are used as medicines in North Africa (Juffe-Bignoli and Darwall, 2012).
Disaster risk reduction including flood control	Regulation of water flow, both by providing places for floods to disperse naturally and by absorbing the impacts of floods with natural vegetation (Russi et al., 2013).
Carbon storage and sequestration	Wetlands, and particularly peatlands, are critically important carbon stores. Destruction of peatlands releases a large amount of carbon, and peat also releases methane; both may contribute to the overall balance of greenhouse gases. PAs play an important role in both protecting and restoring peat (Wetlands International, 2008).
Mental and physical health	An increasing number of health authorities are encouraging people to use nature reserves as "green gyms" to combat a growing physical and mental health crisis. For example, *Moved by Nature* in Finland takes vulnerable groups (e.g., people at risk of type 2 diabetes) into nature for activities including canoeing and ice fishing (ten Brink et al., 2016).

Opportunities for recreation and tourism	Freshwater wetlands offer a wide range of experiences, from sightseeing through boating to adventure sports and nature tourism (Hadwell et al., 2012). PAs maintain these resources and provide important sources of income. Two of the five most visited PAs in the world are freshwater protected areas (Balmford et al., 2015).
Research and education	Freshwater PAs, such as Wicken Fenn in Cambridgeshire, UK (Winston, 2001), provide important research sites and educational centres for both formal and informal education.
Aesthetic values	Many freshwater PAs are protected in part for their scenic values, including lakes, waterfalls, wild rivers and marshes: iconic examples include Victoria Falls on the Zimbabwe/Zambia border (Wynn, 2003).
Spiritual values	Many lakes, springs and rivers hold sacred values for both major and minor faith groups: the entire Ganges River is sacred to Hindus for example. Inclusion within a formal and informal PA can maintain such sites, which are often otherwise under threat as traditions change. For example, Gurudongmar Lake in Sikkim, located above 5,000 metres, is sacred to Hindus, Buddhists and Sikhs and receives over 15,000 visitors annually (Gopal et al., 2008).

There are increasing efforts to put economic values on the ecosystem services and related benefits from freshwater PAs (e.g., Emerton, 2005; de Groot et al., 2006). For example, the value of the Whangamarino Wetland in New Zealand in mitigating a particularly serious flood was estimated at $5.2 million in 2007 dollars (Waugh, 2007). Wetland resources from Stoeng Treng Ramsar Site, Cambodia, provided income averaged at US$3,000 per household (Chong, 2005). The Economics of Ecosystems and Biodiversity (TEEB) process provided a global overview of valuation studies, which shows dramatic variation in estimates (Russi et al., 2013); for example, there are more than a hundred valuation studies for mangroves, with the maximum estimate over a hundred times larger than the minimum. This variety of estimates is caused by differences in the size and/or type of area under investigation and the valuation methods used.

The various benefits from freshwater PAs are not all of equal importance. For a PA to be recognised as such by IUCN the primary role should be nature conservation as stated in the principles that accompany IUCN's definition of a PA (Dudley, 2008); while other roles can be important, in the event of a conflict between them nature conservation comes first. In practice, it has been proven many times that freshwater PAs can, if well-managed, fulfil a wide variety of roles, such as flood prevention, local fishing and ecotourism, without undermining their core conservation purpose (e.g., Murti and Buyck, 2014). There are many different types of freshwater PA and other related conservation strategies that also protect freshwaters: a selection is shown in Table 5.2 below, along with the relevant IUCN PA categories.

Table 5.2 Compatibility of various inland water strategies with IUCN PA categories (adapted from Dudley, 2008).

Type of protected areas or other conservation area	Compatibility with IUCN PA category							Example (with current IUCN category if assigned)
	Ia	Ib	II	III	IV	V	VI	
Designation/recognition under international convention or programme								
World Heritage site								Lake Malawi (Malawi, Category II, World Heritage site)
Ramsar site								Upper Navua Conservation Area (Fiji, Category II)
Biosphere reserve								Dyfi (Wales, UK, Category IV, UNESCO biosphere reserve)
Freshwater place-based protection measures								
Free-flowing river								Upper Delaware River (USA, Category V)
Riparian reserve/buffer								Douglas/Daly River Esplanade Conservation Area (Australia, Category V)
Floodplain reserve								Pacaya-Samiria (Peru, Category VI)
Fishery/harvest reserve								Lubuk Sahab (Indonesia)
Wetland game/hunting reserve								Ndumo Game Reserve (South Africa)
Recreational fishing restricted area								Onon River (Mongolia, Category II)
Protected water supply catchment								Rwenzori National Park (Uganda)
Protected aquifer recharge area								Susupe Wetland (Saipan, Category V)
Indigenous Protected Area/ICCA								Paraku (Western Australia)
Other place-based mechanisms with potential freshwater benefits								
Marine reserve/coastal management reserve								Danube Delta (Romania)
Seasonally closed fishery								Lake Santo Antonio (Brazil)
Forest reserve								Sundarbans Reserved Forest (Bangladesh)
Certified forest area								Upper St John River (USA, Category V)
Sacred river, lake or spring								Gosaikunda Lake and Ramsar site (Nepal)

■ Particularly compatible with the protected area category

□ Not incompatible with the protected area category

This is not to imply that management is inevitably harmonious or trouble-free. PAs are by their definition places where the needs of other species, and the needs of the wider global community for ecosystem services and recreation, tend to be given precedence over the needs of local human communities. The impact of a particular management emphasis on the people who live in and around the PA depends on many aspects of the way in which a PA has been identified, planned and managed; some are strongly supported by local communities while others dispossess people of access or resources and foster resentment and conflict (West et al., 2006). Ways of avoiding PAs becoming contested unhappy places are discussed in the text further below.

Why managing freshwater protected areas is different from managing other protected areas

Freshwater PAs are particularly susceptible to outside influences, from catchment disturbances, diversion of flow or pollution (Arthington et al., Chapter 8, this volume). Unlike most terrestrial PAs, their boundaries are often based around movable features like river banks or shorelines, meaning even their area and location may shift over time (Linke et al., 2011). Their particular roles need different approaches to land-based PAs and their changeable nature also provides some legal challenges in terms of delineation, as outlined below.

While all PAs are influenced by their surroundings, this is particularly the case for those containing freshwaters, where management decisions hundreds or thousands of miles away can create dramatic changes in ecology (Kingsford and Biggs, 2011). Alteration to water flow from abstraction, dams or other forms of diversion have dramatic impacts throughout a river system and on downstream lakes or inland seas (Arthington et al., Chapter 3, this volume). For example, damming of rivers in the Amazon basin has impacts on seasonal flooding, downstream deposition of sediment and the migration of unique and commercially valuable catfish species (Castello and Macedo, 2015); at 6,000 km currently the longest known fish migration on the planet (Barthem and Goulding, 1997). PA managers are often powerless to counter impacts beyond their boundaries, as shown by the effects of water regulation on PAs in the Murray-Darling Basin in Australia (Pittock and Finlayson, 2011). As a result, freshwater PA specialists have often been forced to think at a broader scale, focusing on the connectivity and integration of PAs across landscapes (Finlayson et al., Chapter 12, this volume) including within watershed-scale conservation initiatives (Postel and Thompson, 2005). This complicates management and makes monitoring the success or failure of freshwater PAs particularly problematic (Adams et al., 2015), but means that they are amongst the leaders in adopting a landscape approach (Flitcroft et al., Chapter 10, this volume).

External influences can originate nearby or far away across national boundaries. Long-range transport of sulphur and nitrogen oxides, falling as acidic deposition or "acid rain" altered river and stream chemistry in large areas of southern Scandinavia during the 1980s, including lakes in national parks (Schofield, 1976). More localised pollution, particularly from the leaching of agrochemicals, sewage or discharge of industrial waste, undermines conservation efforts in many PAs. These problems are likely to increase: use of synthetic nitrogen fertilizers has grown nine-fold since the 1950s, with a further increase of 40–50 per cent expected in the next 50 years, and phosphate use has tripled (Sutton et al., 2013).

This interconnectedness also influences the way in which freshwater PAs operate and provides additional planning challenges (Kingsford and Biggs, 2011). For example, if PAs are to be useful in addressing the needs of migratory fish they need to encompass critical life-cycle habitats including spawning and nursery areas, and feeding areas, as well as ensuring that migratory routes stay open (Bower et al., 2014; Arthington et al., 2016). Advances in prioritisation and modelling techniques can ensure the design of cost-effective dam replacement and river barrier planning to maximise ecological gains (Winemiller et al., 2016). For example, King et al. (2016) found that significant increases in species richness can be achieved for modest investment in barrier mitigation in the River Wey in England.

Managers of freshwater PAs have limited options for addressing offsite problems except through lobbying for policy changes; for example, acid deposition was not addressed in Europe until agreement of the Convention on Long-Range Transport of Air Pollutants (Maas and Grennfelt, 2016). The same applies to landscape level planning challenges which often involve several regional, or even national, jurisdictions.

Managing the relationship between freshwater protected areas and people

A handful of freshwater PAs are probably so remote or so inhospitable that they can be declared and managed without any concern about their implications for human societies. Others – a larger group – exist on private land or are purchased from private landowners for the specific purpose of conservation. But these are a small proportion of the total: lakes, marshes, rivers, streams and springs usually play a role in the functioning of human societies and will therefore have official or traditional ownership arrangements, agreements about resource use, cultural or spiritual values or other forms of rights, hopes or expectations.

PAs help to conserve biodiversity and water resources as a common good, but if planned or managed badly can undermine local livelihoods by restricting access to freshwater resources. Designation of a PA by a remote and faceless government, particularly if this results in an exclusionary form of management, can be devastating for people who rely on freshwater resources for their subsistence or livelihoods, with the term "green grabbing" having been coined to describe this kind of expropriation (Fairhead et al., 2012). On the other hand, rising populations, human migration, changing traditions and many forms of development are undermining wetlands to a devastating extent (Gardner et al., 2015). Without active conservation, resources and functions provided by wetlands are degrading and disappearing, and both biodiversity and people are suffering as a result. Therefore, protecting critical freshwater habitats is becoming essential not just to meet legal and ethical imperatives of biodiversity conservation but also to protect essential ecosystem services (Abell et al., 2007). Balancing local and global needs, such as global needs for biodiversity conservation and carbon sequestration versus local needs for fishing and fuelwood collection, is always challenging. The rapid rate of growth of PAs means that governments and other managers are learning as they go along. Governments are increasingly in the position of negotiating trade-offs between different demands for ecosystem services, for example international reporting to UN conventions and local constituents, with their own pressing need for resources. Ways of clearly measuring and articulating the costs and benefits, and of compensating local stakeholders for costs of services provided to a wider group of stakeholders, are important components in this process.

The way that these debates play out depends to a large extent on how they are approached by the various stakeholders involved. Most people are sensitive to the risks of environmental degradation; when this occurs it is often because people are forced by economic or social pressures to exploit resources above their level of sustainability, or because they are dispossessed of resources and in consequence stop treating them with respect. PAs can, if set up correctly, help to maintain landscapes and uses alongside biodiversity conservation, but they work best if the needs of all relevant stakeholders are addressed to the extent possible. PAs that report positive social outcomes also generally report positive biodiversity outcomes as well (Oldekop et al., 2015).

Critical, commonsense steps include adopting a transparent and participatory approach, based around principles of prior informed consent, so that everyone knows what is being proposed and has a chance to make their opinion heard (Borrini-Feyerabend et al., 2013). Ideal processes include reaching consensus, but realistically this is not always possible; however a PA established and managed against the entrenched opposition of the majority of local people is unlikely to succeed in the long term, however laudable its wider aims. Key items of negotiation include not only the existence and location of a freshwater PA but also how it is managed, with a wide portfolio of different management options available as described by the six IUCN categories (Dudley, 2008; Hermoso et al., Chapter 4, this volume).

Perhaps even more important is the issue of governance: who is consulted and who is in charge. Traditional PAs are identified, owned and managed by the state as a resource for the wider community; while this model probably remains dominant in terms of area, at least outside Europe, it is changing fast. The IUCN recognises four governance types: state governance, shared governance, private governance and governance by indigenous peoples and local communities: each has several sub-types as outlined in Table 5.3 (Borrini-Feyerabend et al., 2013).

Table 5.3 IUCN Governance types of protected areas (Dudley, 2008).

Type	Name	Description	Example
A	Governance by government	• Federal or national ministry/agency in charge • Sub-national ministry/agency in charge • Government-delegated management (e.g., to NGO)	Victoria Falls National Park in Zimbabwe is run by the government, attracting large numbers of visitors every year and bringing employment to nearby towns and communities (Wynn, 2003)
B	Shared governance	• Collaborative management (various degrees of influence) • Joint management (pluralist management board) • Transboundary management (various levels over frontiers)	The Varzea area of Brazil, a vast wetland, is increasingly managed through co-management approaches with local communities (WWF Brazil, 2006)
C	Private governance	• By individual owner • By non-profit organisations (NGOs, universities, cooperatives) • By for-profit organsations (individuals or corporate)	In the UK, large areas of the Somerset Levels wetlands are managed by a range of private conservation organisations, aimed at maintaining wildfowl and ecosystem services including flood control (Stolton et al., 2014)
D	Governance by indigenous peoples and local communities	• Indigenous peoples' conserved areas and territories • Community conserved areas – declared and run by local communities	The Toda people in southern India have protected freshwater sources through recognising their sacred values, thus providing important watershed protection missed by official government protected area programmes (Chhabra, 2009)

An increasing number of PAs have some kind of shared governance: either formally or informally; others are entirely controlled by local owners. From a freshwater perspective, this might mean negotiating with local fishing communities about fishing rights; working with recreational users to agree on management of leisure activities that do not undermine conservation aims; or integrating communities' use of sacred sites into management. For example, in Djoudj Biosphere Reserve and World Heritage site, on the border between Senegal and Mauritania, cooperation has included community-run nurseries and alternative energy sources for cooking along with restoration and reconnection of wetlands (Mbaye and Sow, 2013). In Monterrico Biosphere Reserve in Guatemala, local communities work with PA managers and those landscape planners to address increased flooding as a result of activities such as deforestation and drainage beyond the reserve boundaries (Moya et al., 2014). Sometimes cooperation needs to spread across national borders, such as in the Prespa Lake complex in southern Europe, where cooperative management between Greece, Macedonia and Albania is developing (Ivanovski, 2011). Success and failure are relative terms: it may be almost impossible to fulfil the needs of all stakeholders in natural resource management but the skills and experience at avoiding unnecessary and intractable problems are increasing all the time.

The global crisis in freshwater conservation has forced a necessary debate about the role of PAs in this context. New thinking about landscape-scale conservation, participatory approaches and an emphasis on ecosystem services all supply lessons that feed into other PAs and into wider management of the planet's resources. To repeat a point made above: PAs are still a new idea in many places and managers are feeling their way. Over the next few years there is every expectation that more can be learned about integrating freshwater protection into the wider arena of sustainable development with resultant benefits for freshwater PAs, noting the many functions that these serve for ecosystems and for people, as outlined above.

Notes

1 https://sustainabledevelopment.un.org/?menu=1300.
2 www.cbd.int/sp/targets/.

References

Abell, R., Allan, J.D. and Lehner, B. (2007) 'Unlocking the potential of protected areas for freshwaters'. *Biological Conservation*, vol 134, pp48–63.

Adams, V.M., Setterfield, S.A., Douglas, M.M., Kennard, M.J. and Ferdinands, K. (2015) 'Measuring benefits of protected area management: trends across realms and research gaps for freshwater systems'. *Phil. Trans. R. Soc. B*, vol 370, pp20140274.

Arthington, A.H., Dulvy N.K., Gladstone W. and Winfield I.J. (2016) 'Fish conservation in freshwater and marine realms: status, threats and management'. *Aquatic Conservation: Marine and Freshwater Ecosystems*, vol 26, pp838–857.

Balmford, A., Green, J.M.H., Anderson, M., Beresford, J., Huang, C., Naidoo, R., et al. (2015) 'Walk on the wild side: estimating the global magnitude of visits to protected areas'. *PLoS Biol*, vol 13(2): e1002074.doi:10.1371/journal.pbio.1002074.

Barthem, R. and Goulding, M. (1997) The Catfish Connection: Ecology, Migration and Conservation of Amazon Predators. Columbia University Press, New York.

Borrini-Feyerabend, G., Dudley, N., Jaeger, T., Lassen, B., Pathak Broome, N., Phillips, A. and Sandwith, T. (2013) *Governance of Protected Areas: From Understanding to Action*. Best Practice Protected Area Guidelines Series No. 20, Gland, Switzerland.

Bower, S.D., Lennox, R.J. and Cooke, S.J. (2014) 'Is there a role for freshwater protected areas in the conservation of migratory fish?' *Inland Waters*, vol 5, pp1–6.

Brander, L., Brouwer, R. and Wagtendonk, A. (2013) 'Economic valuation of regulating services provided by wetlands in agricultural landscapes: a meta-analysis'. *Ecological Engineering* vol 56, pp 89–96.

Butchart, S.H.M., Clarke, M., Smith, B., Sykes, R., Scharlemann, J.P.W., Harfoot, M., Buchanan, G., Angulo, A., Balmford A., Bertzky, B., et al. (2015) 'Shortfalls and solutions for meeting national and global protected area targets'. *Conservation Letters*, vol 8, pp329–337.

Castello, L. and Macedo, M.N. (2015) 'Large-scale degradation of Amazonian freshwater ecosystems'. *Global Change Biology* doi: 10.1111/gcb.13173.

Chhabra, T. (2012) 'Total sanctification of freshwater sources'. In Martin, V.G. and Shay Sloan, S. (eds) *Protecting Wild Nature on Native Lands: Case studies by native people from around the world, Volume 2*. The Native Lands and Wilderness Council and The Wild Foundation, Boulder, Colorado.

Chong, J. (2005) Valuing the Role of Wetlands in Livelihoods: Constraints and Opportunities for Community Fisheries and Wetland Management in Stoeng Treng Ramsar Site, Cambodia. IUCN Water, Nature and Economics Technical Paper No. 3, IUCN, Ecosystems and Livelihoods Group Asia, Colombo.

Collen, B., McRae, L., Deinet, S., De Palma, A., Carranza, T., Cooper, N., Loh, J. and Baillie, J.E.M. (2011) 'Predicting how populations decline to extinction'. *Philosophical Transactions of the Royal Society B: Biological Sciences*, vol 366 (1577), pp2577–2586, doi: 10.1098/rstb.2011.0015.

Cooperman, M.S., So, N., Arias, M., Cochrane, T.A., Elliott, V., Hand, T., Hannah, L., Holtgrieve, G.W., Kaufman, L., Koning, A.A., Koponen, J., Kum, V., McCann, K.S., McIntyre, P.B., Min, B., Ou, C., Rooney, N., Rose, K.A., Sabo, J.L. and Winemiller, K.O. (2012) 'A watershed moment for the Mekong: newly announced community use and conservation areas for the Tonle Sap Lake may boost sustainability of the world's largest inland fishery'. *Cambodian Journal of Natural History*, pp101–106.

De Groot, R.S., Stuip, M.A.M., Finlayson, C.M. and Davidson, N. (2006) *Valuing Wetlands: Guidance for Valuing the Benefits Derived from Wetland Ecosystem Services*, Ramsar Technical Report No. 3/CBD Technical Series No. 27. Ramsar Convention Secretariat, Gland, Switzerland & Secretariat of the Convention on Biological Diversity, Montreal, Canada.

Dudgeon, D., Arthington, A.H., Gessner, M.O, Kawabata, Z., Knowler, D., Lévêque, C., Naiman, R.J., Prieur-Richard, A.-H., Soto, D., Stiassny, M.L.J. & Sullivan C.A. (2006). Freshwater biodiversity: importance, threats, status, and conservation challenges. *Biological Reviews* 81 (2): 163–182.

Dudley, N. (ed.) (2008) Guidelines for Applying Protected Area Management Categories. IUCN, Gland, Switzerland.

Emerton, L. (ed.) (2005) *Values and Rewards: Counting and Capturing Ecosystem Water Services for Sustainable Development*. IUCN Water, Nature and Economics Technical Paper No. 1, IUCN — The World Conservation Union, Ecosystems and Livelihoods Group Asia.

Fairhead, J., Leach, M. and Scoones, I. (2012) 'Grabbing: a new appropriation of nature?' *The Journal of Peasant Studies*, 39, pp237–261.

Gardner, R.C., Barchiesi, S., Beltrame, C., Finlayson, C.M., Galewski, T., Harrison, I., Paganini, M., Perennou, C., Pritchard, D.E., Rosenqvist, A. and Walpole, M. (2015) *State of the World's Wetlands and their Services to People: A Compilation of Recent Analyses*. Ramsar Convention Secretariat, Ramsar Scientific and Technical Briefing Note No. 7, Gland, Switzerland.

Gopal, B., Chattergee, A. and Gautam, P. (2008) *Sacred Waters of the Himalaya*. WWF India, New Delhi.

Hadwell, W.L., Boon, P.I. and Arthington, A.H. (2012) Aquatic ecosystems in inland Australia: tourism and recreational significance, ecological impacts and imperatives for management. *Marine and Freshwater Research*, vol 63, pp1–16.

Hortle, K.G. (2007) *Consumption and the Yield of Fish and Other Aquatic Animals from the Lower Mekong Basin*. MRC Technical Paper 16, Mekong River Commission: Vientiane, Lao PDR.

Ivanovski, A. (2011) 'Prespa Lakes — where green diplomacy wins'. In Vasilijević, M. and Pezold, T. (eds) (2011). *Crossing Borders for Nature. European Examples of Transboundary Conservation*. IUCN Programme Office for South-Eastern Europe, Gland, Switzerland and Belgrade, Serbia, pp18–21.

Juffe-Bignoli, D. and Darwall, W.R.T (eds) (2012) *Assessment of the Socio-economic Value of Freshwater Species for the Northern African Region*. IUCN, Gland, Switzerland and Málaga, Spain.

Juffe-Bignoli, D., Harrison, I., Butchart, S.H.M., Flitcroft, R., Hermoso, V., Jonas, H., Lukasiewicz, A., Thieme, M., et al. (2016) 'Achieving Aichi Biodiversity Target 11 to improve the performance of protected areas and conserve freshwater biodiversity'. *Aquatic Conservation: Marine and Freshwater Systems*, vol 26 Supplement 1, pp133–151.

Junk, W.J., Brown, M., Campbell, I.C., Finlayson, C.M., Gopal, B., Ramberg, L. and Warner, B.G. (2006) 'Comparative biodiversity of large wetlands: a synthesis'. *Aquatic Sciences* vol 68, pp400–414.

King, S., O'Hanley, J.R., Newbold, L.R., Kemp, P.S. and Diebel, M.W. (2016) 'A toolkit for optimizing fish passage barrier mitigation actions'. *Journal of Applied Ecology*, doi:10.1111/1365-2664.12706.

Kingsford, R.T. and Biggs, H.C. (2011) 'Adaptive management guidelines for effective conservation of freshwater ecosystems in and around protected areas of the world'. IUCN WCPA Freshwater Taskforce, Australian Wetlands and Rivers Centre, Sydney.

Linke, S., Hermoso, V., Nel, J., Swartz, E., Tweddle, D., Turak, E. and Nel, J. (2011) 'Conservation of freshwater ecosystems'. In Darwall, W.R.T., Smith, K.G., Allen, D.J., Holland, R.A, Harrison, I.J. and Brooks, E.G.E. (eds) *The Diversity of Life in African Freshwaters: Under Water, Under Threat. An Analysis of the Status and Distribution of Freshwater Species Throughout Mainland Africa*. IUCN, Cambridge, UK, and Gland, Switzerland.

Macadam, C.R. and Stockan, J.A. (2015) 'More than just fish food: ecosystem services provided by freshwater insects'. *Ecological Entomology*, vol 40 (Supplement 1), pp113–123, doi: 10.1111/een.12245.

Maas, R. and Grennfelt, P. (eds) (2016) *Towards Cleaner Air: Scientific Assessment Report 2016: Summary for Policymakers.* EMEP Steering Body and Working Group on Effects of the Convention on Long-Range Transboundary Air Pollution, Oslo.

Mbaye, K. and Sow, M.S. (2013) 'Community engagement in the Djoudj National Bird Sanctuary and World Heritage site within the governance of a Transboundary Biosphere Reserve'. In Brown, J. and Hay-Edie, T. (eds) *COMPACT: Engaging Local Communities in the Stewardship of World Heritage.* UNDP, New York, pp107–116.

Moya, F.C., Barrios, M., Dávila, V., García Vettorazzi, M., Morales, A.S., Méndez, C. and Fischborn, M. (2014) 'Thinking outside the protected area boundaries for flood risk management: the Monterrico Multiple Use Natural Reserve in Guatemala'. In Murti, R. and Buyck, C. (eds) *Safe Havens: Protected Areas for Disaster Risk Reduction and Climate Change Adaptation.* IUCN, Gland, Switzerland, pp41–48.

Murti, R. and Buyck, C. (eds) (2014) Safe Havens: Protected Areas for Disaster Risk Reduction and Climate Change Adaptation. IUCN, Gland, Switzerland.

Oldekop, J.A., Holmes, G., Harris, W.E. and Evans, K.L. (2015) 'A global assessment of the social and conservation outcomes of protected areas'. *Conservation Biology*, vol 30 (1), pp133–141.

Pittock, J. and Finlayson, C.M. (2011) 'Australia's Murray Darling Basin: freshwater ecosystem conservation options in an era of climate change'. *Marine and Freshwater Research*, vol 62, pp232–243.

Postel, S.L. and Thompson Jnr., B.H. (2005) Watershed protection: Capturing the benefits of nature's water supply services. *Natural Resources Forum*, vol 29, pp98–108.

Ramsar Convention Bureau (2008) *Water Purification: Wetland Values and Functions* leaflet, Ramsar Bureau, Switzerland.

Rhazi, L., Grillas, P., Poulin, B. and Mathevet, R. (2012) 'Case study 4.2: Socio-economic importance of *Phragmites australis* in northern Africa'. In Juffe-Bignoli, D. and Darwall, W.R.T (eds) *Assessment of the Socio-economic Value of Freshwater Species for the Northern African Region.* IUCN, Gland, Switzerland and Málaga, Spain.

Russi, D., ten Brink, P., Farmer, A., Badura, T., Coates, D., Förster, J., Kumar, R. and Davidson, N. (2013) *The Economics of Ecosystems and Biodiversity for Water and Wetlands.* Institute for European Environmental Policy, London and Brussels and Ramsar Secretariat, Gland, Switzerland.

Schofield, C. (1976) 'Acid precipitation: effects on fish'. *Ambio*, vol 5 (5/6), pp228–230.

Stolton, S., Redford, K.H. and Dudley, N. (2014) *The Futures of Privately Protected Areas.* IUCN, Gland, Switzerland.

Sutton, M.A., Bleeker, A., Howard, C.M., Bekunda, M., Grizzetti, B., de Vries, W., van Grinsven, H.J.M., Abrol, Y.P., Adhya, T.K., Billen, G., Davidson, E.A., Datta, A., Diaz, R., Erisman, J.W., Liu, X.J., Oenema, O., Palm, C., Raghuram, N., Reis, S., Scholz, R.W., Sims, T., Westhoek, H. and Zhang, F.S. with contributions from Ayyappan, S., Bouwman, A.F., Bustamante, M., Fowler, D., Galloway, J.N., Gavito, M.E., Garnier, J., Greenwood, S., Hellums, D.T., Holland, M., Hoysall, C., Jaramillo, V.J., Klimont, Z., Ometto, J.P., Pathak, H., Plocq Fichelet, V., Powlson, D., Ramakrishna, K., Roy, A., Sanders, K., Sharma, C., Singh, B., Singh, U., Yan, X.Y. and Zhang, Y. (2013) *Our Nutrient World: The Challenge to Produce more Food and Energy with Less Pollution. Global Overview of Nutrient Management.* Centre for Ecology and Hydrology, Edinburgh on behalf of the Global Partnership on Nutrient Management and the International Nitrogen Initiative.

ten Brink, P., Mutafoglu, K., Schweitzer, J.-P., Kettunen, M., Twigger-Ross, C., Kuipers, Y., Emonts, M., Tyrväinen, L., Hujala, T. and Ojala, A. (2016) *The Health and Social Benefits of Nature and Biodiversity Protection – Executive summary*. A report for the European Commission (ENV.B.3/ETU/2014/0039), Institute for European Environmental Policy, London/Brussels.

UNEP [United Nations Environment Programme] (2014) *Green Infrastructure: Guide for Water Management*. Nairobi.

UNEP-WCMC and IUCN (2016) *Protected Planet Report 2016*. UNEP-WCMC and IUCN: Cambridge UK and Gland, Switzerland.

Watson, J.E.M., Dudley, N., Segan, D.B. and Hockings, M. (2014) 'The performance and potential of protected areas'. *Nature*, vol 515, pp67–73.

Waugh, J. (2007) Report on the Whangamarino Wetland and its Role in Flood Storage on the Lower Waikato River. Department of Conservation.

West, P., Igoe, J. and Brockington, D. (2006) 'Parks and people: the social impact of protected areas'. *Annual Review of Anthropology*, vol 35, pp251–277.

Wetlands International (2008) 'Advice to UNFCCC Parties for COP14 and associated meetings, December'. Wetlands International, Wageningen, Netherlands.

Winemiller, K.O., McIntyre, P.B., Castello, L., Fluet-Chouinard E,Giarrizzo, T., Nam, S., Baird, I.G., Darwall, W., Lujan, N.K. and Harrison, I. (2016) 'Balancing hydropower and biodiversity in the Amazon, Congo and Mekong'. *Science*, vol 351, pp128–129.

Winston, R.P. (2001) 'Discipline and interdiscipline: approaches to study abroad'. *Frontiers: The Interdisciplinary Journal of Study Abroad*, vol 7, pp61–93.

WWF Brazil (2006) Developing Community-Based Management Systems for the Amazon Floodplain: Lessons we are learning. WWF, Brasilia.

Wynn, S. (2003) 'Zambezi River: wilderness and tourism research into visitor perceptions about wilderness and its value. In Watson, A. and Sproull, J. (eds) (2003) *Science and Stewardship to Protect and Sustain Wilderness Values: Seventh World Wilderness Congress symposium; 2001 November 2–8; Port Elizabeth, South Africa*. RMRSP-27. US Department of Agriculture, Forest Service, Rocky Mountain Research Station, Ogden, UT.

Chapter 6

Managing threats to freshwater systems within protected areas

J. Pittock, L. Baumgartner, C.M. Finlayson,
J.D. Thiem, J.P. Forbes, L.G.M. Silva and
A.H. Arthington

Key messages

- Freshwater and estuarine ecosystems are among the most threatened in the world, are under-represented in Protected Area Policies and have the highest portion of species threatened with extinction.
- Freshwater biodiversity is particularly threatened because its conservation depends on: maintaining hydrological processes; retaining longitudinal connectivity of water flows without barriers along rivers; conserving lateral connectivity between a water body and its floodplain; sustaining adequate groundwater–surface water interactions; managing exogenous threats that are propagated across catchments; and integrating governance by multiple management authorities. The impacts of agriculture, aquaculture and fishing need to be managed.
- Where PAs are established particular attention should be given to: minimising impacts of water infrastructure, invasive species incursion control; reducing impacts of visitor facilities and activities; and pollution prevention.
- Floods, droughts and fire are natural processes in many freshwater ecosystems and plants and animals can normally tolerate or recover from them. However, these processes are increasingly exacerbated by climate change leading to changes in ecological character. In different circumstances freshwater ecosystems may attenuate or increase the impacts of natural disasters on people.

Introduction: threats to freshwater ecosystems

Freshwater and estuarine ecosystems are among the most threatened in the world, with the Millennium Ecosystem Assessment describing freshwater ecosystems as being over-used, under-represented in PAs and having the highest portion of species threatened with extinction (MEA, 2005). People and societies worldwide have inextricable links to freshwater ecosystems, and both people and nature benefit by managing risks to the health of these habitats (Dudgeon et al., 2006; Vörösmarty et al., 2010). Primary direct drivers of degradation and loss of riverine and other wetlands include land conversion, infrastructure development, water withdrawal, pollution, over-harvesting and overexploitation of freshwater species, the introduction of invasive alien species, and global climate change (Dudgeon et al., 2006; MEA, 2005). The World Commission on Protected Areas outlines how freshwater biodiversity is particularly threatened because its conservation depends on (Dudley, 2013):

- maintaining hydrological processes;
- retaining longitudinal connectivity of water flows without barriers along rivers;
- conserving lateral connectivity between a water body and its floodplain;
- sustaining adequate groundwater–surface water interactions;
- managing threats that are propagated in catchments and transmitted to inland waters at the lowest points on the landscape;
- integrating governance by multiple management authorities;
- developing policies to establish protected areas capable of sustaining the health and integrity of freshwater ecosystems.

This initial section on threats is focussed on those that managers need to address within PAs, whereas later sections provide advice on managing threats at the landscape scale. A particularly concise source of information for managing wetlands in PAs to avoid or mitigate these threats is *Wetland Management Planning: A Guide for Site Managers* (Chatterjee, Phillips, and Stroud, 2008). The resolutions and guidelines of the Ramsar Convention and the Ramsar handbooks for the wise use of wetlands (Ramsar Convention, 2011) provide excellent advice on good international practices for almost any wetland management challenge. Finlayson (2013) provides an analysis of how the wise use principles of the Convention relate to the threats raised by climate change. Management of a number of specific anthropogenic threats common to freshwater ecosystems within PAs are discussed here.

Water infrastructure and diversions

Freshwater, food and energy are inextricably linked on a global scale. Freshwater is essential for potable supply, producing food and agricultural products, for supporting subsistence communities, and to harness "renewable" energy. Yet human development of freshwater systems has been inequitable. Eleven per cent of the global population are still without access to clean water, sanitation, food resources or modern energy. To redress this inequity, humans are transforming the riverine environment and exploiting natural resources at an unprecedented rate. By 2050, it has been estimated that the world will require 70 per cent more agricultural production (Bruinsma, 2011) and, by 2035, 50 per cent more primary energy (de Fraiture et al., 2007). To meet these demands irrigated agriculture will need to be extended (Döll, 2002) and hydropower development is expanding at 1,500 per cent per annum; not expected to peak until 2030 (Zarfl et al., 2015). Irrigation and hydropower network expansion is therefore inevitable for most freshwater systems throughout the world (Ellis, 2011). The resultant situation will be increased conflict over limited resources, and degradation of freshwater environments, unless appropriate steps are taken to improve outcomes on a global scale.

The global expansion of irrigation and hydropower challenges the long-term sustainability of the historically productive freshwater systems; upon which many people depend for income and food, and that contribute substantial biodiversity globally. Activities in catchments (e.g., vegetation removal, land conversion), water infrastructure, flow regulation (Figures 6.1 and 6.2), diversions and inter-basin transfers (IBTs) alter flows that are vital to maintain freshwater biodiversity, change habitats from riverine to lacustrine and disrupt connectivity to the point where physical and biological processes are substantially disturbed (Poff et al., 1997).

Figure 6.1 Water diversions for irrigation, both small and large-scale, has had a major impact on freshwater wetlands (photograph © C.M. Finlayson).

Figure 6.2a & b In-stream infrastructure used to divert water for irrigated agriculture (photograph © C.M. Finlayson).

Catchment-wide planning is essential to sustain water for the environment (see upcoming section; Arthington et al., Chapter 8, this volume). The worst impacts of dam developments on freshwater may be reduced through appropriate basin-wide planning that seeks to locate new water infrastructure at sites that minimise environmental and social impacts while providing the desired economic benefits (Winemiller et al., 2016). For example, it may be possible to fully develop dams on already damaged river tributaries while retaining connectivity and natural flow variability on other tributaries or the main stem of a river. This requires assessment to be undertaken at a large scale before individual dams are approved. Freshwater PA managers should promote such processes by their governments where new developments are proposed. In the Mekong River basin there are examples of strategic environmental assessment (ICEM, 2010) and a sustainability assessment tool (MRC, 2010) to inform such decisions, which sadly, in that case, have not been used to date. The guidelines of the World Commission on Dams (WCD, 2000) and the industry endorsed Hydropower Sustainability Assessment Protocol (IHA, 2010) provide further tools that can be used to advocate for better environmental outcomes from water infrastructure development.

Major water storage dams are often located within nature reserves. Many of these were built over the objections of protected area managers. For example, the Hetch Hetchy Reservoir was built in Yosemite National Park in 1913–23 (Righter, 2005). In an era of increasing water scarcity more proposals to exploit water resources within nature reserves are certain. These developments should be resisted to prevent damage to freshwater biodiversity, but if imposed on protected area managers the mitigation measures described below should become mandatory. While this discussion has focussed on surface waters, the conservation of groundwater systems and dependant ecosystems requires similar vigilance in the face of increasing water extraction.

Where surface water diversions or infrastructure are proposed or in place, five key measures will reduce but not fully compensate for the impact on freshwater ecosystems and the conservation of aquatic species (Davies et al., 2010; Pittock and Hartmann, 2011): restoration of fish passage around dams, especially for migratory fish species and other mobile taxa; provision for release of environmental flows that mimic natural river flows (see upcoming section, and Arthington et al. (Chapter 8, this volume); building dam outlet structures that can eliminate any thermal pollution (Rolls et al., 2013); and conservation and restoration of the river corridor below the dam, for example, by restoring riparian vegetation and strategically placing infrastructure in areas that will minimise impact from the outset. Screening water diversion intakes to prevent loss of fish and other aquatic wildlife may also help (Baumgartner et al., 2009; Ghassemi and White, 2007).

Many protected area managers have installed small dams, either to supply PA managers and visitors with a potable water supply or to enhance water security for wildlife and wildlife viewing, as in Kruger National Park (Brits, et al. 2002). To reduce the ecological impacts of water supply infrastructure for people in PAs, it would be preferable to access groundwater sustainably or rely on off-river storage tanks or small dams. Establishing water storages along river corridors for wildlife is a misguided notion that should only be considered in exceptional circumstances, such as part of a targeted threatened species recovery plan. Even small dams across streams can block the passage of aquatic wildlife and reduce their populations. There are also negative impacts on terrestrial and riparian ecosystems, for instance, by enabling concentration and overgrazing by herbivores.

Wherever possible, redundant water storages in PAs should be decommissioned, as is occurring in Kruger National Park. There are a number of manuals available to guide removal of dams (Bowman, 2002; Lindloff, 2000). For example, in the United States two large dams are being removed on the Elwha River to enable migratory salmon to recolonize habitat within the Olympic National Park (Howard, 2012). This dam removal project, the largest in US history, reopens >100 km of spawning and rearing habitat for five species of salmon in the Elwha River and its tributaries. "Elwha chinook – the largest salmon in the river and unique in Puget Sound – cruised right past the former Glines Canyon Dam site just three days after it was blown out of their way" (Mapes, 2016).

Water infrastructure does not last forever, as it progressively degrades and needs to be removed, repaired or replaced (Krchnak, et al. 2009). As climate change alters the hydrological parameters that water infrastructure was designed to operate within a lot of these structures need to be re-operated or even decommissioned (Pittock and Hartmann, 2011). Re-operating water infrastructure is an opportunity to apply higher environmental standards that may have been adopted since first construction, such as by adding sediment flushing, thermal pollution control, fish passage and environmental flow release valves. Beyond water infrastructure, the additional threats from agriculture are now considered.

Agriculture

Agriculture and aquaculture are often excluded from IUCN category I to IV PAs since they are regarded as intensive commercial activities inimical to biodiversity conservation. These industries are discussed in this book on conservation and management of freshwater PAs for two reasons. First, human settlements are often focussed on freshwater ecosystems which have and continue to be used by people. Many freshwater ecosystems are cultural landscapes shaped by the activities of people over millennia such that maintaining traditional practices, including agriculture and aquaculture, is required to sustain biological and other values. Further, conservation of representative areas of freshwater ecosystems in PAs requires use of a full range of IUCN PA categories, to include those areas that remain heavily used by people. Second, agriculture and aquaculture occupy increasing areas around freshwater PAs, so their impacts need to be managed to maintain freshwater ecosystem health.

Agricultural practices over the past few centuries have caused wide-scale changes in land cover, watercourses, and aquifers, contributing to the loss and degradation of wetlands and undermining the ecological processes that support the provision of a wide range of ecosystem services (Boelee et al., 2013; Dixon et al., 2008; MEA, 2005). Among these are reductions in provisioning ecosystem services such as the supply of fresh water and fisheries, reduced regulating services such as storm protection and nutrient retention, and the loss of cultural services such as spiritual and recreational values. This has occurred as many agricultural systems have been managed as if they were not connected with the wider landscape, including the rivers and floodplains that have been so important for maintaining the ecological processes that have underpinned their sustainability (Falkenmark et al., 2007; Gordon et al., 2010). Irrigation and drainage of landscapes, the extensive clearing of vegetation, and the addition of agro-chemicals (fertilizers and pesticides) have altered the quantity and quality of water in the landscape, with modifications of water flows and water quality having major ecological, economic, and social consequences, as well as impacts on human well-being, for example, through insect-borne disease or through changes in nutrition (Horwitz and Finlayson, 2011).

Agriculture threatens freshwater biodiversity in many ways, including: conversion of habitat, degradation of riparian vegetation, development of water infrastructure, diversion and regulation of water, grazing of vegetation by livestock, soil and bank erosion, pollution from farm chemicals, and introduction of alien species. Examples of the cropping systems that generate numerous such impacts include cotton and sugar production in Australia (Arthington, 1996; Arthington et al., 1997). Riparian degradation has many implications for stream ecosystems, including alterations to shading and the thermal characteristics of streams, the failure of diminished vegetation to intercept runoff and filter sediments and nutrients, loss of bank stability, erosion and sedimentation, degraded aquatic habitats and reduced energy subsidies (Pusey and Arthington, 2003). The measures described above for conserving riparian corridors, providing environmental flows, screening water intakes and preventing pest species introductions are important to mitigate the many impacts of agriculture.

All too often the consequences of modifying agricultural landscapes have not been fully considered nor adequately monitored. In some instances agricultural practices have caused some ecosystems, such as inland lakes, to pass ecological tipping points, leading to a regime change in the ecosystem and the loss of important ecosystem services and benefits for local people. Although strong evidence lines to support many claims of changes of wetlands passing tipping points have not been supported by clear evidence and long-term monitoring (Capon et al., 2015). Despite issues with evidence and understanding of the consequences of changes in wetlands due to agriculture, Falkenmark et al. (2007) have questioned whether or not we have gone too far in developing agriculture at the expense of the many benefits that wetland ecosystems can provide, especially in terms of how we have changed water flows and disposed of waste products. The answer to this question is dependent on many factors and also on local circumstances and is influenced by history as well as changing climates and changes in agricultural practices and the demands for food.

Rebelo et al. (2010) in an overview of wetland distribution, type and condition across Sub-Saharan Africa highlighted the reliance of local communities on wetland agriculture (Figures 6.3 and 6.4) and showed that the nature of household dependence varied significantly from place to place and with socio-economic status. Consequently, measures to manage agriculture in wetlands will need to differ markedly from one location to another and across socio-economic groups. Rebelo et al. (2009) also reported that 78 per cent of wetlands listed as internationally important in 2006 under the Ramsar Convention supported agricultural activities, with 80–90 per cent of sites in forest and savanna biomes containing agriculture; and more than 50 per cent in other biomes. Nagabhatla et al. (2010), noting the limits of the available data, reported an increase in wetlands under cultivation from 25 per cent in 1926 to 43 per cent in 2006. Such figures demonstrate the importance of wetlands for agriculture and highlight how agriculture has very likely shaped the character of many wetlands, in some cases over long time periods. The duality of agriculture being a major cause of wetland loss and degradation as well as shaping the very character of remaining wetlands provides a dilemma for future wetland management.

Land clearing and more intense land uses as well as increased water regulation and diversion of water away from riverine systems is now widespread and prompted Davis et al. (2015) to examine "the challenge of protecting freshwater ecosystems under multiple land use and hydrological intensification scenarios" in Australia's Murray-Darling Basin. They highlighted the importance of managing and improving water quality, the value of providing environmental flows within

Figure 6.3 Grazing of stock is common in freshwater wetlands (photograph ©
C.M. Finlayson).

Figure 6.4 Recreational use of freshwater wetlands is increasing (photograph ©
C.M. Finlayson).

a watershed framework, and the important role that innovative science and adaptive management must play in developing proactive and robust responses to intensification. They also suggested the following research priorities to support improved systemic governance: including: i) determining the relative contributions of surface water and groundwater in supporting freshwater ecosystems in agricultural landscapes; ii) identifying and protecting freshwater biodiversity hotspots and refuges; iii) improving our capacity to model hydro-ecological relationships and predict ecological outcomes from land-use intensification and climate change; iv) developing an understanding of long-term behaviour of riverine ecosystems; and v) exploring systemic approaches to enhancing governance systems, including planning and management systems affecting freshwater outcomes. They also saw the integration of land and water management as essential.

The dilemma is all the more important as it has been estimated that by 2050 food demand will double with an expected increase in the demand for water for irrigated agriculture. In a simple sense the increased demand for water could be met through increased water use on current agricultural lands, or an expansion of agricultural lands, or through increased water productivity, each with different implications for wetland ecosystems. While all are plausible outcomes it is expected that a mix of solutions is likely (Falkenmark et al., 2007). Dependent on local conditions and the further adoption of existing and new technologies both agricultural and wetland management practices will need to be substantially improved to reduce the impacts from agriculture, whether within wetlands or on other lands. Further intensification of agriculture could lead to further water pollution, while expanding agriculture will require careful management to prevent further degradation and loss of ecosystem services. In some cases there have been efforts to reverse the loss and degradation of wetlands through rehabilitation and, in some cases, full restoration, although this is likely to be expensive, if possible at all. The latter is an important consideration as some changes can be nearly irreversible, for example, changes in ecological regimes in inland freshwater lakes (Capon et al., 2015). These changes can occur suddenly, although they often represent the outcome of a slow decline in the wetland (lake) and reduced ecological resilience that undermines the ability of the ecosystem to retain the same function, structure and feedbacks even when the pressure is reduced.

Failure to tackle the loss and degradation of wetlands, including that caused by the development and management of agriculture and related water resources, will undoubtedly undermine progress toward achieving the UN Sustainable Development Goals of reducing poverty, and hunger, and increasing wetland sustainability. An integrated approach to wetland management at a catchment scale is needed to halt past trends and restore, where possible, essential ecosystem services, especially where local communities are dependent on these services. This includes making decisions on the trade-offs that have occurred where agriculture has affected wetlands and occurs within wetlands. These should be based on a set of alternative scientifically informed scenarios, supported by social and economic analyses, as outlined in the guidance provided by the Ramsar Convention for the wise use of wetlands (Finlayson et al., 2011). To date such guidance has proven difficult to implement given political decisions, such as those outlined in plans to restore wetlands in Australia's Murray-Darling Basin, that are unlikely to sufficiently address the past degradation caused by ill-informed agricultural practices and misguided agricultural policies (Pittock and Finlayson, 2011). As well as sound science, ongoing attention is needed to communicate ecological process understanding across disciplinary and sectoral boundaries and to relevant policy and decision-makers.

In view of the huge scale of future demands on agriculture to feed people and eradicate hunger, and the past undermining of the ecological functions on which agriculture depends, Falkenmark et al. (2007) have highlighted the essential need to change the way we have been doing business. To achieve this, we need to:

- address social and environmental inequities and failures in governance and policy as well as on-ground management;
- rehabilitate degraded ecosystems, and, where possible, restore lost ecosystems;
- develop institutional and economic measures to prevent further loss and to encourage further changes in the way we do business;
- increase transparency in decision-making about agriculture-related water management and increase the exchange of knowledge about the consequences of these decisions.

The growing number of agricultural sustainability certification programs, including the Water Stewardship Standard (AWS, 2014), offer the prospect of recognising and rewarding better practices in agriculture that reduce the industry's impacts on freshwater ecosystems.

Over-fishing and over-harvesting

Although PAs are usually designated to protect wildlife, in many societies fish and other harvested aquatic species are not afforded the same level of legal protection as a mammal might receive. Further, many fish species are migratory at some point in their life cycles and are thus both vulnerable to harvesting outside PAs and may have life stages dependent on accessing critical habitat elsewhere (Arthington et al., 2016). For these reasons managing the threats from over-harvesting aquatic species are discussed here.

Over-harvest is one of the biggest threats to wild freshwater fish stocks globally, especially considering that freshwater fish contribute 15.3 per cent of total animal protein consumed by people globally (Tacon and Metian, 2008). It is further estimated that 1 billion people are involved in freshwater fisheries supply chains – processing, packing, transport and retailing (Allan et al., 2005). Some freshwater fisheries are highly productive. For example, the annual freshwater fish harvest from the Lower Mekong Basin comprises 2 per cent of the total global fish harvest (Hortle, 2007). But for almost all fisheries worldwide, harvest rates are unsustainable. An increasing demand for resources linked to an expanding global population and improvements in the efficiencies of fishing gears have led to a situation where over-fishing is a major cause of global freshwater fisheries extinctions (Allan et al., 2005).

Several types of fishing activity that are commonly deployed include commercial, subsistence (or artisanal), recreational (or sport) and illegal (Abell et al., 2007). Commercial fishing is the catch and re-sale for economic gain. It is generally perceived to have greater impacts than other forms because high yields can quickly deplete fish stocks and threaten extinction (Pauly et al., 2002). Subsistence fishing refers to harvest primarily for consumptive purposes. Subsistence fishing is often unregulated because of household reliance as a source of income, protein and micronutrients (Panayotou, 1982). Recreational fishing occurs globally for either harvest, or as catch and release. It can occupy many forms, deploying many gear types and is often species-specific (Cooke et al., 2015). Illegal fishing may comprise any form but in general refers to any fishing that occurs in contravention to established laws, cultures and regulations (Sullivan, 2002). Each method or approach has different impacts, and can rapidly deplete the resource base if left unregulated or unmanaged.

The impacts of fishing are managed in a variety of different ways. Commercial fisheries may have gear restrictions or harvest quotas (Pauly et al., 2002). Subsistence fisheries may be regulated by community co-management frameworks (Panayotou, 1982). Recreational fisheries are controlled via harvest restrictions such as size or bag limits as well as seasonal closures to protect sensitive life history events such as spawning aggregations, parental care, etc. (Forbes et al., 2015). Illegal fishing is generally combated using strong compliance frameworks, although there is evidence that non-mandated regulations can also be effectively applied (Cooke et al., 2013). Irrespective of the control method, the overall goal of fisheries management is to reduce stress on the resource and ensure harvest occurs at sustainable levels over the long term (Pauly et al., 2002). Thus, in the context of fisheries' sustainability, management strategies in PAs should focus on establishing zones or locations where one or more types of fishing are restricted or outlawed.

There are many examples of freshwater PAs that have been specifically established to protect fish from exploitation. But a major deficiency in PA policy is that regulations are restricted, often permitting fishing by a sub-set of methods or by imposing temporary restrictions. For instance, special rights are often granted to recreational fishers in many areas of the world under the assumption of minimal or no impact to the target or non-target species. For example, downstream of Yarrawonga on the Murray River, Australia, the nationally endangered trout cod (*Maccullochella macquariensis*) is protected by a complete ban on harvest by any method, but fishing for other species is permitted in the protected zone. Trout cod often comprise unwanted by-catch and therefore endure unquantified post-release mortality or physiological impairment. Similarly, the Pacaya-Samiria National Reserve (Peru, South America) was established to protect migratory fish (specifically *Arapaima gigas*). The freshwater reserve is hailed as a conservation success as netting is completely prohibited in the region. However, the ban has given rise to a strong sports fishery. Many thousands of tourists fish the region annually in hope of catching an Arapaima, which can exceed 100 kg. From a long-term perspective, large-bodied freshwater species are at the biggest risk of extinction from over-harvest due to small numbers and lower growth rates compared with smaller fish (Allan et al., 2005). However, this is seldom formally recognised in any PA planning processes. The broader conservation impacts of sustained harvest are often not considered significant in the context of PA planning.

A key to preventing overharvest of fish in PAs is recognising that the place of capture may only represent a snapshot of a fish's life and the habitats it occupies (Fausch et al., 2002). It is important to recognise that the majority of fish are migratory (Schlosser and Angermeier, 1995). For long-lived species, the requirement to move long distances risks exposure to multiple threats over many years (Baird, 2006). The challenge for freshwater PAs is to recognise this diversity of life-history strategies and ensure that fishing regulations are implemented simultaneously with a range of other management interventions, such as habitat protection, sustainable infrastructure and water management, over longer time-frames.

Threats from aquaculture

Fish, and other freshwater organisms, account for 72.4 per cent of all capture harvest in developing countries (Diana, 2009). Increasing global human populations are therefore driving a sharp decline in freshwater resources (Tidwell and Allan, 2001). Options that can augment wild production, such as aquaculture or aquaponics, are becoming increasingly seen as a mechanism to reduce pressure on wild stocks. Aquaculture is generally defined as the farming of fish, shellfish, or aquatic plants and production practices vary accordingly. Aquaculture techniques are being rapidly developed for fish, molluscs, crustaceans and plants. It is a fast-growing industry aimed at addressing predicted global food shortages (Naylor et al., 2001). In 2008, 60 per cent of the world aquaculture production was occurring in freshwater systems (Bostock et al., 2010). It is expected the aquaculture industry will continue to grow significantly until 2025 (Diana, 2009).

Most aquaculture ventures are either pond-based, where animals are bred and grow in isolated systems, or cage-based, where animals are reared and fed in pens in open waterways. Growing freshwater aquaculture trends include methods to improve intensification and cost-effectiveness with the development of new bio-engineering technologies and genetic fish strains with high growth rates and meat yield (Bostock et al., 2010). High intensity production of economically important species can alleviate stress on wild fish stocks, provide employment opportunities for local people (Diana, 2009) and, in a few cases, with species introduction providing desirable ecosystem functions (Schlaepfer et al., 2011).

However, a precautionary approach is required to analyse any possible benefits of species introduction (Vitule et al., 2012) and can also provide a mechanism for conservation re-stocking of wild stocks under enormous pressure. Under such scenarios, especially where aquaculture is a cultural activity that has been permitted for many generations, encouraging such practices within protected areas may provide substantial benefits to communities.

In general, however, the more intense the operation is, the higher the negative impact on freshwater systems. The predominant negative impacts include water quality deterioration which arises from uneaten food, surplus chemical supplements and waste discharge (Diana, 2009). Freshwater finfish, such as tilapias, are often grown in closed systems such as inland ponds and increasingly in floating cages in open water bodies from which escapes are inevitable (McCrary et al., 2001). Aquaculture species frequently establish reproducing populations when they escape from the aquaculture system into suitable habitats or are introduced into the wild (Arthington and Blühdorn, 1996), and many have a history of rapid spread (Canonico et al., 2005). Escape from cages is a significant issue, especially when culture of non-indigenous species occurs on large scales (Gozlan et al., 2010). In many areas, escape from aquaculture has led to the proliferation of many non-native species, especially carp (*Cyprinus carpio*) and tilapia (*Oreochromis* spp) (Cooke and Cowx, 2006; Weber and Fausch, 2003). High-density production can also lead to substantial disease outbreaks (McGinnity et al., 2003).

The main means to minimise aquaculture impacts are either mechanical or operational. For instance, improved production techniques and holding technology, as well as sound management programs, can generally minimise risk. Treating aquaculture ponds prior to, and post, harvest can substantially improve water quality outcomes (Lin and Yi, 2003). The development of reproductively-inert triploid-strains of species can minimise the chance of aquaculture fish breeding with wild populations (Cotter et al., 2000). The use of wild fish to feed farmed fish can be reduced by developing alternative and novel food sources (Naylor et al., 2000).

Aquaculture impacts can also be mitigated through strong policy development (Naylor et al., 2000). Negative examples of policy development for aquaculture have been described in Brazil (Pelicice et al., 2014). Obviously the most effective way to mitigate aquaculture impacts is to disallow it completely, but then positive benefits will not be realised. A pragmatic way forward is to establish oversight groups which can regulate or provide stewardship for the industry (Bush et al., 2013). For example, in Australia, the establishment of an Aquaculture Stewardship Council led to improved industry standards. The Council was instrumental in helping to develop a hatchery quality assurance scheme (HQAS) which clearly specifies the obligations of aquaculture operators and outlines acceptable procedures; all of which were supported by strong research and development (Department of Primary Industries, 2008). Advisory groups can also help to create aquaculture development zones which set boundaries for the production of certain species in the context of potential economic returns and likelihood to minimise impacts on the surrounding environment (Tidwell and Allan, 2001).

Aquaculture will be a dominant feature of the freshwater landscape in coming years. The future challenge is to find a balance between the negative and positive impacts of aquaculture (Diana, 2009). Where aquaculture is permitted in PAs, the combination of sound operational practices, and the development and implementation of robust policy, is needed to ensure impacts are minimised and positive outcomes are generated.

Invasive species

The growth in global transport and communication vectors is increasing the ability of alien species to move to new habitats and become invasive (Canning-Clode, 2015). Vectors include movement of species by air, ships' ballast water, overland transport and internet mail order deliberately or incidentally to agriculture, aquaculture escape, international development, inter-basin water transfer, tourism, gardening, aquarium pet escapes, biological control and scientific research endeavours. Once species have invaded they may significantly change ecological processes, including the generation of ecosystem services for people (Vilà and Hulme, 2017). Both the Convention on Biological Diversity and Ramsar Convention on Wetlands have adopted resolutions urging all governments to adopt a precautionary approach to movement of alien species (CBD, 2002; Ramsar, 2002).

Alien animal and plant species once introduced into water bodies are particularly difficult to eliminate or control. Unintentional introductions of aquatic species occur through ballast-water discharge from shipping, bait-bucket releases by anglers, and escapes from the ornamental fish trade, fish farms and ornamental ponds (Canonico et al., 2005). Invasive fish species threaten native taxa by predation, competition, habitat alteration, hybridization and the transfer of parasites and diseases (Strayer, 2010). Several species of the mosquitofish *Gambusia* (Poeciliidae), introduced deliberately for biocontrol of pest mosquitoes in many countries, threaten native fish species in numerous freshwater habitats, including PAs, by preying on eggs, competition for food and aggressive behaviour (Pyke, 2008). Deliberate sport fish introductions threaten native species, especially when the alien species are top predators, such as bass (Centrarchidae) and trout (Salmonidae). The zebra mussel (*Dreissena polymorpha*), a prohibited invasive species native to Eastern Europe and Western Russia, entered the US Great Lakes in ships' ballast, with devastating consequences for native American unionid clams, ecosystem functions (e.g., foodweb structure), and human uses of these waterbodies for fishing and recreation (Schloesser and Nalepa, 1994; Strayer, 2010).

Where invasive species are present in a country, distribution modelling can guide the design of efficient species-specific monitoring programs to forecast future distributions and prevent invasions from adjacent river basins. These outputs can also be used in developing regional freshwater conservation plans (e.g., (Esselman and Allan, 2011).

A stepwise process is needed to prevent alien introductions and translocations and control those that do occur (Chatterjee et al., 2008):

- identify local vectors for introduction of species (e.g., aquaculture farms, ornamental gardens) and seek voluntary or regulatory measures to prevent pest releases;
- monitor freshwater ecosystems to identify new problem species, drawing on information on pest species in your country or region that may invade;
- eliminate newly observed populations of threat species – incursion management;
- prevent the spread of pest species (this may be a case where a barrier dam in a stream serves to protect upstream populations of indigenous species from alien species spreading from downstream);
- institute control measures where they are feasible.

Confounding the challenge of managing invasive species is the impact of climate change that is leading to changes in indigenous freshwater species distributions (Daufresne and Boet, 2007). This raises questions for freshwater PA managers of which species are considered acceptable and which ones are invasive. Novel ecosystems are emerging and will require new approaches to identify, value and conserve freshwater biodiversity (Catford et al., 2012; Finlayson et al., 2017). There is a particular threat of expansion of invasive freshwater species due to climate change, resulting in a decline in populations of native species (Rahel et al., 2008). Greater control over vectors such as inter-basin water transfers and recreational use of water bodies are two measures that are required to limit the spread of alien species in a changing climate.

Recreational use of water bodies

Freshwater ecosystems are a magnet for a wide range of recreational users and activities. International tourism represented 7 per cent of global exports in 2015 and tourist arrivals are expected to increase by 3.3 per cent per year to 2030 (UNWTO, 2016). While this section focuses on the management of the impacts of recreation on freshwater PAs, access to healthy freshwater ecosystems is also critical driver for tourism that sustains PA systems. For example, research at Kruger National Park in South Africa found that if rivers were degraded approximately 30 per cent of tourism business would be lost (Turpie and Joubert, 2001).

Most PA managers face decisions in trading off the environmental impacts of tourists versus the benefits that they bring in terms of revenue for park management, livelihoods for local people, and the health and cultural benefits for visitors. Fundamental for managing this trade-off is a decision as to what level of environmental impact is acceptable. The IUCN has provided guidance on the full range of issues in managing visitors in PAs (Spenceley and Goodwin, 2007), and the Ramsar Convention has adopted guidance on managing tourism and recreation in wetlands (Ramsar, 2012). Here we focus on the impacts of visitors on freshwater ecosystems.

Freshwater ecosystems are a major focus of visitor activities and in most PAs require trade-offs between freedoms of visitor use and biodiversity conservation (Hadwen et al., 2012). Riparian areas often provide a biodiverse corridor of moisture-loving vegetation running through drier regions, vegetation which creates its own moist micro-climate and habitat for many terrestrial species and the terrestrial life stages of aquatic species (e.g., insects). Fragmentation and trampling of this vegetation can significantly impact on the freshwater ecosystem. For example, in the Wet Tropics World Heritage Area in Australia, visitors damaged riparian vegetation and reduced water quality at water holes used for swimming along the rivers (Turton, 2005). Canoeing, kayaking and rafting may result in impacts along a river corridor, such as the impacts on wildlife and with pollution from human waste observed in the Ganga River gorge in India (Farooquee et al., 2008).

Sediment-laden runoff from roads and tracks into water bodies can seriously harm aquatic biota, for example, by reducing filter feeding and prey visibility, and by smothering rocky substrates used for fish spawning and insect development (Pusey and Arthington, 2003). The smallest 'jump' up to or over a causeway or culvert across a waterbody may be a barrier to migration of aquatic species like fish and invertebrates, with implications for population exchange, recruitment and upstream–downstream biodiversity patterns (Olden, 2016).

The IUCN advises on four generic strategies for reducing the impacts of large numbers of visitors, namely managing the (Spenceley and Goodwin, 2007: 737):

1   supply of tourism or visitor opportunities, for instance, by having quotas for river rafting groups;
2   demand for visitation, such as by limiting the length of stay or restricting more damaging activities like fishing;
3   environment's capability of handling high use, for example, hardening sites with infrastructure like board walks;
4   impact of use, for instance, by requiring river rafting groups to carry away all their wastes or by distributing visitors over a broader area.

Key management responses for freshwater ecosystems should include: zoning land access, siting visitor facilities away from water bodies, fencing visitors out of riparian areas, creating hardened board walks and access points to the water, and regulating use of motorised vehicles and boating (Chatterjee et al., 2008; Mosisch and Arthington, 1998). Roads and tracks should be located to drain runoff away from water bodies and onto land (Sheridan and Noske, 2007). Crossings should be built as bridges or broad culverts sunk into the stream bed so as to maintain passage for aquatic fauna. Avoiding contaminated discharge and treating sewage are particularly important in preventing pollution of water bodies. Toilet facilities should be sited well away from water bodies and provided with drainage facilities that avoid pollution of wetlands and lakes.

Pollution

Pollution of freshwater ecosystems in protected areas may originate externally or within a PA. Verhoeven (2014) advocates that wetland managers assess threats to water quality in the following four ways:

1   determine the condition of wetlands in terms of water chemistry compared with least disturbed reference sites;
2   establish the nutrient status of wetlands and try and establish the thresholds for nutrient loadings that, if exceeded, would lead to change in their structure and functioning;
3   assess the capability of wetlands to remove or retain nutrients or sediments from through-flowing water without changing the ecological character of the wetland;
4   evaluate the risk of wetlands receiving toxic substances, human pathogens or other hazardous materials from inflows.

The Ramsar Convention has adopted guidance on risk assessment and selection of early warning indicators that can be used to proactively manage pollution threats (Ramsar, 1999). Earlier sections discussed management of the risks of cold deoxygenated water releases from dams and other pollution from agriculture, aquaculture and recreational use. An upcoming chapter focusses on the importance of management at catchment scales, in part to reduce the threats from pollution generated from outside protected areas (Flitcroft et al., Chapter 10, this volume).

We now focus on management of other pollution threats arising within PAs. A particular challenge for freshwater PAs is the management of water quality for environmental flows, to ensure that water releases optimize ecological responses. For example, along the River Murray, attempts to use minimal volumes of environmental water to inundate large floodplain wetlands in warmer months can lead to anoxic 'blackwater' events that kill aquatic fauna (Howitt et al., 2007). Attempts to engineer freshwater PAs to conserve biodiversity with less water may end up exacerbating salinity (Pittock et al., 2012). Consequently PA managers need to ensure that decisions on water management consider more than water volumes such that the water quality is adequate to sustain biota.

Conservation management requires use of chemicals such as fuels and herbicides that would have negative impacts if discharged into water bodies. Spills should be prevented wherever possible through good occupational health and safety practices, including siting chemicals away from water bodies, and securing and labelling stored chemicals. Potential pollutants should be stored and used on hard, internally draining surfaces that can contain accidental spills. Materials for soaking up any spills such as hay, sawdust or cat litter should be available on site, plus tools and bags for removing them for treatment. Spills into waterways require urgent advice to downstream authorities to close water diversions and prevent use of polluted water by people, wildlife and livestock wherever possible. Training of PA staff and development of pollution response plans for incident management are critical to ensure that spills are prevented where possible and well managed if they do occur (Worboys, 2015). Erstwhile 'disasters' such as floods are inevitable such that freshwater PAs need to be managed to minimise the obvious risks of resulting pollution.

Wetlands and disasters

Floods, droughts and fire are natural processes in many freshwater ecosystems and plants and animals can normally tolerate or recover from them (Bond et al., 2008). In particular, many freshwater species and ecosystems are adapted to variability in water volumes and timing of flows and require variability to thrive, such that regulated water bodies should not be managed with unnatural, permanent or stable flows (Postel and Richter, 2003). However, climate change may exacerbate the frequency and intensity of flood and drought events (Field and Van Aalst, 2014), and has many adverse consequences for freshwater ecosystems and aquatic species. Adaptation options are considered further in the upcoming chapter on climate change (Finlayson and Pittock, Chapter 13, this volume).

Some freshwater ecosystems are adapted to fire, such as floodplain forests in southern Australia, whereas others are destroyed and should be protected from fire, for example, peat swamp forests in Borneo. Riparian forests are often naturally fire resistant even where situated among other, flammable vegetation types. Traditional practices of local and indigenous peoples of cool patch burns around these ecosystems may conserve them from the effects of intense hot wildfires (Pittock et al., 2015).

A key question is whether wetland conservation can contribute to disaster risk reduction, a concept often described as ecosystem based conservation (UNEP, 2012). Wetland conservation has been proposed as helping to mitigate the impacts of natural hazards such as dust and sand storms, floods, droughts, fires, landslides, coastal erosion, tsunamis, hurricanes, storms, and storm surges, and also accelerated sea level rise (Ramsar, 2015). There are examples of wetland ecosystem restoration reducing the impacts of some natural hazards. For instance, restoration of floodplain wetlands is being used to provide room to hold and safely release flood peaks, such as along the Danube and Yangtze rivers (Ebert et al., 2009; Yu et al., 2009). Further, in South Africa, programs to remove invasive alien trees and restore eroding peat wetlands are increasing the base flow of rivers and streams (Ellery et al., 2011; Gorgens and Wilgen, 2004; Le Maitre et al., 2002).

However other kinds of wetlands may not offer such benefits. In upland areas wetlands may promote rather than attenuate flood flows by raising groundwater levels, limiting water storage (McCartney et al., 2013; Ramchunder et al., 2009). For instance, some kinds of wetlands may increase evaporation and reduce downstream flows (Bullock and Acreman, 2003). Consequently different kinds of wetlands play different roles in the hydrological cycle. Freshwater PA managers need to assess how their wetlands may aid disaster risk reduction, establishing a case for broader societal support for conservation, or require a more nuanced approach to ensure recognition of their biodiversity conservation values while managing any disservices.

While this brief section on threats cannot detail all mitigation measures, the Ramsar Convention materials and guides for site managers (Chatterjee et al., 2008; Ramsar, 2011) are a valuable source of information for dealing with most threats and challenges to freshwater ecosystem conservation. Threats to particular aquatic ecosystems in PAs, and management options, are addressed in other chapters in this volume (Arthington et al., Chapter 8, this volume; Finlayson et al., Chapter 12, this volume; Flitcroft et al., Chapter 10, this volume). In the next chapter we turn to conservation of freshwater species and protected area design options that involve mitigating threats and maximising biodiversity protection (Turak and Pittock, Chapter 7, this volume).

References

Abell, R., Allan, J. D. and Lehner, B. (2007). 'Unlocking the potential of protected areas for freshwaters', *Biological Conservation* 134:48–63.

Allan, J. D., Abell, R., Hogan, Z. E. B., Revenga, C., Taylor, B. W., Welcomme, R. L. and Winemiller, K. (2005). 'Overfishing of inland waters', *BioScience* 55:1041–1051.

Arthington, A. H. (1996). 'The effects of agricultural land use and cotton production on tributaries of the Darling River, Australia', *GeoJournal* 40:115–125.

Arthington, A. H. and Blühdorn, D. R. (1996). 'The effects of species interactions resulting from aquaculture operations', *Aquaculture and Water Resource Management* 114–139.

Arthington, A. H., Dulvy, N. K., Gladstone, W. and Winfield, I. J. (2016). 'Fish conservation in freshwater and marine realms: status, threats and management', *Aquatic Conservation: Marine and Freshwater Ecosystems* 26:838–857.

Arthington, A. H., Marshall, J., Rayment, G., Hunter, H. and Bunn, S. (1997). 'Potential impacts of sugarcane production on the riparian and freshwater environment', in B. A. Keating and J. R. Wilson (eds), *Intensive Sugar Cane Production: Meeting the Challenges Beyond 2000*, pp. 403–421, CAB International, Wallingford.

AWS (2014). *The AWS International Water Stewardship Standard*, Alliance for Water Stewardship, Edinburgh, allianceforwaterstewardship.org/assets/documents/AWS-Standard-v-1-Abbreviated-print.pdf.

Baird, I. G. (2006). '*Probarbus jullieni* and *Probarbus labeamajor*: The management and conservation of two of the largest fish species in the Mekong River in southern Laos', *Aquatic Conservation: Marine and Freshwater Ecosystems*, 16:517–532.

Baumgartner, L. J., Reynoldson, N. K., Cameron, L. and Stanger, J. G. (2009). 'Effects of irrigation pumps on riverine fish', *Fisheries Management and Ecology* 16:429–437.

Boelee, E., Scherr, S. J., Pert, P. L., Barron, J., Finlayson, M., Descheemaeker, K., Milder, J. C., Fleiner, R., Nguyen-Khoa, S., Barchiesi, S., Buntin, S. W., Tharme, R. E., Khaka, E., Coates, D., Solowey, E. M., Lloyd, G. J., Molden, D. and Cook, S. (2013). 'Management of water and agroecosystems in landscapes for sustainable food security', in E. Boelee (ed.) *Managing Water and Agroecosysems for Food Security*, pp. 156–170, CAB International, Wallingford.

Bond, N., Lake, P. and Arthington, A. (2008). 'The impacts of drought on freshwater ecosystems: an Australian perspective', *Hydrobiologia* 600, 3–16.

Bostock, J., McAndrew, B., Richards, R., Jauncey, K., Telfer, T., Lorenzen, K., Little, D., Ross, L., Handisyde, N., Gatward, I. and Corner, R. (2010). 'Aquaculture: global status and trends', *Philosophical Transactions of the Royal Society of London B: Biological Sciences* 365:2897–2912.

Bowman, M. B. (2002). 'Legal perspectives on dam removal', *BioScience* 52:739–747.

Brits, J., Van Rooyen, M. W. and Van Rooyen, N. (2002). 'Ecological impact of large herbivores on the woody vegetation at selected watering points on the eastern basaltic soils in the Kruger National Park', *African Journal of Ecolog*, 40:53–60.

Bruinsma, J. (2011). 'The resource outlook to 2050: By how much do land, water use and crop yields need to increase by 2050?', in P. Conforti (ed.), *Looking ahead in World Food and Agriculture: Perspectives to 2050*, pp. 1–33, FAO, Rome.

Bullock, A. and Acreman, M. (2003). 'The role of wetlands in the hydrological cycle', *Hydrology and Earth System Sciences Discussions* 7:358–389.

Bush, S. R., Belton, B., Hall, D., Vandergeest, P., Murray, F. J., Ponte, S., Oosterveer, P., Islam, M. S., Mol, A. P. J., Hatanaka, M., Kruijssen, F., Ha, T. T. T., Little, D. C. and Kusumawati, R. (2013). Certify Sustainable Aquaculture? *Science* 341:1067–1068.

Canning-Clode, J. (2015). *Biological Invasions in Changing Ecosystems*, De Gruyter, Warschau/Berlin.

Canonico, G. C., Arthington, A., McCrary, J. K. and Thieme, M. L. (2005). 'The effects of introduced tilapias on native biodiversity', *Aquatic Conservation: Marine and Freshwater Ecosystems* 15:463–483.

Capon, S. J., Lynch, A. J. J., Bond, N., Chessman, B. C., Davis, J., Davidson, N., Finlayson, M., Gell, P. A., Hohnberg, D. and Humphrey, C. (2015). 'Regime shifts, thresholds and multiple stable states in freshwater ecosystems: A critical appraisal of the evidence', *Science of The Total Environment* 534:122–130.

Catford, J., Naiman, R., Chambers, L., Roberts, J., Douglas, M. and Davies, P. (2012). 'Predicting novel riparian ecosystems in a changing climate', *Ecosystems* 16:382–400.

CBD (2002). *COP 6 Decision VI/23 Alien species that threaten ecosystems, habitats or species*, Convention on Biological Diversity, Monreal, www.cbd.int/decision/cop/default.shtml?id=7197.

Chatterjee, A., Phillips, B. and Stroud, D. (2008). *Wetland Management Planning. A Guide for Site Managers*, WWF, Wetlands International, IUCN and Ramsar Convention, Gland.

Cooke, S. J., Arlinghaus, R., Johnson, B. M. and Cowx, I. G. (2015). 'Recreational fisheries in inland waters', in J. F. Craig (ed.), *Freshwater Fisheries Ecology*, pp. 449–465, Wiley Blackwell, Hoboken.

Cooke, S. J. and Cowx, I. G. (2006). 'Contrasting recreational and commercial fishing: searching for common issues to promote unified conservation of fisheries resources and aquatic environments', *Biological Conservation* 128:93–108.

Cooke, S. J., Suski, C. D., Arlinghaus, R. and Danylchuk, A. J. (2013). 'Voluntary institutions and behaviours as alternatives to formal regulations in recreational fisheries management', *Fish and Fisheries* 14:439–457.

Cotter, D., O'Donovan, V., O'Maoiléidigh, N., Rogan, G., Roche, N. and Wilkins, N. P. (2000). 'An evaluation of the use of triploid Atlantic salmon (*Salmo salar* L.) in minimising the impact of escaped farmed salmon on wild populations', *Aquaculture* 186:61–75.

Daufresne, M. and Boet, P. (2007). 'Climate change impacts on structure and diversity of fish communities in rivers', *Global Change Biology* 13:2467–2478.

Davies, P. E., Harris, J. H., Hillman, T. J. and Walker, K. F. (2010). 'The Sustainable Rivers Audit: Assessing river ecosystem health in the Murray–Darling Basin, Australia', *Marine and Freshwater Research* 61:764–777.

Davis, J., O'Grady, A. P., Dale, A., Arthington, A. H., Gell, P. A., Driver, P. D., Bond, N., Casanova, M., Finlayson, M. and Watts, R. J. (2015). 'When trends intersect: The challenge of protecting freshwater ecosystems under multiple land use and hydrological intensification scenarios', *Science of The Total Environment* 534:65–78.

de Fraiture, C., Wichelns, D., Rockstrom, J., Kemp-Benedict, E., Eriyagama, N., Gordon, L. J., Hanjra, M. A., Hoogeveen, J., Huber-Lee, A. and Karlberg, L. (2007). 'Looking ahead to 2050: scenarios of alternative investment approaches', in D. Molden (ed.), *Water for Food, Water for Life: A Comprehensive Assessment of Water Management in Agriculture*, pp. 91–145, Earthscan, London; International Water Management Institute, Colombo.

Department of Primary Industries (2008). *The NSW Hatchery Quality Assurance Scheme*, New South Wales Department of Primary Industries, Sydney.

Diana, J. S. (2009). 'Aquaculture production and biodiversity conservation', *BioScience* 59:27–38.

Dixon, A., Wood, A., Finlayson, M. and van Halsema, G. E. (2008). 'Exploring agriculture – wetland interactions: a framework for analysis', in A. Wood and G. E. van Halsema (eds), *Scoping Agriculture –Wetland Interaction: Towards a Sustainable Multiple-Response Strategy*, FAO Water Report 33, pp. 5–27, FAO, Rome.

Döll, P. (2002). 'Impact of climate change and variability on irrigation requirements: a global perspective', *Climatic Change* 54:269–293.

Dudgeon, D., Arthington, A. H., Gessner, M. O., Kawabata, Z. I., Knowler, D. J., Lévêque, C., Naiman, R. J., Prieur-Richard, A. H., Soto, D., Stiassny, M. L. and Sullivan, C. A. (2006). 'Freshwater biodiversity: importance, threats, status and conservation challenges', *Biological Reviews* 81:163–182.

Dudley, N. (ed.) (2013). *Guidelines for Applying Protected Area Management Categories*, International Union for the Conservation of Nature, Gland.

Ebert, S., Hulea, O. and Strobel, D. (2009). 'Floodplain restoration along the Lower Danube: a climate change adaptation case study', *Climate and Development* 1: 212–219.

Ellery, W., Grenfell, M., Grenfell, S., Kotze, D., McCarthy, T., Tooth, S., Grundling, P. L., Beckedahl, H., Maitre, D. le and Ramsay, L. (2011). *WET-origins: Controls on the Distribution and Dynamics of Wetlands in South Africa*, WRC Report (334/09), Water Research Commission, Pretoria.

Ellis, E. C. (2011). 'Anthropogenic transformation of the terrestrial biosphere', *Philosophical Transactions of the Royal Society of London A: Mathematical, Physical and Engineering Sciences* 369:1010–1035.

Esselman, P. C. and Allan, J. (2011). 'Application of species distribution models and conservation planning software to the design of a reserve network for the riverine fishes of northeastern Mesoamerica', *Freshwater Biology* 56:71–88.

Falkenmark, M., Finlayson, C. M. and Gordon, L. (2007). 'Agriculture, water, and ecosystems: Avoiding the costs of going too far', in Molden, D. (ed.), *Water for Food, Water for Life: A Comprehensive Assessment of Water Management in Agriculture*, pp. 234–277, Earthscan, London.

Farooquee, N. A., Budal, T. K. and Maikhuri, R. (2008). 'Environmental and socio-cultural impacts of river rafting and camping on Ganga in Uttarakhand Himalaya', *Current Science – Bangalore* 94:587–594.

Fausch, K. D., Torgersen, C. E., Baxter, C. V. and Li, H. W. (2002). 'Landscapes to riverscapes: Bridging the gap between research and conservation of stream fishes: A continuous view of the river is needed to understand how processes interacting among scales set the context for stream fishes and their habitat', *BioScience* 52:483–498.

Field, C. and Van Aalst, M. (2014). *Climate Change 2014: Impacts, Adaptation, and Vulnerability*, Inter-governmental Panel on Climate Change, Geneva.

Finlayson, C. M. (2013). 'Climate change and the wise use of wetlands: information from Australian wetlands', *Hydrobiologia* 708:145–152.

Finlayson, C. M., Capon, S. J., Rissik, D., Pittock, J., Fisk, G., Davidson, N. C., Bodmin K. A., Papas P., Robertson H. A., Schallenberg M., Saintilan N., Edyvane K. and Bino, G. (2017). 'Policy considerations for managing wetlands under a changing climate', *Marine and Freshwater Research* [online early] doi:https://doi.org/10.1071/MF16244.

Finlayson, C. M., Davidson, N., Pritchard, D., Milton, G. R. and MacKay, H. (2011). 'The Ramsar Convention and ecosystem-based approaches to the wise use and sustainable development of wetlands', *Journal of International Wildlife Law and Policy* 14:176–198.

Forbes, J. P., Watts, R. J., Robinson, W. A., Baumgartner, L. J., Allen, M. S., McGuffie, P., Cameron, L. and Crook, D. A. (2015). 'System-specific variability in Murray cod and golden perch maturation and growth influences fisheries management options', *North American Journal of Fisheries Management* 35:1226–1238.

Ghassemi, F. and White, I. (2007). *Inter-basin Water Transfer: Case Studies from Australia, United States, Canada, China and India*, Cambridge University Press, Cambridge.

Gordon, L. J., Finlayson, C. M. and Falkenmark, M. (2010). 'Managing water in agriculture for food production and other ecosystem services', *Agricultural Water Management* 97:512–519.

Gorgens, A. H. M. and Wilgen, B. W. (2004). 'Invasive alien plants and water resources in South Africa: current understanding, predictive ability and research challenges', *South African Journal of Science* 100:27–33.

Gozlan, R. E., Britton, J. R., Cowx, I. and Copp, G. H. (2010). 'Current knowledge on non-native freshwater fish introductions', *Journal of Fish Biology* 76:751–786.

Hadwen, W. L., Boon, P. I. and Arthington, A. H. (2012). 'Aquatic ecosystems in inland Australia: tourism and recreational significance, ecological impacts and imperatives for management', *Marine and Freshwater Research* 63:325–340.

Hortle, K. G. (2007). *Consumption and the Yield of Fish and Other Aquatic Animals from the Lower Mekong Basin*. MRC Technical Paper, Mekong Research Commission, Vientiane.

Horwitz, P. and Finlayson, C. M. (2011). 'Wetlands as settings: Ecosystem services and health impact assessment for wetland and water resource management', *BioScience* 61:678–688.

Howard, B. C. (2012). Salmon re-enter Olympic National Park river thanks to Elwha Dam removal. *National Geographic NewsWatch*, newswatch.nationalgeographic.com/2012/08/21/salmon-enter-olympic-national-park-for-the-first-time-thanks-to-elwha-dam-removal/.

Howitt, J. A., Baldwin, D. S., Rees, G. N. and Williams, J. L. (2007). 'Modelling blackwater: Predicting water quality during flooding of lowland river forests', *Ecological Modelling* 203:229–242.

ICEM (2010). *MRC Strategic Environmental Assessment (SEA) of Hydropower on the Mekong Mainstream: Final Report*, International Center for Environmental Management, Hanoi.

IHA (2010). *Hydropower Sustainability Assessment Protocol*, International Hydropower Association, Sutton.

Krchnak, K., Richter, B. and Thomas, G. (2009). *Integrating Environmental Flows into Hydropower Dam Planning, Design, and Operations*, World Bank Group, Washington.

Le Maitre, D. C., van Wilgen, B. W., Gelderblom, C. M., Bailey, C., Chapman, R. A. and Nel, J. A. (2002). 'Invasive alien trees and water resources in South Africa: Case studies of the costs and benefits of management', *Forest Ecology and Management* 160:143–159.

Lin, C. K. and Yi, Y. (2003). 'Minimizing environmental impacts of freshwater aquaculture and reuse of pond effluents and mud', *Aquaculture* 226:57–68.

Lindloff, S. (2000). *Dam Removal: A Citizen's Guide to Restoring Rivers*, River Alliance of Wisconsin and Trout Unlimited, Madison and Arlington.

Mapes, L. V. (2016). 'Elwha: Roaring back to life', *The Seattle Times* 13 February 2016, projects.seattletimes.com/2016/elwha/.

McCartney, M., Cai, X. and Smakhtin, V. (2013). *Evaluating the Flow Regulating Functions of Natural Ecosystems in the Zambezi River Basin*, IWMI Research Report 148, International Water Management Institute, Colombo.

McCrary, J. K., van den Berghe, E. P., McKaye, K. R. and Lopez Perez, L. J. (2001). 'Tilapia cultivation: a threat to native fish species in Nicaragua', *Encuentro* 58:3–19.

McGinnity, P., Prodöhl, P., Ferguson, A., Hynes, R., Maoiléidigh, N., Baker, N., Cotter, D., O'Hea, B., Cooke, D., Rogan, G. and Taggart, J. (2003). 'Fitness reduction and potential extinction of wild populations of Atlantic salmon, *Salmo salar*, as a result of interactions with escaped farm salmon', *Proceedings of the Royal Society of London B: Biological Sciences* 270:2443–2450.

MEA (Millennium Ecosystem Assessment) (2005). *Ecosystems and Human Well-being: Wetlands and Water Synthesis*, World Resources Institute, Washington.

Mosisch, T. D. and Arthington, A. H. (1998). 'The impacts of power boating and water skiing on lakes and reservoirs', *Lakes and Reservoirs: Research and Management* 3:1–17.

MRC (2010). *Basin-wide Rapid Sustainability Assessment Tool*, Mekong River Commission, Vientiene.

Nagabhatla, N., Wickramasuriya, R., Prasad, N. and Finlayson, C. M. (2010). 'A multi-scale geospatial study of wetlands distribution and agricultural zones, and the case of India', *Tropical Conservation Science* 3:344–360.

Naylor, R. L., Goldburg, R. J., Primavera, J. H., Kautsky, N., Beveridge, M. C., Clay, J., Folke, C., Lubchenco, J., Mooney, H. and Troell, M. (2000). 'Effect of aquaculture on world fish supplies', *Nature* 405:1017–1024.

Naylor, R. L., Williams, S. L. and Strong, D. R. (2001). 'Aquaculture – A gateway for exotic species', *Science* 294:1655–1656.

Olden, J. D. (2016). *Challenges and Opportunities for Fish Conservation in Dam-impacted Waters*, Cambridge University Press, Cambridge.

Panayotou, T. (1982). *Management Concepts for Small-scale Fisheries: Economic and Social Aspects*, FAO Fisheries Technical Papers No. 228, Food and Agriculture Organisation of the United Nations, Rome.

Pauly, D., Christensen, V., Guenette, S., Pitcher, T. J., Sumaila, U. R., Walters, C. J., Watson, R. and Zeller, D. (2002). 'Towards sustainability in world fisheries', *Nature* 418:689–695.

Pelicice, F. M., Vitule, J. R. S., Lima Junior, D. P., Orsi, M. L. and Agostinho, A. A. (2014). 'A serious new threat to Brazilian freshwater ecosystems: the naturalization of nonnative fish by decree', *Conservation Letters* 22:55–60.

Pittock, J. and Finlayson, C. M. (2011). 'Australia's Murray-Darling Basin: freshwater ecosystem conservation options in an era of climate change', *Marine and Freshwater Research* 62:232–243.

Pittock, J., Finlayson, C. M. and Howitt, J. A. (2012). 'Beguiling and risky: "Environmental works and measures" for wetlands conservation under a changing climate', *Hydrobiologia* 708:111–131.

Pittock, J., Finlayson, M., Arthington, A. H., Roux, D., Matthews, J. H., Biggs, H., Harrison, I., Blom, E., Flitcroft, R., Froend, R., Hermoso, V., Junk, W., Kumar, R., Linke, S., Nel, J., Nunes da Cunha, C., Pattnaik, A., Pollard, S., Rast, W., Thieme, M., Turak, E., Turpie, J., van Niekerk, L., Willems, D. and Viers, J. (2015). 'Managing freshwater, river, wetland and estuarine protected areas', in G. L. Worboys, M. Lockwood, A. Kothari, S. Feary, and I. Pulsford (eds), *Protected Area Governance and Management*, pp. 569–608, ANU Press, Canberra.

Pittock, J. and Hartmann, J. (2011). 'Taking a second look: climate change, periodic relicensing and improved management of dams', *Marine and Freshwater Research* 62:312–320.

Poff, N. L., Allan, J. D., Bain, M. B., Karr, J. R., Prestegaard, K. L., Richter, B. D., Sparks, R. E. and Stromberg, J. C. (1997). 'The natural flow regime', *BioScience* 47:769–784.

Postel, S. and Richter, B. (2003). *Rivers for Life: Managing Water for People and Nature*, Island Press, Washington.

Pusey, B. J. and Arthington, A. H. (2003). 'Importance of the riparian zone to the conservation and management of freshwater fish: a review', *Marine and Freshwater Research* 54:1–16.

Pyke, G. H. (2008). 'Plague minnow or mosquito fish? A review of the biology and impacts of introduced Gambusia species', *Annual Review of Ecology, Evolution, and Systematics* 39:171–191.

Rahel, F. J., Bierwagen, B. and Taniguchi, Y. (2008). 'Managing aquatic species of conservation concern in the face of climate change and invasive species', *Conservation Biology* 22:551–61.

Ramchunder, S., Brown, L. and Holden, J. (2009). 'Environmental effects of drainage, drain-blocking and prescribed vegetation burning in UK upland peatlands', *Progress in Physical Geography* 33:49–79.

Ramsar (1999). *Resolution VII.10 Wetlands Risk Assessment Framework*, Ramsar Convention on Wetlands, Gland.

Ramsar (2002). *Resolution VIII:18 Invasive Species and Wetlands*, Ramsar Convention on Wetlands, Gland.

Ramsar (2011). *The Ramsar Handbooks for the Wise Use of Wetlands, 4th Edition*, Ramsar Convention on Wetlands, Gland.

Ramsar (2012). *Resolution XI.7 Tourism, Recreation and Wetlands*, Ramsar Convention on Wetlands, Gland.

Ramsar (2015). *Resolution XII:13 Wetlands and Disaster Risk Reduction*, Ramsar Convention on Wetlands, Gland.

Rebelo, L.-M., Finlayson, C. M. and Nagabhatla, N. (2009). 'Remote sensing and GIS for wetland inventory, mapping and change analysis', *Journal of Environmental Management* 90:2144–2153.

Rebelo, L.-M., McCartney, M. P. and Finlayson, C. M. (2010). 'Wetlands of Sub-Saharan Africa: Distribution and Contribution of Agriculture to Livelihoods', *Wetlands Ecology and Management* 18:557–572.

Righter, R. W. (2005). *The Battle over Hetch Hetchy: America's Most Controversial Dam and the Birth of Modern Environmentalism*, Oxford University Press, Oxford.

Rolls, R. J., Growns, I. O., Khan, T. A., Wilson, G. G., Ellison, T. L., Prior, A. and Waring, C. C. (2013). 'Fish recruitment in rivers with modified discharge depends on the interacting effects of flow and thermal regimes', *Freshwater Biology* 58:1804–1819.

Schlaepfer, M. A., Sax, D. F. and Olden, J. D. (2011). 'The potential conservation value of non-native species', *Conservation Biology* 25:428–437.

Schloesser, D. W. and Nalepa, T. F. (1994). 'Dramatic decline of unionid bivalves in offshore waters of western Lake Erie after infestation by the zebra mussel, *Dreissena polymorpha*', *Canadian Journal of Fisheries and Aquatic Sciences* 51:2234–2242.

Schlosser, I. J. and Angermeier, P. L. (1995). 'Spatial variation in demographic processes in lotic fishes: Conceptual models, empirical evidence, and implications for conservation', *American Fisheries Society Symposium* 17:360–370.

Sheridan, G. J. and Noske, P. J. (2007). 'A quantitative study of sediment delivery and stream pollution from different forest road types', *Hydrological Processes* 21:387–398.

Spenceley, A. and Goodwin, H. (2007). 'Nature-based tourism and poverty alleviation: Impacts of private sector and parastatal enterprises in and around Kruger National Park, South Africa', *Current Issues in Tourism* 10:255–277.

Strayer, D. L. (2010). 'Alien species in fresh waters: Ecological effects, interactions with other stressors, and prospects for the future', *Freshwater Biology* 55(s1):152–174.

Sullivan, M. G. (2002). 'Illegal angling harvest of walleyes protected by length limits in Alberta', *North American Journal of Fisheries Management* 22:1053–1063.

Tacon, A. G. and Metian, M. (2008). 'Global overview on the use of fish meal and fish oil in industrially compounded aquafeeds: Trends and future prospects', *Aquaculture* 285:146–158.

Tidwell, J. H., and Allan, G. L. (2001). 'Fish as food: aquaculture's contribution', *EMBO reports* 2:958–963.

Turpie, J. and Joubert, A. (2001). 'Estimating potential impacts of a change in river quality on the tourism value of Kruger National Park: An application of travel cost, contingent, and conjoint valuation methods', *WaterSA* 27:387–398.

Turton, S. M. (2005). 'Managing environmental impacts of recreation and tourism in rainforests of the wet tropics of Queensland World Heritage Area', *Geographical Research* 43:140–151.

UNEP (2012). *Ecosystem-based Adaptation Guideance: Moving from Principles to Practice. Working Document April 2012*, United Nations Environment Program, Nairobi. www.unep.org/climatechange/adaptation/Ecosystem-BasedAdaptation/EBADecisionSupportFramework/tabid/102163/Default.aspx.

UNWTO (2016). *UNWTO Tourism Highlights 2016 Edition*, United Nations World Tourism Organisation, Madrid.

Verhoeven, J. T. A. (2014). 'Water-quality issues in Ramsar wetlands', *Marine and Freshwater Research* 65:604–611.

Vilà, M. and Hulme, P. E. (eds) (2017). *Impact of Biological Invasions on Ecosystem Services*, Springer International Publishing, Basel.

Vitule, J. R. S., Freire, C. A., Vazquez, D. P., Nuñez, M. A. and Simberloff, D. (2012). 'Revisiting the potential conservation value of non-native species', *Conservation Biology* 26:1153–1155.

Vörösmarty, C. J., McIntyre, P. B., Gessner, M. O., Dudgeon, D., Prusevich, A., Green, P., Glidden, S., Bunn, S. E., Sullivan, C. A., Liermann, C. R. and Davies, P. M. (2010). 'Global threats to human water security and river biodiversity', *Nature* 467:555–561.

WCD (2000). *Dams and Development: A New Framework for Decision-making. The Report of the World Commission on Dams*, Earthscan, London.

Weber, E. D. and Fausch, K. D. (2003). 'Interactions between hatchery and wild salmonids in streams: differences in biology and evidence for competition', *Canadian Journal of Fisheries and Aquatic Sciences* 60:1018–1036.

Winemiller, K., McIntyre, P., Castello, L., Fluet-Chouinard, E., Giarrizzo, T., Nam, S., Baird, I. G., Darwall, W., Lujan, N. K., Harrison, I., Stiassny, M. L. J., Silvano, R. A. M., Fitzgerald, D. B., Pelicice, F. M., Agostinho, A. A., Gomes, L. C., Albert, J. S., Baran, E., Petrere, M., Zarfl, C., Mulligan, M., Sullivan, J. P., Arantes, C. C., Sousa, L. M., Koning, A. A., Hoeinghaus, D. J., Sabaj, M., Lundberg, J. G., Armbruster, J., Thieme, M. L., Petry, P., Zuanon, J., Vilara, G. T., Snoeks, J., Ou, C., Rainboth, W., Pavanelli, C. S., Akama, A., Soesbergen, A. v. and Sáenz, L. (2016). 'Balancing hydropower and biodiversity in the Amazon, Congo, and Mekong', *Science* 351: 128–129.

Worboys, G. L. (2015). 'Managing incidents', in G. L. Worboys, M. Lockwood, A. Kothari, S. Feary and I. Pulsford (eds), *Protected area governance and management*, pp. 569–608, ANU Press, Canberra.

Yu, X., Jiang, L., Wang, J., Wang, L., Lei, G. and Pittock, J. (2009). 'Freshwater management and climate change adaptation: Experiences from the central Yangtze in China', *Climate and Development* 1:241–248.

Zarfl, C., Lumsdon, A. E., Berlekamp, J., Tydecks, L. and Tockner, K. (2015). 'A global boom in hydropower dam construction', *Aquatic Sciences* 77:161–170.

Chapter 7

Conserving freshwater species in protected areas

E. Turak and J. Pittock

Key messages

- Freshwater species are those species that would disappear if inland (non-marine) habitats, disappeared or were severely degraded. These habitats include rivers, streams, lakes, ponds, swamps, marshes, bogs, fens, and aquifers. Animals dependent on these habitats account for close to 10 per cent of all animal species. Hence they are a large part of biodiversity on earth and of most Protected Areas (PAs). However, determining what is a freshwater species is difficult for some species groups because on the large variation in the degrees and types of dependence on freshwaters within the group.
- Evidence on the effectiveness of actions that may help conserve freshwater species is growing rapidly. This evidence can help conserving freshwater species in PAs but it needs to be integrated into a system-level approach to species conservation which includes the identification of synergies and conflicts between freshwater species conservation and other management objectives for PAs.
- Dependence on freshwater can often make a species more vulnerable than similar marine or terrestrial species because freshwater habitats are often also sites of multiple and high levels of human activity. Partly for this reason, however, the mechanisms by which human activities impact upon freshwater species are often quite direct and easy to describe. This can lead to more community awareness about how their actions affect freshwater species. Hence it can also translate into greater support for the conservation of freshwater species.
- Steps towards integrating freshwater species in PA management include: documenting freshwater species and ecosystems in the PA; setting goals

for freshwater biodiversity conservation; using conceptual models to explore how these goals relate to other management goals including the conservation of marine and terrestrial species and ecosystems; and efficiently implementing conservation actions that optimise multiple goals and benefits.

Introduction

Most PAs contain a variety of inland water (non-marine) habitats supporting species that would disappear if those habitats were lost or severely degraded. Here we refer to those species as "freshwater species", describe their diversity and explore what needs to be done to conserve these species within PAs. We show that many actions taken to protect individual freshwater species will also help in meeting other management goals. We expand on guidance given in the IUCN's *Protected Area Governance and Management* manual (Worboys et al., 2015) which emphasized the need to "strike the right balance between actions targeted at the level of ecosystems and landscapes and those that target individual freshwater species" (Pittock et al., 2015). To illustrate some of the options when targeting individual freshwater species, here we examine the conservation profile of six freshwater species in different taxonomic groups and across the world in the text boxes. These high-profile, relatively data-rich species provide real-life examples that illustrate successful strategies and actions for conserving freshwater species in PAs and in the wider landscape.

The information in previous chapters of this book allows us to focus this chapter on what needs to be done to optimize freshwater species conservation. The identification and mapping of freshwater ecosystems as defined by Milton and Finlayson (Chapter 2, this volume) is an essential first step towards conserving freshwater species in a given PA. Freshwater PAs have a special role in the conservation of freshwater species and some have been specifically established primarily for this purpose (Hermoso et al., Chapter 4, this volume). Regardless of its primary objectives and designation, however, almost every PA can contribute to freshwater species conservation. The application of freshwater ecological principles in PA management will enhance such contribution (Arthington et al., Chapter 3, this volume; Flitcroft et al., Chapter 10, this volume). Measures of success in protected area management include how well conflicting goals are reconciled and multiple conservation opportunities utilized. The social, cultural and economic benefits of a PA (Dudley et al., Chapter 5, this volume) can be as important as its role in conserving species and often success in the former is needed for success in the latter. The approach we have taken here strongly links species conservation in PAs to the totality of benefits provided to society by a protected area. The benefits are reliant on the effective management of threats (Arthington et al., Chapter 8, this volume; Pittock et al., Chapter 6, this volume) to the values which the PA aims to protect or enhance.

Our focus here is primarily on what can be done within the boundaries of PAs to help ensure the persistence of species and to reduce the risks of their local or global extinction. This information will support, and must be complemented by, what must be done outside PAs to ensure the persistence of the species, which includes actions planned and implemented at the scale of entire catchment, country, continent or the globe (Finlayson et al., Chapter 12, this volume). Some of these actions planned at broader scales are described in the next chapters of this book, for example, establishment and protection of corridors and water use (Pittock et al., Chapter 9, this volume), and climate change adaptation and mitigation (Finlayson and Pittock, Chapter 13, this volume).

What are freshwater species?

A majority of species on earth need freshwater to survive, but this dependence represents a continuum from needing to be immersed in freshwater in all stages of the life cycle to having very little direct contact with freshwater habitats. Here, we define freshwater biodiversity as the diversity of life of inland (non-marine) waters, even though these include some aquatic ecosystems with salinities well above those of marine waters. This definition is congruent with determinations made for the Freshwater Animal Diversity Assessment (FADA), which provided the first overview of global genus- and species-level diversity of selected animal groups and macrophytes in freshwater (Balian et al., 2008a, b). The FADA generally included species that accomplish all or parts of their lifecycles in or on water (i.e., 'real' aquatic species) and 'water-dependent' or 'paraquatic' species such as amphibians and water birds, which depend on inland water habitats during at least parts of their lives. These habitats include: flowing waters (rivers and streams); lacustrine wetlands (lakes, ponds, etc.); palustrine wetlands (swamps, marshes, fens, bogs); and groundwater systems (e.g., karstic systems, aquifers). Despite many obvious taxonomic and geographic gaps, and hence a need to collect more data, the FADA described the diversity of life that depends on inland water habitats in much more detail than had been available previously, and generated essential statistics such as the species richness of major organism groups. In particular, the disproportionate richness of global freshwaters is striking: the total number of freshwater animal species was estimated at 125,531 species, representing 9.5 per cent of 1,324,000 animal species described thus far, even though fresh waters occupy less than 1 per cent of the Earth's surface (Balian et al., 2008c).

Locating information about freshwater species

Current lists of freshwater species

The diversity of animal and plant groups included in the FADA are given in Tables 7.1 and 7.2. For many groups of organisms, whether a species is freshwater dependent or not is not easy to establish. For the FADA, experts on each taxonomic group formulated criteria specific to their group and this formed the basis of the 59 manuscripts published as the first major set of outputs of FADA (Balian et al., 2008a). Checklists are available for freshwater species in about 28 of the 65 major groups included in the FADA (Tables 7.1. and 7.2). The publication of these checklists commenced in August 2009. As of December 2016 the checklists for four more groups (nematodes, tardigrades, leeches and oligochaetes) are currently under review before public release. No checklists were produced for the other (33) groups but global assessments were subsequently completed for some of them, e.g., freshwater crabs (Kawai and Cumberlidge, 2016). The FADA also included macrophytes or aquatic plants that are large enough to be seen with the naked eye.

Obtaining information about freshwater species is particularly difficult for some species groups because of the large variation in the degrees and types of dependence on freshwaters within the group. Birds are a good example of this. Below we examine the terms and categories used in the relationship between bird species and freshwater habitats.

Freshwater birds: multiple lists and definitions

For many bird species, substantial information may be needed to place a species along each of the two major gradients relevant to their dependence on freshwater: the terrestrial–aquatic gradient and the marine–freshwater gradient. There are multiple terms used to describe different groups of bird species that represent parts of those gradients in different combinations not always reflected in the names given to those groups, e.g., waders. Importantly other factors are then added to define groups of birds. For example, the value of the species as game together with dependence on wetlands describe wildfowl, a term commonly used to refer to ducks, geese and swans. This terminology has influenced legislation, international conventions (Ramsar Secretariat, 2008), and has been widely used to communicate information about conservation of species (Williams et al., 2017).

For the purposes of FADA, freshwater birds included species that would disappear if the freshwater habitats disappeared because they would have little chance to adapt (Dehorter and Guillemain, 2008). However, in a departure from the FADA criteria for most other groups (Balian et al., 2008a), Dehorter and Guillemain (2008) adopted a narrow definition of freshwater habitats by only including inland wetlands where salinity was below 0.5 g/l. On this basis 566 freshwater bird species in 45 of the 183 known bird families were assigned to the category of freshwater species (Table 7.1). The FADA output for birds includes the number of species in each bird

Table 7.1 Diversity of vascular plants and vertebrates in freshwaters. The number of freshwater species in each major animal or plant group and species groups are given in brackets in column 2 based on manuscripts in Balian et al., 2008a). The last column describes conservation actions relevant to conservation of freshwater species in protected areas. The information on birds and amphibians are from Williams et al (2017) and Smith et al (2017) respectively. These provide evidence of effectiveness of actions relevant to protected areas management, using the methods in Sutherland et al (2017). All other examples of conservation actions based on individual publications cited there.

Group	Species diversity within the group	Conservation actions
Vascular plants (2,614)	Ferns and allies (171). Flowering plants (2,443). Fully freshwater genera with >50 species: water clover *Marsilae* (60); pondweed, *Potamogeton* (99); riverweed *Apinagia* (57); water trumpet *Cryptocoryne* (56); cape pondweed *Aponogeton* (54); water milfoil *Myriophyllum* (54); water lily *Nymphaea* (53); floating heart *Nymphoides* (53). Other genera with >50 freshwater species: quillwort *Ioetes* (70 of ~150); spikerush *Eleocharis* (70 of ~200); Sedges *Cyperus* (53 of ~900); bladderwort *Utricularia* (52 of 216).	Critically endangered *Nymphoides sivarajanii* (Kerala, India) is easily propagated from runners so it can be introduced to suitable locations (Sadasivaiah and Rao, 2011). *Aiphanes argos* is a new species of critically endangered rheophytic palm known only from the middle basin of the Samaná Norte River in Colombia (Bernal et al., 2017). Artisanal mining is becoming a threat and would need to be controlled but the biggest threat is a proposed dam which will extirpate this population if it proceeds. The biggest threat to the critically endangered quillwort *Isoëtes malinverniana* in Italy is declining water quality in small streams caused by the cultivation of rice. A predictive model for the preliminary evaluation of potential reintroduction sites uses pH, conductivity, and clarity as predictors allows comprehensive exploration of possible sites by limiting expensive chemical analysis to a few sites (Abeli et al., 2012).

Mammals (>122)	Dolphins and porpoises (6); platypus (1); water opossum (1); manatees (3); hippopotamuses (2); fishing bats (2); order insectivora (20 species includes water shrews, otter shrews, and the web-footed tenrec); rabbits (3); rodents (>60 species, including beavers, water rats, swamp rats, capybara, nutrias, aquatic genet, otter civet, and fish-eating mice); order carnivora (25 species, including seals, otters, marsh mangoose, fishing cat, and minks).	Fishing restrictions at critical times of the year reduced the threat to young in Saimaa ringed seals (Box 7.6) from fish traps, gill nets and fish-baited hooks (Niemi, 2013; Sipilä, 2016). Construction of artificial snow drifts during mild winters enabled the seals to build birthing lairs thus reducing reproductive failure (Auttila, 2015; Sipilä, 2016). Instalment of manatee protection devices at water control structures reduced deaths of the Florida manatee *Trichechus manatus latirostris* (Deutsch et al., 2008). Dolphin rescue operations that commenced in 1992 in Pakistan have been successful in reducing deaths of the Indus River dolphin *Platanista gangetica minor* stranded in irrigation canals (Waqas et al., 2012). Intensive control measures of the introduced American mink most likely saved the Spanish population of the European mink *Mustela lutreola* (Maran et al., 2016).
Birds (566)	Ducks, geese, and swans (134); screamers (3); nightjars (1); plovers (13); coursers and pratincoles (4); jacanas (8); gulls, terns and skimmers (19); avocets and stilts (9); painted snipes (2); sandpipers, snipes and phalaropes (43); herons (31); Shoebill (1); storks (7); Hamerkop (1); ibises and spoonbills (17); kingfishers (14); cuckoos (1); Hoatzin (1); hawks and eagles (10); pheasants (4); divers/loons (5); the Limpkin (1); cranes (10); finfoots (2); rails, gallinules and coots (65); dippers (5); buntings (15); waxbills (3); ovenbirds (5); swallows and martins (7); the Marsh Tchagra (1); wagtails and pipits (3); chats and old world flycatchers (11); tits (2); weavers (17); bulbuls (1); old world warblers (36); antbirds (2); babblers (3); wrens(4); tryant flycatchers(10); darters(4); pelicans (2); cormorants (7); grebes (19); owls (3).	**Beneficial**: restoring or creating wetlands (11); artificial nests for waterbirds (27). **Likely beneficial**: restoring or creating traditional water meadows (4); artificial nest for divers/loons (4); artificial/floating islands for ducks, geese and swans (3); foster ibis eggs or chicks with wild non-conspecifics (1); supplementary food to increase adult survival in cranes (1); translocate water birds (3). **Trade-off between benefit and harms**: managing water level in wetlands (6). **Limited evidence**: environmentally sensitive flood management (2); artificial nests for ibises and flamingos (2); grebes (1), moor hen and coot (1): foster eggs or chicks of cranes with wild conspecifics (1); supplementary food to increase adult survival in ducks and geese (2); supplementary food to increase reproductive success in ibises (1), kingfishers (1); snow geese (1) supplementary water to increase survival or reproductive success in ibises (1). **Likely to be ineffective or harmful**: treating wetlands with herbicides to create open water (4) (Williams et al 2017).

(continued)

Table 7.1 (continued)

Group	Species diversity within the group	Conservation actions
Amphibians (4,117)	Frogs (3,978); salamanders (237); and caecilians (79). Detailed information about each species can be found in AmphibiaWeb 2017 http://amphibiaweb.org.	**Beneficial:** create ponds (28); regulate water levels (3); remove invasive fish by drying out ponds (1); deepen, de-silt or re-profile ponds (4); replant vegetation (4). **Likely to be beneficial:** wetland creation (15) and restoration (17); restore ponds (15); manage ditches (1); remove or control fish population by catching (6); use temperature treatment to reduce chytridiomycosis (5); clear vegetation (7); create artificial hibernacula or aestivation sites (2); create refuges (2); translocate amphibians (4); engage volunteers to collect amphibian data (5); education programmes (5); raise awareness amongst the general public (2). **Trade-off between benefit and harm:** install barrier fencing along roads (8); install culverts or tunnels as road crossings (32); remove or control fish using rotenone (3); use antifungal treatment to reduce chytridiomycosis (16). **Limited evidence:** add salt to ponds to control chytridiomycosis (1). **Unlikely to be beneficial:** use prescribed fire or modifications to burning regime in forests (15).
Reptiles (500)	Lizards (73); snakes (153); crocodiles (14); alligators (2); caimans (5); gharials (2); turtles (257). The reptile database (Uetz, 2006, http://reptile-database.org/) has detailed information on each species.	Constructing cages to protect the nests of Blanding's turtle *Emydoidea blandingii* from nest predators was effective (Standing et al., 2000). However, wetland translocation of adults is likely to often be unsuccessful (Congdon et al., 2011). Protection of waterfowl habitats and improvement of water quality may allow the remaining populations of the vulnerable giant garter snake *Thamnophis gigas* to survive (Hammerson, 2007). Community-based conservation management, including enforcement patrols, was successful in reducing poaching of the Siamese Crocodile *Crocodylus siamensis* in Cambodia (Oum et al., 2009). Artificially breeding crocodile lizard *Shinisaurus crocodilurus* and then releasing the young back into nature is likely to be effective in recovering wild populations (Huang et al., 2008).

Fish (15,062) (15,750 in Darwall and Freyhof, 2016)	FADA checklists include species that are strictly freshwater (12,740) and brackish water (2,322). These include: ray finned fishes (14,922); lampreys (57); sharks and rays (74); chimaeras (1). Most of the ray-finned fishes are in five orders: Characiformes (1,801) in 17 families; Cypriniformes (3,664) in 7 families including carps, minnows, loaches and relatives; Siluriformes (2,992) in 34 families, catfishes; Perciformes (3,368) in 34 families, including cichlids; Cyprinodontiformes (1,096) in 9 families of mostly small fish including live bearers. The *FishBase* database (Froese and Pauly, 2017, www.fishbase.org/home.htm) has information about the biology, distribution and conservation status of each species. Trait information on 654 taxa in Europe, e.g., occurrence and migration, five habitat preference attributes, and 14 life-history attributes (Schmidt-Kloiber and Hering, 2015).	Conservation reserves are vital to protect species-rich habitats, important radiations, and endemic species (Arthington et al., 2016). Secific recovery programs are also needed and may involve individual species or entire basin fauna (Arthington et al., 2016). Conservation actions for migratory fishes should include removing barriers or making barriers more passable, using fish passage technologies that address a wide range of fish species and background hydrologies allowing both upstream access and downstream passage (McIntyre et al., 2016). In some situations barriers will need to be kept to prevent unwanted spread of non-native fishes, diseases and contaminants carried by hatchery and wild fishes (McLaughlin et al., 2013). Riparian habitat should be considered biologically critical for most species of freshwater fish, unless the habitat requirements of individual species indicate insensitivity to the ecological functions associated with riparian zones (Richardson et al 2010). Reconnecting isolated habitats is often successful in restoring fish populations (Roni et al., 2012). Managed introductions will be needed for restoration and in response to climate change (Olden et al., 2011). Assessment of conservation status is an important prerequisite for conservation. Darwall and Freyhof (2016) conservatively estimated that 7,000 freshwater fish species are awaiting IUCN Redlist assessments.

Table 7.2 Diversity of freshwater invertebrates of inland water habitats. The number of species in each group is given in brackets. These are based on the Freshwater Animal Diversity Assessment (FADA) in manuscripts (Balian et al., 2008a).

Group	Diversity within group	FADA checklists and other references
Molluscs (3,998)	Freshwater gastropods (3,792) are found in nearly all aquatic habitats. A great majority of the 206 species of bivalves have an obligate parasitic larval stage on fish.	Checklists: bivalves No checklists: gastropods
Crustaceans (11,990)	Conspicuous species include: crabs (1,306); crayfish (640); shrimps (655); aeglid crabs (63). Smaller bodied species include: amphipods (1,866); water fleas (620); mussel shrimps (ostracods 1,936); copepods (2,814); isopods (942). Fairy shrimps (307), clam shrimps (~116), and tadpole shrimps (15) are characteristic of temporary waters in arid, semi-arid zones. Syncarids (240) are confined to caves, springs, and wells. Fishlice (113), and mysids (72) are less ubiquitous. Highly specialised species include: spelaeogriphaceans (4); thermosbaenaceans (18) are eyeless and unpigmented crustaceans from subterranean waters; as well Cumacea (21), and Tanaidacea (4).	Checklist: spelaeogriphaceans, thermosbaenaceans, cumaceans-tanaiceans, mussel shrimps, copepods, water fleas No checklists: fishlice, mysids isopods, aeglid crabs, amphipods, syncarids, crabs, shrimps, crayfish, fairy shrimps, tadpole shrimps, and clam shrimps
Arachnids (>6,146)	Water mites (>6,000) show high specialization to particular microhabitats. Halacarid mites (56) are found in springs, wells, the hyporheic zone of rivers, sandy deposits. Oribatid mites (90) may be present in very large numbers.	Checklists: halacarid mites No checklists: water mites, oribatid mites
Springtails	Springtails (414) are not truly aquatic but exploit water surfaces in all continents.	No checklists: springtails

Aquatic insects (75,874)	Mayflies (3,046); dragonlies (5,680); stoneflies (3,497); heteroptera (truebugs) 801; caddisflies (11,532); alderflies and dobsonflies (328); lacewings (118); beetles (12,604); non-biting midges (4,147); craneflies (15,270); black flies (2,000); mosquitos (3,492); march flies (~5,000). Other true fly families ~13,454; butterflies (737); wasps (150); grasshoppers (188); scorpionflies (8). Trait information is available on 8,586 macroinvertebrate taxa (most of them insects) in Europe (Schmidt-Kloiber and Hering, 2015).	Checklists: mayflies, Chironomidae – Tanytarsini, stoneflies, caddisflies, mosquitos, blackflies, beetles-selected families. No checklists: dragonflies, grasshoppers, alder flies, dobsonflies, fishflies and lacewings, scorpion flies and hanging flies, true bugs, wasps, butterflies, true flies, craneflies, beetles, chironomidae.
Other invertebrates	These include: sponges (219); jellyfish and hydra (18); flatworms (1,303); rotifers (1948); nematodes (2,586); hairworms (2,000); gastrotrichs (318); bryozoans (88); tardigrades (1,100); polychaetes (62); oligochaetes (168); leeches (482). Most or all species in some of these groups (sponges, bryozoans, polychaetes, hydra) are sessile, attached to substratum. Sponges can cover large areas and help water purification. Rotifers are found in every continent including Antarctica and in every type of freshwater environment.	Checklists: Cnidaria–Hydroida rotifers, flatworms, nemertea, nematomorpha, Gastrothricha, Bryozoa, Polychaetes. Checklists in preparation: Oligochaetes, tardigrades, leeches. No checklists: Sponge, nematoda.

family that fits in with the definition of freshwater birds (Table 7.1) but does not identify individual species with very few exceptions. The open access online database of BirdLife International (BirdLife International, 2017) does not explicitly contain a freshwater species attribution. However, it has two fields that may be used to arrive at such determination: "habitat type" and "species type". One of the habitat types is "Inland wetland habitats". Globally 2,189 species of birds are attributed to this habitat. In the species type field the category "waterbird" is used interchangeably with "waterfowl", referring to this species type as "relevant to the Ramsar Convention" (BirdLife International, 2017). Globally 896 species of birds fall into the waterbird category and 671 of these are listed as being associated with the inland water habitat type. This list does not include 159 species from 23 of the bird families that were assessed in the FADA as having at least one freshwater species (Table 7.3). On the other hand, the Birdlife International list of inland water birds includes many species not listed by FADA as freshwater species (Table 7.4). The most notable difference at the order level are flamingos, which were not considered as freshwater species in the FADA. There are also major differences in species numbers from large waterbird families such as Ardeidae (herons), Laridae (gulls, skimmers and terns), Ciconiidae (storks) and Charadriidae (plovers). Of the species of these groups listed in Birdlife as waterbirds associated with inland waters only a small proportion were included in the FADA as freshwater species (Table 7.4). The omission of flamingos from the FADA list is readily explained by the water chemistry of their preferred inland wetlands. However, there are more than 259 other bird species from 23 families identified by BirdLife (2017) as waterbirds associated with inlands that are not included in the FADA (Tables 7.4), presumably because their dependence on freshwater habitats was not strong enough to qualify them as freshwater species.

Table 7.3 Bird families which were included as having freshwater species in the Freshwater Animal Diversity Assessment (FADA; Dehorter and Guillemain, 2008) but not included in the definition of 'waterbird' in the BirdLife database (BirdLife International, 2017).

Bird family	Number of FADA species
Accipitridae (hawks, eagles)	10
Alcedinidae (kingfishers)	14
Caprimulgidae (nightjars)	1
Cinclidae (dippers)	5
Cuculidae (cuckoos)	1
Emberizidae (buntings, American sparrows and allies)	15
Estrildidae (waxbills, grass finches, munias and allies)	3
Furnariidae (ovenbirds)	5
Hirundinidae (swallows and martins)	7
Malaconotidae (helmetshrikes, bushshrikes, puffbacks)	1
Motacillidae (wagtails and pipits)	3
Muscicapidae (chats and old world flycatchers)	11
Opisthocomidae (Hoatzin)	1

Paridae (tits and chickadees)	2
Phasianidae (pheasants, partridges, turkeys, grouse)	4
Ploceidae (weavers and allies)	17
Pycnonotidae (bulbuls)	1
Strigidae (typical owls)	3
Sylviidae (old world warblers)	36
Thamnophilidae (antbirds)	2
Timaliidae (babblers and parrotbills)	3
Troglodytidae (wrens)	4
Tyrannidae (tyrant-flycatchers)	10
Total	159

Table 7.4 Differences between the numbers in FADA and Birdlife International (2017) in number of species in bird families BirdLife International (2017) identifies as having at least one species of waterbird associated with inland wetlands. The bird families that have matching numbers of species in the two lists were excluded from this table.

Family	Birdlife inland waterbirds	FADA
Anatidae (ducks, geese, swans)	157	134
Anseranatidae (magpie goose)	1	0
Ardeidae (herons)	65	31
Burhinidae (thick-knees)	3	0
Charadriidae (plovers)	55	13
Ciconiidae (storks)	19	7
Eurypygidae (sunbittern)	1	0
Glareolidae (coursers, pratincoles)	9	4
Gruidae (cranes)	15	10
Haematopodidae (oystercatchers)	2	0
Heliornithidae (finfoots)	3	2
Ibidorhynchidae (ibisbill)	1	0
Laridae (gulls, terns, skimmers)	48	19
Pelecanidae (pelicans)	6	2
Phalacrocoracidae (cormorants)	10	7
Phoenicopteridae (flamingos)	5	0
Pluvianellidae (Magellanic plover)	1	0
Pluvianidae (Egyptian plover)	1	0
Podicipedidae (grebes)	23	19
Rallidae (rails, gallinules, coots)	115	65
Recurvirostridae (avocets, stilts)	7	9
Scolopacidae (sandpipers, snipes, phalaropes)	68	43
Thinocoridae (seedsnipes)	3	0
Threskiornithidae (ibises, spoonbills)	28	17
Total	646	382

An authoritative list of freshwater birds would help to better link management of freshwater habitats to the conservation of some bird species. Official documents and websites of both BirdLife International and the Ramsar Convention on Wetlands currently use the term "waterbird" interchangeably with the term "waterfowl" (BirdLife 2017; Ramsar Secretariat, 2008). This could lead to inadequate attention to the needs of some freshwater bird species because the term waterfowl is generally understood to refer to a subset of water dependent species: "ducks, geese, or other large aquatic birds, especially when regarded as game" (*Oxford English Dictionary* https://en.oxforddictionaries.com/). Official Ramsar documents do allow for a broader use of the term water bird to include "birds ecologically dependent on wetlands" (Ramsar Secretariat, 2008). In the same document bird groups are listed at the order level and wetland dependent raptors, owls, coucals and the Hoatzin are specifically mentioned (Ramsar Secretariat, 2008). However, these specifically mentioned groups and species are not included in the list of waterbirds available from BirdLife International (2017). More precise information on the dependence of each species on wetland extent and condition would help in placing a species in the terrestrial–aquatic gradient but information about the marine–inland gradient is also needed. Local and expert knowledge, including indigenous knowledge, could help to decide on the reliance of bird species on the condition and extent of freshwater habitats in a given protected area. The list in Table 7.3 could be a useful starting point, alerting PA managers, ground staff and practitioners that some groups of birds not traditionally considered as waterbirds may require management of freshwater habitats to persist in the PA. The numbers of waterbird species for each bird family associated with inland wetlands in BirdLife International (2017) in relation to the numbers in the FADA (Table 7.4) may also help to determine which of the species traditionally known as waterbirds are not strictly dependent on freshwater habitats.

What must be done to conserve freshwater species?

Knowledge of the extent of and nature of the dependence of a species on inland water habitats along with other information on its biology are critical for making good decisions about freshwater species conservation. However, it is also important to use the vast conservation experience from across the world in determining management priorities and optimizing conservation outcomes. Resources like the database of conservation actions documenting and summarizing published evidence of the effectiveness of a large number of conservation actions (Sutherland et al., 2017; Table 7.1) can be very helpful. Sutherland et al (2017) used published evidence to assign a given conservation action to one of the following categories of effectiveness: beneficial; likely to be beneficial; trade-off between benefit and harm; unknown effectiveness (limited evidence); unlikely to be beneficial; likely to be ineffective or harmful. Effectiveness assessments of a selection of the actions that are relevant to the conservation of birds and amphibians is given in the final column of Table 7.1 along with the numbers of studies upon which these assessments were based. This database still only covers a limited taxonomic range (birds, bats and amphibians). Further, in assessing each group, it is biased toward particular subsets of the species in a group. For example, under almost every broad category of actions there are multiple effectiveness assessments for wildfowl but there is only one assessment that is relevant to kingfishers (Table 7.1), probably reflecting the general absence of kingfishers in lists of waterbirds (Table 7.3).

Even with a comprehensive freshwater-focussed database of conservation evidence, the complex requirements of successful conservation of individual species would require more information. For example, the social–political setting, history of conservation and resource use and the surrounding landscape are also important. Below in boxes 7.1–7.6 we present the conservation profile of six species from different continents and across a wide phylogenetic range: plants, crustaceans, mammals, fish, reptiles and amphibians. A mere six species is clearly insufficient to illustrate the challenges in conserving freshwater species but these profiles can help illustrate how freshwater species conservation can be integrated in to the business of managing protected areas.

Conservation status, international profile and community awareness

One of the main challenges of freshwater species conservation in PAs is finding the resources needed for conservation and, in particular, justifying the investment in conserving freshwater species over other management priorities. Credible assessments of the conservation status of a species can play an important role in mobilizing resources for species conservation. The IUCN *Red Lists of Threatened Species* (IUCN, 2016) classify threatened species using clearly defined categories and criteria (IUCN, 2012). The placement of a species in critically endangered, endangered, vulnerable, near threatened, or least concern categories in the IUCN red list can greatly help obtain public support and funds to conserve freshwater species. For example, the listing of the Singapore freshwater crab (Box 7.1) as critically endangered by IUCN was followed by its adoption as an icon of national heritage in Singapore, which facilitated the development of an effective multi-faceted conservation strategy for the species. Community awareness and support was particularly important in the case of the Singapore freshwater crab, where a strict PA in a very highly urbanized country was unable to protect a species from a threat like stream acidification. However, credible assessments of the conservation status of a species cannot be made without some data about its distribution, population size and the threats it faces. This applies as much to national or jurisdictional listings as it does to IUCN red lists. Each of the jurisdictions responsible for species conservation for the six species in Boxes 7.1–6 have rigorous processes to evaluate the conservation status of species which often overlap with the IUCN processes to different degrees, at the very least by using the best available data and best experts. PA managers and ground staff can help this process by collecting or facilitating the collection of high quality data. It would also help if they can ensure that these data are readily available. The list of taxa and species numbers given in Tables 7.1 and 7.2 are global summaries of freshwater biodiversity but they also provide an insight into the potential richness of life in freshwater habitats within any PA. In an ideal world each PA would have a complete inventory of species from the groups listed in Tables 7.1 and 7.2.

Box 7.1   Conservation of the Singapore freshwater crab: *Johora singaporensis*

The Singapore freshwater crab *Johora singaporensis* (Figure 7.1) is regarded as an icon of Singapore's national as well as natural heritage. Out of the five known populations, the population at the site from which the species was first described has declined significantly, possibly as a result of anthropogenic acidification while the other populations remain robust (Ng et al., 2015). In 2008, it was assessed by IUCN to be critically endangered (Esser et al., 2008), and in 2012 it was declared by the IUCN as one of the world's 100 most threatened species (Baillie and Butcher, 2012).

Figure 7.1 Singapore freshwater crab: *Johora singaporensis*. Photo: Daniel Ng.

Saving the Singapore freshwater crab required multiple actions involving many stakeholders. The National Parks Board, Singapore, commenced working closely with National University of Singapore and Wildlife Reserves Singapore and the IUCN Species Survival Commission. Other stakeholders were subsequently brought into a process that led to the development of a conservation strategy (Figure 7.2) focusing on the broad

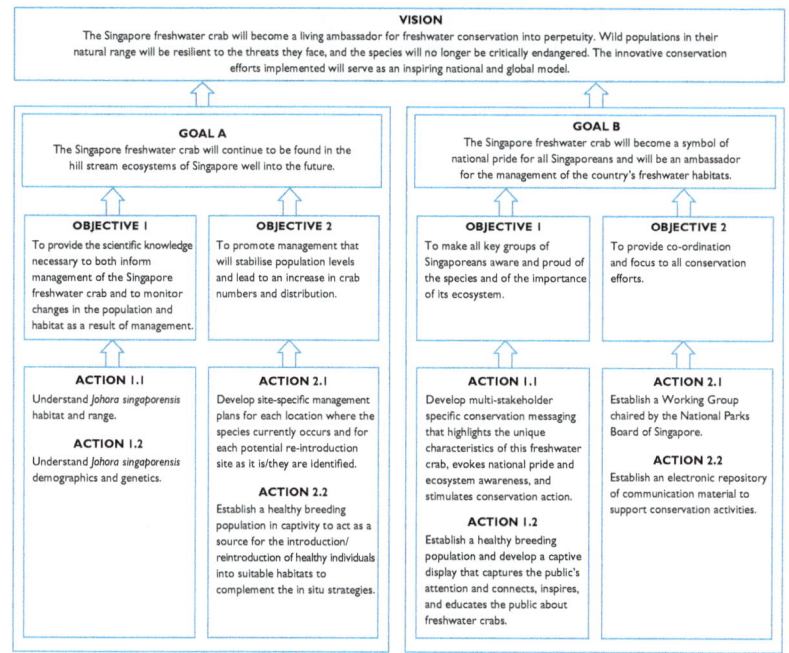

Figure 7.2 The conservation strategy for Singapore freshwater crab, *Johora singaporensis*. (This figure was adapted from Yeo et al., 2016).

aspects of ecological research, site management, captive breeding, and education and communication (Yeo et al., 2016).

The *J. singaporensis* conservation strategy has important broader global implications. It can serve as an example of what can be done in even a highly urbanized country with multiple competing economic and land use priorities, first and foremost for other similarly threatened crab species, but also for other freshwater invertebrate species (Yeo et al., 2016).

Acknowledgement

Daniel Ng, National Parks Board, Singapore Botanic Gardens, provided helpful comments on a draft of the text in Box 7.1.

International and national recognition of the high extinction risk of a species may not be sufficient for people to change behaviour that threatens a species. For example, the Central American river turtle (Box 7.2) is listed by the IUCN as Critically Endangered and there are major initiatives in Mexico, Guatemala and Belize aimed at protecting it, including a specially designated PA in Mexico which is a Ramsar site. Yet the species still faces extinction, mainly because of overexploitation as a food source. National or jurisdictional legislation and international agreements can also help to protect species threatened by overexploitation. One such international agreement is the Convention on International Trade in Endangered Species of Wild Fauna and Flora (CITES, https://cites.org/eng). The Central American Turtle is listed in CITES Appendix II which means international trade in it may be authorized by a permit or export certificate and its sustainable use may be possible under certain circumstances, such as captive breeding following a strict management plan in authorized management units (https://cites.org/eng). Allowing controlled trade under CITES Appendix II may contribute to saving species like the Central American river turtle by helping to meet the demand for the species though controlled farming and hence removing or reducing the threat to wild populations.

Box 7.2   Conservation of the Central American river turtle: *Dermatemys mawii*

The Central American river turtle *Dermatemys mawii* (Figure 7.3) is one of the most endangered species of turtles and the only living species in the family Dermatemydidae. It is listed as critically endangered by IUCN (Vogt et al., 2006). *D. mawii* has been listed in CITES appendix II since 1981. Currently the greatest threat to this species is human over-consumption. Anthropogenic habitat disturbance, mainly caused by cattle ranching or changes on river flows, also threaten this species (Guichard, 2006). In

Figure 7.3 The Central American river turtle *Dermatemys mawii*. Photo: Gerardo Ceballos Gonzáles/CONABIO.

Mexico *D. mawii* was included in the Mexican Red List data lists as endangered (SEMARNAT, 2010). In Belize there is national legislation designed to control the level of harvest and establish some protected populations. *D. mawii* is protected under Guatemalan law and is categorized as a highly protected species under the US Endangered Species Act.

In addition to the legal protection, there have been significant conservation actions to help protect *D mawii*. Laguna La Popotera, Veracruz, Mexico, was designated a Ramsar site in 2005 with specific hopes of creating the first extensively managed wildlife reserve specifically for this species. The species is present in Laguna del Tigre N.P. in Peten, Guatemala, and was recommended as a focal species for park management (CONABIO, 2014). In 2004 a group for *D. mawii* conservation was established, proposing a program that includes raising the authorities' and peoples' awareness of the species, its captive conservation that includes the genetic management of the captive populations, evaluating the wild populations and their habitat, improved enforcement of existing legal protection, promotion of harvest management measures for sustainable use, and the creation of a reintroduction program. One of the possibilities proposed is meeting some, if not all, of the intensive commercial demand for the species by sustainable production from farms (CONABIO, 2014).

Acknowledgements

Esther Quintero and Rafael Ramirez from CONABIO, Mexico, provided helpful comments on a draft of the text in Box 7.2.

Conflicts and synergies between conservation of freshwater species and other objectives

Some management practices intended for maintaining healthy terrestrial ecosystems may harm freshwater species. For example, spraying of invasive plants with herbicides, prescribed burning, and the use of fire retardants in managing protected areas on the coastal fringe regions in Eastern Australia could affect a small threatened fish species, the Oxleyan pygmy perch (Box 7.3), inhabiting streams in the region. In addition, some of the benefits offered by the PA to visitors such as camping and four-wheel driving may also be causing harm to the species. However, communicating the needs of individual species to users of protected areas and ensuring compliance with species-specific rules can be difficult. Focusing conservation efforts at the level of habitats, ecosystems or ecological communities may be an effective way of conserving freshwater species, especially if these ecosystems, habitats or ecosystems are easy to recognize and describe. This approach is particularly advantageous if there is a possibility that, because of their distinctiveness, these ecosystems or ecological communities are likely to harbour undiscovered species not found elsewhere.

Box 7.3   Conservation of the Oxleyan pygmy perch: *Nannoperca oxleyana*

The Oxleyan Pygmy Perch *Nannoperca oxleyana* (Figure 7.4) is a small fish found mostly in small, low gradient, low pH coastal streams in Eastern Australia (Arthington and Knight, 2012; Knight et al., 2012). Its natural freshwater habitat is highly fragmented and has increasingly been degraded and lost since European settlement, leading to a major reduction in its distribution (Knight and Arthington, 2008). In 1996 it was listed by the IUCN as endangered (Wager, 1996). Although the majority of extant populations of *N. oxleyana* are located in national parks and other protected areas, the species remains threatened by degradation of aquatic habitats downstream from reserves, disruption to fish passage, water pollution and riparian degradation associated with public access, fire management activities and the introduction of alien (e.g., *Gambusia holbrooki*) and translocated species.

The current recovery plan for *N. oxleyana* details the need to investigate and implement options for providing increased protection for key areas of *N. oxleyana* habitat (NSW DPI, 2005a, b). Another important

Figure 7.4 **The Oxleyan pygmy perch** *Nannoperca oxleyana*. **Photo: Gunther Schmida.**

aspect of the recovery program for *N. oxleyana* is engagement and communication with public authorities responsible for managing *N. oxleyana* habitat. Engaging with PA managers is fundamental to managing threats within the PAs including run-off originating from unsealed roads, herbicide use, fire management activities and disturbance by recreational users such as four-wheel-drive vehicles and campers. The range of measures adopted to lessen impacts on *N. oxleyana* include, for example: avoiding prescribed burning activities during the main breeding season (October to April); utilising a mosaic burning pattern where only a third to a quarter of the wetlands are burnt in any given year; avoiding the pumping of water from *N. oxleyana* habitats; and avoiding the use of fire-fighting foam and retardants within 50 m of *N. oxleyana* habitat.

The recovery plan also promotes the reporting and monitoring of introduced fish in environments inhabited by *N oxleyana* and supports measures to reduce their impacts. Emphasis has been placed on documenting and monitoring the distribution and range expansions of species such as *Gambusia holbrooki* relative to *N. oxleyana*, and gathering evidence of its impact.

The recovery plan recognizes the need for a strategically focused monitoring program to enable the effectiveness of the recovery objectives to be evaluated. A long-term monitoring program has been developed to assess the ongoing status of populations (NSW DPI, 2015). To date, monitoring surveys in NSW and Queensland have found many populations persisting where previously recorded (NSW DPI, 2015; 2016).

Acknowledgement

Angela Arthington, Emeritus Professor, Griffith University, Australia, provided helpful feedback on earlier drafts of the text in Box 7.3.

Focussing on protecting and restoring the ecological community as a step towards recovery of a threatened species is particularly useful when recovery of the species is a long-term goal. For example, recovery of the water grass (Box 7.4) in the modified lowland wetlands of the Cape Peninsula may not be feasible in the short term, or even in the medium term, but the knowledge of the hydrological regime, water chemistry and other environmental requirements of such species, derived from both historical and present-day data on wetlands where the species are found, can be particularly helpful in setting benchmarks and targets for ecosystem restoration.

Box 7.4 Conservation of water grass: *Cotula myriophylloides*

The water grass *Cotula myriophylloides* (Figure 7.5) is endemic to the Western Cape region of South Africa. It was thought to be extinct until three small isolated populations were discovered in 2008 on the southern Cape coast, scattered over an area of over 400 kilometres, and it is currently listed as Critically Endangered in the IUCN red lists (Helme and Raimondo, 2010) and the Red List of South African Endangered Plants (Powell et al., 2013). Herbarium records indicate that this species was common on the Cape Peninsula in the past, but it is now locally extinct, probably because previously seasonal wetlands are now mostly permanent eutrophic wetlands. Important ongoing threats include urbanization and associated eutrophication, alteration to hydrology and invasive alien species.

Figure 7.5 Water grass *Cotula myriophylloides*. Photo: Nick Helme.

The vegetation types associated with this species persist as remnants in various protected areas in the Cape peninsula as well as further along the southern Cape coast. These reserves, which include Table Bay Reserve surrounded by heavy urban development, are being managed to protect and restore these ecological communities (Retief, 2011), possibly creating the conditions for *C. myriophylloides* to re-establish in the wetlands where it was once abundant.

Acknowledgement

Jeanne Nel from Nelson Mandela Metropolitan University, South Africa, provided helpful feedback on the text in Box 7.4.

Understanding ecological processes and prioritising management responses

Successful conservation of species requires reliable assessments of the magnitude of threats and realistic prioritization of conservation actions. The Patagonia Frog (Box 7.5) is potentially facing threats from disease, habitat destruction and eutrophication by livestock, desiccation of ponds and predation plus competition for food from introduced fish. The research conducted on this and other frog species in similar landscapes has provided insight into the likely causes of the disappearance of the species from the main water body of Laguna Blanca as well as the resilience of remaining populations in the face of ongoing multiple threats.

Box 7.5   Conservation of the Patagonia frog: *Atelognathus patagonicus*

The Patagonia frog *Atelognathus patagonicus* (Figure 7.6) inhabits permanent and temporary water bodies in a system of endorheic ponds on the basaltic plateau in and around Laguna Blanca National Park in northwestern Patagonia, Argentina (Fox et al., 2005; Cuello et al., 2009a). The species was named in 1962 from Laguna Blanca (1,670 ha), the largest lake in the region. *A. patagonicus* was very abundant in Laguna Blanca until the late 1970s but the numbers declined sharply in the late 1980s and the species is assumed to have been subsequently extirpated from that lake (Fox et al., 2005). Since then the species has been recorded at 23 small (0.5–135 ha), water bodies in the vicinity of Laguna Blanca which were characterized by the absence of *Percichthys colhuapiensis* (Cuello et al., 2009a), which is an invasive predatory fish native to Patagonia but not to the Laguna Blanca system. It was deliberately introduced into Laguna Blanca (Fox et al., 2005). The small water bodies containing populations of *A. patagonicus* are not connected with Laguna Blanca, which was fishless until the introduction and translocation of fish species in the 1940s to 1960s (Úbeda et al., 2010). These introductions are thought to have caused significant negative changes to the native aquatic biota (Ortubay et al., 2006).

(continued)

(*continued*)

Figure 7.6 Patagonia frog *Atelognathus patagonicus*. Photo: Richard Sage.

Currently A. *patagonicus* is listed by the IUCN as Endangered (Úbeda et al., 2010). A revision of the amphibians of Argentina undertaken by the Argentinian Herpetological Society also concluded that this species is Endangered and recommended several actions to help protect the existing populations (Vaira et al., 2012; Basso et al., 2012).

The main cause of the extirpation of the population of A. *patagonicus* in Laguna Blanca is thought to be the presence of a large population of *Percichthys colhuapiensis* in the lake (Fox et al., 2005; Ortubay et al., 2006). The mechanisms causing this extirpation may be complex and probably include predation by P. *colhuapiensis* on the amphipod Hyalella sp. which is the preferred food of A. *patagonicus* (Cuello et al., 2006; 2009b). Also P. *colhuapiensis* is thought to be responsible for the destruction of the macrophytes in the lake, which provided shelter for tadpoles and frogs and their prey (Ortubay et al., 2006; Cuello et al., 2009a). Fox et al (2006) proposed infectious disease as a possible contributor to the disappearance of A. *Patagonicus* from Laguna Blanca. In 2005 populations of A. *patagonicus* in other lakes were observed to be affected by an infectious disease but they were not decimated by it (Fox et al., 2005). Since that time there have been no reports of disease-related mortality in A. *patagonicus* (Carmen Úbeda pers.com.). There was also some evidence that destruction of aquatic macrophytes and marginal vegetation on the shores of other ponds by livestock may be depriving the frogs of shelter and causing decline in A *Patagonicus* populations (Úbeda, et al., 2010; Cuello et al., 2014).

The populations of A. *patagonicus* at greatest risk are those found in water bodies currently not protected by Laguna Blanca National Park (Cuello et al., 2009a). Conservation actions recommended by Basso et al., (2012) include: rigorous implementation of protocols developed for preventing the spread of the chytrid fungus; limiting livestock access to the water bodies where this species is currently found; preventing any changes to stream channels that may allow fish to reach water bodies that currently have populations of A. *patagonicus*; and strengthening and enforcing the existing legislation that prohibits new fish introductions to water bodies in the region. The last point is particularly important because of a strong push towards the development of recreational fishing in the region (Basso et al., 2012). The feasibility and benefits of limiting livestock access to the ponds inhabited by A. *patagonicus* was demonstrated when National Parks Administration authorities fenced one of the ponds as a pilot experiment. Marginal vegetation on the shores recovered rapidly, providing additional shelter for A. *patagonicus* metamorphs (Cuello et al., 2014). It was also shown that A. *patagonicus* can employ different life-history strategies to optimize the utilization of both temporary and permanent ponds (Cuello et al., 2008, 2014). Based on this, Cuello et al (2014) recommended that temporary ponds with hydroperiods of at least four months should be protected because they are suitable environments for the breeding and development of A. *patagonicus*.

Acknowledgements

Carmen Úbeda and Maria Elena Cuello, from Comahue University, Argentina, provided helpful feedback and guidance on earlier drafts of the text in Box 7.5.

This research has demonstrated that populations of the A. *Patagonicus* are able to respond to changes in hydrology, utilize a wide range of food sources, survive disease and disturbance as well as predation by fish. Having access to the results of such research is very helpful for PA managers in conserving species because it helps direct management efforts towards actions that are likely to have multiple benefits and where there is greater certainty of success.

A system-level response and where people fit in

Prioritizing conservation actions for species facing major multiple threats is particularly difficult when there are large uncertainties about many of these threats. Climate change clearly puts the future of species like the Saimaa ringed seal in the balance (Box 7.6). Despite the recent increase in population size, many factors make increased reproductive failure and juvenile mortality a real possibility in the near future: extremely low genetic variability in the population; the decline in availability of snow to build lairs; shorter periods of ice cover; increased runoff from the forest floor caused by higher precipitation and the consequent likely increase of higher mercury concentrations in seal tissues; and emergence of avian predators as a major threat to seal pups because of their earlier arrival to Lake Saimaa caused by climate (Box 7.6). The challenges in conserving Saimaa ringed seal (Box 7.6) exemplify many of the disadvantages of being dependent on freshwater. Like the Florida manatee and Indus River dolphin (Table 7.1), the dependence of the Saimaa ringed seal on freshwater places them at greater proximity to human activity and greater danger, compared with their relatives that live in marine environments.

Box 7.6   Conservation of the Saimaa ringed seal:
Pusa hispida saimensis

The Saimaa ringed seal *Pusa hispida saimensis* (Figure 7.7) is a subspecies of ringed seal that has a very small isolated population landlocked within Lake Saimaa in Finland, at the southern edge of the global range of the species. The size of the population is currently estimated to be around 360 individuals (Tero Sipilä, pers. comm.). Although the numbers are slowly growing, the population is not stable (Sipilä et al., 2013) and the subspecies is listed as endangered by the IUCN (Sipilä, 2016).

Figure 7.7 A new-born pup of the Saimaa ringed *seal Pusa hispida saimensis*. Photo: Timo Seppäläinen.

The populations of the Saimaa ringed seal declined severely until the 1980s. Some major causes of this decline are no longer considered threats. For example, early last century seals were shot to make oil or because they were seen as a threat to commercial and recreational fishing (Kunnasranta et al., 2016), but the subspecies has been legally protected in Finland since 1955 (Sipilä, 2016). Another past threat which stopped in 1991 was the manipulation of the lake water-levels which caused untimely breaks in the ice sheet near the shorelines (Sipilä, 2016). Mercury pollution, which was considered to be one of the causes of decline in the Saimaa ringed seal populations last century (Hyvärinen et al., 1998), is not regarded currently to be a major threat (Sipilä, 2016). However, high mercury levels in the tissues of the Saimaa ringed seal may return as a threat in the future because of increased runoff from the land following higher precipitation brought about by climate change (Sipilä, 2016). Climate change could also bring about new threats, e.g., predation of pups by avian predators, or exacerbate currently minor threats, e.g., predation by red foxes (Sipilä, 2016). Today the most serious ongoing threats to the population of the Saimaa ringed seal are fishing by gill nets, large fish traps and fish baited hooks; and the warming climate which is already having adverse effects on pupping habitat via poor snow and ice conditions (Sipilä, 2016). Other ongoing threats include use of the lake for snowmobiling, ice skating, cross-country skiing, tourism; and building and use of cottages on the lake shoreline (Sipilä, 2016).

Current conservation interventions to protect the Saimaa ringed seal primarily address the threats from fishing and climate warming. Fishing restrictions extending over 60 per cent of the area of Lake Saimaa during the most critical times of the year are probably the most important reason that population sizes of the seal have been growing in the past three decades (Kunnastranta et al., 2016). Responding to the threat of climate warming has required some extraordinary actions. For example, the winter of 2013–2014 was extremely mild and there was no wind-drifted snow at the seals' lair sites. A project of large-scale construction of the human-made snowdrifts was implemented. Altogether 225 snowdrifts were piled all around Lake Saimaa through the efforts of over 100 volunteers. Over 90 per cent of pups observed in the annual lair census were born in the piled drifts and the perinatal mortality remained at a much lower level than it was expected to be without this intervention (Autilla et al., 2016). Effective PA management within conservation in Kolovesi and Linnansari national parks and at sites in the Natura 2000 are considered to have a critical role in reducing harmful disturbance of the habitat and, in particular, disturbance of breeding grounds during recreational activities (Sipilä, 2016).

Conservation of the Saimaa ringed seal is supported by rigorous scientific research. Valtonen (2014) showed recently that genetic diversity of

(continued)

(continued)

the Saimaa ringed seal was extremely low and that it was likely to continue to decrease unless the seal's population size can be increased substantially. Research using novel non-invasive monitoring methods and a novel method of building artificial snowdrifts (Autilla, 2015) is now used successfully to improve the seals' breeding success during winters with poor snow conditions (Autilla et al., 2016; Kunnasranta et al., 2016). Research into the behavioural ecology of the Saimaa ringed seal generated precise maps of the distribution and breeding areas and revealed new information on site fidelity, the timing and extent of the movements of the seals and their behavioural responses to various human activities (Niemi, 2013). These research findings were critical for determining effective specific conservation actions (Kunnasranta et al., 2016). These findings also underpin the overall conservation strategy for the Saimaa ringed seal, for example by providing justification for investing in efforts to increase the reproductive rate and pup survival so that the population size can quickly reach levels that would make the subspecies less vulnerable to stochastic events.

Acknowledgements

Tero Sipilä, Metsähallitus, and Marja Niemi, the University of Eastern Finland, provided helpful feedback on a draft of the text in Box 7.6.

The very attributes that make freshwater species particularly vulnerable to disturbances in a watershed can also be a plus for their conservation. The pathways through which human activities impact upon freshwater species are often easier to elucidate and interrupt than in terrestrial and marine ecosystems. For example, the largemouth perch *Perchictys colhuapiensis*, implicated in the extirpation of the main population of the Patagonia Frog in Laguna Blanca, is native to Patagonia and found in a drainage system that is at a distance of only 10 km from Laguna Blanca. Yet this species did not reach Laguna Blanca until it was deliberately introduced, possibly in the1960s, and at least until now the species has been successfully kept out of the 23 ponds that have populations of Patagonia frog, including populations that are separated from a very large population of *P. colhuapiensis* by only several hundred meters of land (Box 7.5).

The capacity for human activities anywhere within a watershed to cause direct harm to freshwater species also gives local communities more opportunities and capacity to make a direct positive contribution to conserving threatened freshwater species compared with what they can do directly for threatened marine or terrestrial species. There is little doubt that ongoing support from the communities of Lake Saimaa is essential for success in the conservation of the Saimaa ringed seal. Jaakkola et al (2014) argued that the national conservation programs of the 1980s and 1990s might have contributed to the feeling among rural inhabitants of the Saimaa area that conservation decisions are made "somewhere else". This perception may have led to weaker local support for the conservation of the Saimaa ringed seal compared with the strong support in other parts of the country. It was suggested that the adoption of more participatory planning and decision-making procedures would probably increase support for Saimaa seal protection in the local community (Jaakkola et al., 2014). In a similar way, success in the conservation of the Patagonia frog may depend, at least partly, on the support and engagement of local subsistence pastoralists whose livelihoods depend on their livestock having easy access to freshwater (Box 7.5).

An integrative approach to freshwater species conservation in Protected Areas

The examples we looked at in this chapter illustrate the value of approaching freshwater species conservation in PAs at the system level, with people being a part of that system. Useful tools to use in working out these system-level responses include conceptual models of how these systems work and where people, freshwater species and other components of biodiversity fit in. For PA management it would be particularly useful to develop such conceptual models for entire PAs or at least for ecological features such as wetland types within a PA (Turak et al., 2016). Ideally multiple interlinked conceptual models would be developed to guide, coordinate and evaluate conservation actions locally, regionally and nationally. At the scale of an individual PA the steps include: documenting freshwater species and ecosystems in the PA; setting conservation goals for freshwater species and ecosystems; using conceptual models to explore how these goals relate to other management goals including the conservation of marine and terrestrial species and ecosystems; and efficiently implementing conservation actions that optimise multiple goals and benefits (Figure 7.8).

As part of the process of defining objectives and determining how to meet these, traditional ecological knowledge of Indigenous and other local peoples should be used to identify cultural keystone species (Noble et al., 2016). These are fish, molluscs, crustaceans and other aquatic species that perform key functions in or are sensitive to changes in the ecosystem, and are also of great cultural importance. This approach engenders greater support across many parts of society offering the prospects of catalysing more resources for conservation.

Once freshwater conservation objectives are determined, they can be considered alongside management objectives specific to marine and terrestrial features and social, cultural or economic objectives that are not necessarily specific to any natural feature. The clarity with which freshwater-related management objectives are defined will have a major impact on the success of the PA in conserving freshwater species.

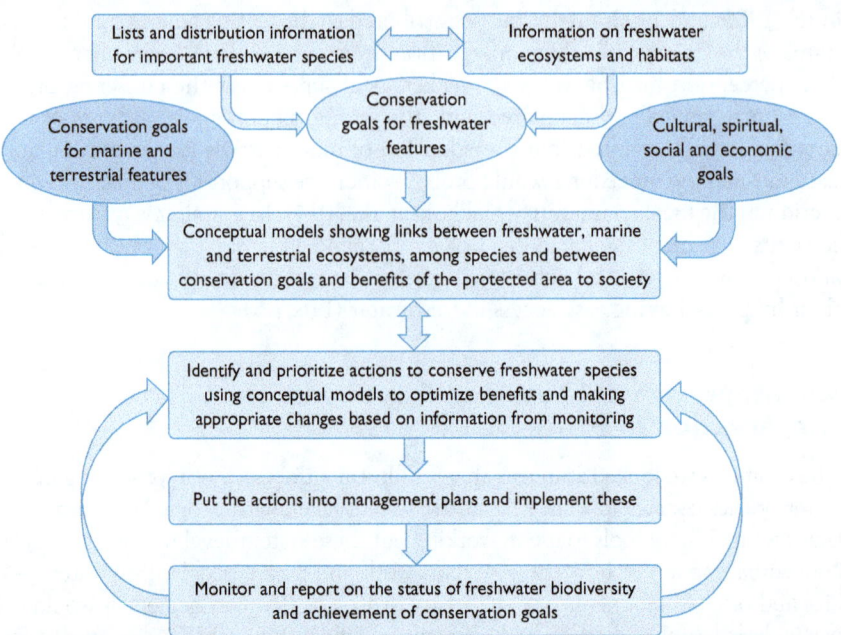

Figure 7.8 A schematic representation of steps that PA managers can take in integrating freshwater species conservation into the management of protected areas. These steps are applicable to any PA. This process is focussed on freshwater biodiversity but assumes that similar steps may be taken for marine and terrestrial biodiversity.

References

Abell, R., Thieme, M.L., Revenga, C., Bryer, M., Kottelat, M., Bogutskaya, N., Coad, B., Mandrak, N., Balderas, S.C., Bussing, W. and Stiassny, M.L. (2008) 'Freshwater ecoregions of the world: a new map of biogeographic units for freshwater biodiversity conservation', *BioScience*, vol 58, pp403–414.

Abeli, T., Barni, E., Siniscalco, C., Amosso, C. and Rossi, G. (2012) 'A cost-effective model for preliminary site evaluation for the reintroduction of a threatened quillwort', *Aquatic Conservation: Marine and Freshwater Ecosystems*, vol 22(1), pp66–73.

AmphibiaWeb (2017) http://amphibiaweb.org, University of California, Berkeley, CA, USA. Accessed 19 Mar. 2017.

Arthington, A.H., Dulvy, N.K., Gladstone, W. and Winfield, I.J. (2016) 'Fish conservation in freshwater and marine realms: status, threats and management', *Aquatic Conservation: Marine and Freshwater Ecosystems*, vol 26(5), pp838–857.

Arthington, A.H. and Knight, J.T. (2012) 'Oxleyan pygmy perch *Nannoperca oxleyana* Whitley'. In Curtis, L.K., Dennis, A.J., McDonald, K.R., Kyne, P.M. and Debus, S.J.S. (eds), *Queensland's Threatened Animals*. CSIRO Publishing, Melbourne, pp123–124.

Auttila, M. (2015) 'The endangered Saimaa ringed seal in a changing climate – challenges for conservation and monitoring'. PhD thesis, University of Eastern Finland.

Auttila, M., Heikkilä, P., Koskela, J., Kunnasranta, M., Marttinen, I., Niemi, M., Tiilikainen, R. and Sipilä, T. (2016) *New Methods Improve Conservation and Monitoring of the Saimaa Ringed Seal in a Changing Climate*. Metsähallitus publications, Series A 223,Vantaa.

Baillie, J.E.M. and Butcher, E.R. (2012) *Priceless or Worthless? The World's Most Threatened Species*. London, UK: Zoological Society of London.

Balian, E.V., Lévêque, C., Segers, H. and Martens, K. (eds) (2008a) *Freshwater Animal Diversity Assessment* (Vol. 198). Springer Science & Business Media.

Balian, E.V., Lévêque, C., Segers, H. and Martens, K. (2008b) 'An introduction to the freshwater animal diversity assessment (FADA) project', *Hydrobiologia*, vol 595, pp3–8.

Balian, E.V., Segers, H., Lévéque, C. and Martens, K. (2008c). 'The freshwater animal diversity assessment: An overview of the results', *Hydrobiologia*, vol 595, pp627–637.

Basso N.G., Úbeda C.A., Martinazzo, L. and Cuello, M.E. (2012) 'Atelognathus patagonicus (Gallardo, 1962). Rana de Laguna Blanca, Categorización del estado de conservación de la herpetofauna de la República Argentina. Ficha de los Taxones. Anfibios', *Cuadernos de herpetología*, vol 26 (Supl 1): p173.

Bernal, R., Hoyos-Gomez, S.E. and Borchsenius, F. (2017) 'A new, critically endangered species of Aiphanes (Arecaceae) from Colombia', *Phytotaxa*, vol 298(1), pp65–70.

BirdLife International (2017) http://datazone.birdlife.org/species/search. Accessed 28 May 2017.

CONABIO (2014) 'Tortuga blanca (*Dermatemys mawii*)', http://bios.conabio.gob.mx/especies/8001027.pdf. Accessed 28 May 2017.

Cuello, M.E., Bello, M.T., Kun, M. and Úbeda, C.A. (2006) 'Feeding habits and their implications for the conservation of the endangered semiaquatic frog *Atelognathus patagonicus* (Anura, Neobatrachia) in a northwestern Patagonian pond', *Phyllomedusa*, vol 5(1), pp67–76.

Cuello, M.E., Úbeda C.A. and Bello M.T. (2008) 'Relationship between morphotypes of *Atelognathus patagonicus* (Anura, Neobatrachia) and environmental conditions: evidence and possible explanation', *Phyllomedusa*, vol 7, pp35–44.

Cuello, M.E., Perotti, M.G. and Iglesias, G.J. (2009a) 'Dramatic decline and range contraction of the endangered patagonian frog, *Atelognathus patagonicus* (Anura, Leptodactylidae)', *Oryx*, vol 43, pp443–446.

Cuello M.E., Úbeda, C.A., Bello, M.T. and Kun, M. (2009b) 'Seasonal trophic activity of the aquatic morphotype of *Atelognathus patagonicus* (Anura, Neobatrachia) and prey availability in the littoral benthos of a permanent pond in Argentinean Patagonia', *Phyllomedusa*, vol 8, pp135–146.

Cuello, M.E., Úbeda, C.A., Bello, M.T. and Perotti, M.G. (2014) 'Plastic patterns in larval development of Endangered endemic *Atelognathus patagonicus*: implications for conservation strategies', *Endangered Species Research*, vol 23(1), pp83–92.

Darwall, W.R.T. and Freyhof, J. (2016) 'Lost fishes, who is counting? The extent of the threat to freshwater fish biodiversity'. In *Conservation of Freshwater Fishes*, Closs, G.P., Krkosek, M. and Olden, J.D. (eds). Cambridge University Press: Cambridge, pp1–36.

Dehorter, O. and Guillemain, M. (2008) 'Global diversity of freshwater birds (Aves)', *Hydrobiologia*, vol 595(1), pp619–626.

Deutsch, C.J., Self-Sullivan, C. and Mignucci-Giannoni, A. (2008) '*Trichechus manatus*'. In *The IUCN Red List of Threatened Species*: e.T22103A9356917, http://dx.doi.org/10.2305/IUCN.UK.2008.RLTS.T22103A9356917.en. Downloaded on 20 May 2017.

Esser, L., Cumberlidge, N. and Yeo, D.C.J. (2008) 'Johora singaporensis'. In The IUCN Red List of Threatened Species: e.T134219A3921290, http://dx.doi.org/10.2305/IUCN.UK.2008.RLTS.T134219A3921290.en. Downloaded on 12 March 2017.

Fox, S.F., Yoshioka, J.H., Cuello, M.E. and Úbeda, C. (2005) 'Status, distribution, and ecology of an endangered semi-aquatic frog (Atelognathus patagonicus) of northwestern Patagonia, Argentina', Copeia, vol 4, pp921–929.

Fox, S.F., Greer, A.L., Torres-Cervantes, R. and Collins, J.P. (2006) 'First case of ranavirus-associated morbidity and mortality in natural populations of the South American frog Atelognathus patagonicus', Diseases of aquatic organisms, vol 72(1), pp87–92.

Froese, R. and D. Pauly, D. (eds) (2017) FishBase, www.fishbase.org, version (02/2017). Accessed 20 May 2017.

Guichard Romero, C.A. (2006) Situación actual de las poblaciones de tortuga blanca (Dematemys mawii) en el sureste de México. México, D.F.

Helme, N. and Raimondo, D. (2010) 'Cotula myriophylloides'. In The IUCN Red List of Threatened Species: e.T185419A8407927, http://dx.doi.org/10.2305/IUCN.UK.2010-3.RLTS.T185419A8407927.en. Downloaded on 17 March 2017.

Hammerson, G.A. (2007) 'Thamnophis gigas'. In The IUCN Red List of Threatened Species: e.T21706A9310655, http://dx.doi.org/10.2305/IUCN.UK.2007.RLTS.T21706A9310655.en. Downloaded on 20 May 2017.

Huang, C.M., Yu, H., Wu, Z.J., Li, Y.B., Wei, F.W. and Gong, M.H. (2008) 'Population and conservation strategies for the Chinese crocodile lizard (Shinisaurus crocodilurus) in China', Animal Biodiversity and Conservation, vol 31(2), pp63–70.

Hughes, J.M., Ponniah, M.H., Hurwood, D.A., Chenoweth, S. and Arthington, A.H. (1999) 'Strong genetic structuring in a habitat specialist, the Oxleyan Pygmy Perch Nannoperca oxleyana', Heredity, vol 83, pp5–14.

Hyvärinen, H., Sipilä, T., Kunnasranta, M. and Koskela, J.T. (1998) 'Mercury pollution and the Saimaa ringed seal (Phoca hispida saimensis)', Marine Pollution Bulletin, vol 36(1), pp76–81.

IUCN (2012) IUCN Red List Categories and Criteria: Version 3.1, Second edition. Gland, Switzerland and Cambridge, UK: IUCN, iv + 32pp.

IUCN (2016) The IUCN Red List of Threatened Species, Version 2016–3, www.iucnredlist.org, downloaded on 7 December 2016.

Jaakkola, M., Laasonen, S. and Vuorisalo, T. (2014) 'Safeguarding the Saimaa ringed seal. Development and present status of the attitudes and atmosphere for Saimaa ringed seal conservation'. LIFE12 NAT/FI/000367 LIFE. Project Delivery 5/2014 Saimaa Seal. University of Turku. Accessed on 9 May 2017 at: www.utu.fi/fi/yksikot/ffrc/tutkimus/hankkeet/Sivut/life.aspx.

Kawai, T. and Cumberlidge, N. (2016) A Global Overview of the Conservation of Freshwater Decapod Crustaceans. Springer.

Knight, J.T. and Arthington, A.H. (2008) 'Distribution and habitat associations of the endangered Oxleyan pygmy perch, Nannoperca oxleyana Whitley, in eastern Australia'. Aquatic Conservation: Marine and Freshwater Ecosystems, vol 18 (7): pp1240–1254.

Knight, J.T., Arthington, A.H., Holder, G.S. and Talbot, R.B. (2012) 'Conservation biology and management of the endangered Oxleyan pygmy perch Nannoperca oxleyana in Australia', Endangered Species Research, vol 17(2), pp169–178.

Kunnasranta, M., Niemi, M. and Auttila, M. (2016) 'aimaanorpan suojelubiologiaa: tutkimuksista toimenpiteisiin', Suomen Riista, vol 62, pp71–82.

Maran, T., Skumatov, D., Gomez, A., Põdra, M., Abramov, A.V. and Dinets, V. (2016) 'Mustela lutreola'. In The IUCN Red List of Threatened Species: e.T14018A45199861, http://dx.doi.org/10.2305/IUCN.UK.2016-1.RLTS.T14018A45199861.en. Downloaded on 20 May 2017.

McLaughlin, R.L., Smyth, E.R., Castro-Santos, T., Jones, M.L., Koops, M.A., Pratt, T.C. and Vélez-Espino, L.A. (2013) 'Unintended consequences and trade-offs of fish passage', Fish and Fisheries, vol 14(4), pp580–604.

Mcintyre, P.B., Liermann, C.R., Childress, E., Hamann, E.J., Hogan, J.D., Januchowski-Hartley, S.R., Koning, A.A., Neeson, T.M., Oele, D.L. and Pracheil, B.M. (2016) Conservation of Migratory Fishes in Freshwater Ecosystems. Cambridge University Press: Cambridge, pp324–360.

Niemi, M. (2013) Behavioural Ecology of the Saimaa Ringed Seal, PhD Thesis. University of Eastern Finland.

Ng, D.J.J., Yeo, D.C.J., Sivasothi, N. and Ng, P.K.L. (2015) 'Conservation challenges and action for the critically endangered Singapore freshwater crab Johora singaporensis', Oryx, vol 49, pp345–351.

Noble, M., Duncan, P., Perry, D., Prosper, K., Rose, D., Schnierer, S., Tipa, G., Williams, E., Woods, R. and Pittock, J. (2016). 'Culturally significant fisheries: keystones for management of freshwater social-ecological systems', Ecology and Society, vol 21(2).

NSW DPI (2016) 'Oxleyan Pygmy Perch – Nannoperca oxleyana', October, Primefact 181, Second Edition. Threatened Species Unit, Port Stephens Fisheries Institute. www.dpi.nsw.gov.au/__data/assets/pdf_file/0004/635422/Oxleyan-pygmy-perch-nannoperca-oxleyana.pdf.

NSW DPI (2015) Review of the Oxleyan Pygmy Perch Recovery Plan. www.dpi.nsw.gov.au/__data/assets/pdf_file/0010/635455/Review-of-the-oxleyan-pygmy-perch-recovery-plan.pdf.

NSW DPI (2005a) Oxleyan Pygmy Perch: Recovery Plan. NSW Department of Primary Industries Fisheries Management Branch, Port Stephens Fisheries Centre, Taylors Beach, www.environment.gov.au/system/files/resources/fde7b65a-446e-4aa3-b3c6-e9e3b4b9a41e/files/n-oxleyana.pdf. Accessed on 3 April 2017.

NSW DPI (2005b) Oxleyan Pygmy Perch: Background Paper. NSW Department of Primary Indus- tries Fisheries Management Branch, Port Stephens Fisheries Centre, Taylors Beach, www.environment.gov.au/system/files/resources/fde7b65a-446e-4aa3-b3c6-e9e3b4b9a41e/files/n-oxleyana-background.pdf. Accessed on 3 April 2017.

Olden, J.D., Kennard, M.J., Lawler, J.J. and Poff, N.L. (2011) 'Challenges and opportunities in implementing managed relocation for conservation of freshwater species', Conservation Biology, vol 25(1), pp40–47.

Ortubay, S., Cussac, V., Battini, M., Barriga, J., Aigo, J., Alonso, M., Macchi, P., Reissig, M., Yoshioka, J. and Fox, S. (2006) 'Is the decline of birds and amphibians in a steppe lake of northern Patagonia a consequence of limnological changes following fish introduction?', Aquatic Conservation, Marine and Freshwater, vol 16, pp93–105.

Oum, S., Hor, L., Sam, H., Sonn, P., Simpson, B. and Daltry, J.C. (2009) 'A comparative study of incentive-based schemes for Siamese crocodile Crocodylus siamensis conservation in the Cardamom Mountains, Cambodia', Cambodian Journal of Natural History, pp40–57.

Pittock, J., Finlayson, M., Arthington, A.H., Roux, D., Matthews, J.H., Biggs, H., Harrison, I., Blom, E., Flitcroft, R., Froend, R., Hermoso, V., Junk, W., Kumar, R., Linke, S., Nel, J.,

Nunes da Cunha, C., Pattnaik, A., Pollard, S., Rast, W., Thieme, M., Turak, E., Turpie, J., van Niekerk, L., Willems, D. and Viers, J. (2015) 'Managing freshwater, river, wetland and estuarine protected areas'. In Worboys, G.L., Lockwood, M., Kothari, A., Feary, S. and Pulsford, I. (eds), *Protected Area Governance and Management*. ANU Press, Canberra, pp569–608.

Powell, R.F., Helme, N.A., Raimondo, D. and von Staden, L. (2013) '*Cotula myriophylloides*', *Harv. National Assessment: Red List of South African Plants, Version 2017.1*, http://redlist.sanbi.org/species.php?species=3162-30. Accessed on 17 March 2017.

Ramsar Secretariat (2008) *Strategic Framework and Guidelines for the Future Development of the List of Wetlands of International Importance of the Convention on Wetlands* (Ramsar, Iran, 1971). Secretariat of the Convention on Wetlands of International Importance, Gland, Switzerland.

Retief, J.J. (2011) *Integrated Reserve Management Plan*. Table Bay Nature Reserve, June. Biodiversity Management Branch, Environmental Resource Management Department, City of Cape Town, https://studylib.net/download/10504532. Accessed 17 March 2017.

Richardson, J.S., Taylor, E., Schluter, D., Pearson, M. and Hatfield, T. (2010) 'Do riparian zones qualify as critical habitat for endangered freshwater fishes?' *Canadian Journal of Fisheries and Aquatic Sciences*, 67(7), pp1197–1204.

Roni, P., Beechie, T., Schmutz, S. and Muhar, S. (2012) *Prioritization of Watersheds and Restoration Projects. Stream and Watershed Restoration: A Guide to Restoring Riverine Processes and Habitats*, John Wiley & Sons, Ltd, Chichester, UK, pp189–214.

Sadasivaiah, B. and Rao, M.L.V. (2011) '*Nymphoides sivarajanii*'. In *The IUCN Red List of Threatened Species*: e.T194156A8884548, http://dx.doi.org/10.2305/IUCN.UK.2011-1.RLTS.T194156A8884548.en. Downloaded on 23 May 2017.

Schmidt-Kloiber, A. and Hering, D. (2015) 'www.freshwaterecology.info—an online tool that unifies, standardises and codifies more than 20,000 European freshwater organisms and their ecological preferences', *Ecological Indicators*, vol 53, pp271–282.

SEMARNAT (2010) *Norma Oficial Mexicana NOM-059-SEMARNAT-2010, Protección ambiental – Especies nativas de México de flora y fauna silvestres -Categorías de riesgo y especificaciones para su inclusión, exclusión o cambio – Lista de especies en riesgo*. s.l.: Secretaría de Medio Ambiente y Recursos Naturales. Accessed on 15 April 2017 at: www.profepa.gob.mx/innovaportal/file/435/1/NOM_059_SEMARNAT_2010.pdf.

Sipilä, T. (2003) 'Conservation biology of Saimaa ringed seal (*Phoca hispida saimensis*) with reference to other European seal populations', PhD thesis, University of Helsinki, Finland.

Sipilä, T., Kokkonen, T. and Koskela, J. (2013) 'The growth of the saimaa ringed seal population is unstable'. Factsheet, Metsähallitus, Vantaa.

Sipilä, T. (2016) '*Pusa hispida ssp. Saimensis*'. In *The IUCN Red List of Threatened Species*: e.T41675A66991678, http://dx.doi.org/10.2305/IUCN.UK.2016-1.RLTS.T41675A66991678.en. Downloaded on 19 March 2017.

Smith, R.K., Meredith, H. and Sutherland, W.J. (2017) 'Amphibian Conservation'. In Sutherland, W.J., Dicks, L.V., Ockendon, N. and Smith, R.K. (eds), *What Works in Conservation*. Open Book Publishers: Cambridge, UK.

Standing, K.L., Herman, T.B., Shallow, M., Power, T. and Morrison, I.P. (2000) 'Results of the nest protection program for Blanding's turtle in Kejimkujik National Park, Canada: 1987–1997', *Chelonian Conservation and Biology*, 3, pp637–642.

Sutherland, W.J., Dicks, L.V., Ockendon, N. and Smith, R.K. (2017) *What Works in Conservation*. Open Book Publishers: Cambridge, UK, http://dx.doi.org/10.11647/OBP.0109.

Turak, E., Brazill-Boast, J., Cooney, T., Drielsma, M., DelaCruz, J., Dunkerley, G., Fernandez, M., Ferrier, S., Gill, M. and Jones, H. (2016) 'Using the Essential Biodiversity Variables Framework to measure biodiversity change at national scale', *Biological Conservation*, http://dx.doi.org/10.1016/j.biocon.2016.08.019.

Úbeda, C., Lavilla, E. and Basso, N. (2010) '*Atelognathus patagonicus*'. In *The IUCN Red List of Threatened Species*: e.T56323A11459931, http://dx.doi.org/10.2305/IUCN.UK.2010-2.RLTS.T56323A11459931.en. Downloaded on 18 March 2017.

Uetz, P. (ed) (2006) *The Reptile Database*, www.reptile-database.org, accessed 23 May 2006.

Vaira, M., Akmentins, M., Attademo, A., Baldo, D., Barrasso, D., Barrionuevo, S., Basso, N., Blotto, B., Cairo, S., Cajade, R., Céspedez, J., Corbalán, V., Chilote, P., Duré, M., Falcione, C., Ferraro, D., Gutierrez, F., Junges, P., Lajmanovich, R., Lescano, J., Marangoni, F., Martinazzo, L., Marti, L., Moreno, L., Natale, G., Pérez Iglesias, J., Peltzer, P., Quiroga, L., Rosset, S., Sanabria, E., Sánchez, P., Schaefer, E., Úbeda, C. and Zaracho, V. (2012) 'Categorización del estado de conservación de los Anfibios de la República Argentina', *Cuadernos de Herpetología*, vol 26(Suppl. 1), pp131–159.

Valtonen, M. (2014) 'Conservation genetics of the Saimaa ringed seal: Insights into the history of a critically endangered population'. PhD thesis, University of Eastern Finland, Joensuu, Finland, 61 pp.

Vogt, R.C., Gonzalez-Porter, G.P. and Van Dijk, P.P. (2006) '*Dermatemys mawii*' (errata version published in 2016). In *The IUCN Red List of Threatened Species*: e.T6493A97409830. Downloaded on 12 March 2017.

Wager, R. (1996) '*Nannoperca oxleyana*'. In *The IUCN Red List of Threatened Species*: e.T14321A4431441,http://dx.doi.org/10.2305/IUCN.UK.1996.RLTS.T14321A4431441.en. Downloaded on 12 March 2017.

Waqas, U., Malik, M.I. and Khokhar, L.A. (2012) 'Conservation of Indus River Dolphin (*Platanista gangetica minor*) in the Indus River system, Pakistan: an overview', *Rec. Zool. Surv. Pak.*, vol 21, pp82–85.

Williams, D.R., Child, M.F., Dicks, L.V., Ockendon, N., Pople, R.G., Showler, D.A., Walsh, J.C., zu Ermgassen, E.K.H.J. and Sutherland, W.J. (2017) 'Bird conservation'. In Sutherland, W.J., Dicks, L.V., Ockendon, N. and Smith, R.K. (eds), *What Works in Conservation*. Open Book Publishers, Cambridge, UK.

Worboys, G.L., Lockwood, M., Kothari, A., Feary, S. and Pulsford, I. (eds) (2015) *Protected Area Governance and Management*. ANU Press.

Yeo, D.C.J., Luz, S., Cai, Y., Cumberlidge, N., McGowan, P.J., Ng, D.J.J., Raghavan, R. and Davison, G.W.H. (2016) 'Conservation first: strategic planning to save the critically endangered Singapore freshwater crab, *Johora singaporensis*'. In *A Global Overview of the Conservation of Freshwater Decapod Crustaceans*. Springer International Publishing, pp359–372.

Chapter 8

Managing specific freshwater ecosystems

A.H. Arthington, C.M. Finlayson, D.J. Roux, J.L. Nel, W. Rast, R. Froend, J. Turpie and L. van Niekerk

Key messages

- A number of relatively simple changes to the way Protected Areas (PAs) are designed and managed can help to further improve their conservation benefits for surface- and ground-water-dependent freshwater ecosystems and estuaries. These include: avoid using a river as the boundary of a PA; incorporate natural large-scale catchment processes into PAs; ensure that the water regimes of rivers, lakes, peatlands and groundwater-dependent ecosystems, as well as their linkages and interactions, are recognized and well managed within PAs and their catchments; avoid development of visitor infrastructure on priority freshwater ecosystems in PAs; encourage expansion of existing PAs to incorporate biodiversity hotspots, functional processes and connectivity; and promote new PAs for the last remaining free-flowing rivers and other high priority freshwater ecosystems.

- PA on their own are unlikely to fully protect freshwater ecosystems, especially where large-scale catchment threats impinge on the PA and its ecosystems. Therefore, conservation efforts should not stop at the boundaries of PAs, but extend into developing cooperative relationships and activities among various entities with overlapping water management mandates. Integrated and adaptive management approaches that facilitate engagement and empowerment of all stakeholders, inclusive and iterative learning, and purposeful action amidst inherent complexities are recommended.

- Establishing environmental flows and maintaining the water regimes of rivers, lakes and Groundwater Dependent Ecosystems (GDEs) are critical for sustaining biodiversity and ecosystem services inside freshwater and estuarine PAs. Conservation managers should aim to ensure that the natural water regimes of specific freshwater ecosystems are protected. Setting a limit on hydrologic alteration remains the most challenging aspect of implementing environmental flows, although methods and guidelines are available to assist in this task.

- Effective management of GDEs requires integration of associated surface and groundwater resources and necessitates an understanding of the origins, pathways and storages of water. Some GDEs are entirely dependent on continuous groundwater discharges, whilst others are maintained by minor but critical groundwater inflows restricted to particular seasons or inter-annual episodes.

- The International Lake Environment Committee (ILEC 2005) has identified six major pillars for lake governance: 1) policies, which represent the "rules of the game"; 2) institutions, representing the entities responsible for carrying out the rules of the game; 3) stakeholder participation in implementing effective management plans; 4) technology, involving selection of "hard" versus "soft" management approaches; 5) knowledge and information, focusing on obtaining the most accurate information and data; and 6) finances, including identifying and ensuring sustainable sources. These pillars are encompassed within the concept of Integrated Lake Basin Management (ILBM), which represents a major complement to Integrated Water Resources Management (IWRM) for addressing lakes, reservoirs, wetlands and other lentic water systems.

Freshwater ecosystems in protected areas

Relatively few PAs have been established specifically to protect freshwater habitats and biota. The norm is for terrestrial biodiversity features to drive the design of PAs with the inclusion of freshwater features being a secondary or neglected consideration (Abell et al., 2007; Allan et al., 2010; Herbert et al., 2010). As an example, the iconic Kruger National Park in South Africa was proclaimed in 1926 in the north-eastern corner of the country as a last local refuge for large savannah mammals. The use of modern-day conservation planning tools shows that Kruger's inherited design (as a belt of land across the middle of five river catchments) caters poorly for freshwater conservation, both in terms of representing the diversity of freshwater features in the associated eco-region and supporting persistence of those features (Roux et al., 2008). By contrast, Kakadu National Park in Australia largely incorporates the catchments of two major rivers from their watersheds to the sea.

Notwithstanding their terrestrial bias, PAs do extend critical protection to freshwater systems; for example, the ecological state of rivers in formally declared

PAs of South Africa is significantly better than of their counterparts outside PAs (Nel et al., 2007). A number of relatively simple changes to the way PAs are designed or expanded can help to further improve their conservation benefits for freshwater ecosystems, namely to: avoid using a river as the boundary of a protected area; encourage expansion of existing PAs to incorporate natural large-scale catchment processes into PAs where possible; ensure that the water regimes of rivers, lakes, peatlands and groundwater-dependent ecosystems are well managed within PAs and their catchments, enabling them to recover from the impact of activities upstream as they flow through the protected area; avoid development of visitor infrastructure on priority freshwater ecosystems in PAs; encourage expansion of existing PAs to incorporate biodiversity hotspots, functional processes and connectivity pathways at relevant scales; and promote new PAs for the last remaining free-flowing rivers and other high priority freshwater ecosystems (Nel et al., 2009).

No matter how considerate their designs, PAs on their own are unlikely to ever fully protect freshwater ecosystems because of their connected nature across longitudinal and lateral drainage lines spanning large-scale landscapes (Dudgeon et al., 2006). Therefore, conservation efforts should not stop at the boundaries of PAs but extend into developing cooperative relationships among various entities with overlapping water management mandates, including policy sectors for water, biodiversity, land-use planning and agriculture at catchment scale. Indeed, PAs can be powerful catalysts for freshwater ecosystem management outside their boundaries (Pollard et al., 2003). Ultimately, different legislative frameworks and place-based mechanisms (Abell et al., 2007; Allan et al., 2010) need to work in better harmony to enable integrated management of rivers at basin scales and bring effective freshwater conservation strategies to bear. Moreover, active cooperation across all levels and spheres of government is required to arrest the main threats to global freshwater biodiversity, notably overexploitation, water pollution, flow modification, destruction or degradation of habitat, invasion by alien species and, increasingly, climate change (Dudgeon et al., 2006; Pittock et al., Chapter 6, this volume).

The dual imperative of having to cooperate with multiple management agencies and stakeholders, and deal with multiple interacting and constantly changing threats (with uncertain and contested outcomes) add both a social and a biophysical layer of complexity to the task of conserving freshwaters. Under these circumstances, adaptive management seems to be gaining traction as an approach to management. Such an approach essentially enables a process of social learning among scientists, managers and stakeholders (Pahl-Wostl et al., 2007), leading towards a shared understanding of issues, consensus around a vision of the desired state of a catchment (Arthington, 2012) and a hierarchy of objectives that connects the high-level vision with measurable and implementable end points (McLoughlin et al., 2015). Through engagement and empowerment of stakeholders, as well as inclusive and iterative learning, adaptive management facilitates purposeful action amidst inherent complexities and uncertainties (Kingsford et al., 2011a; Pollard and Du Toit, 2011). In Australia the water plan for the Murray-Darling Basin is built around an adaptive management framework with five-year reviews of the environmental watering, water quality and salinity management plans, and after ten years a review of the overall plan (Neave et al., 2015; Swirepik et al., 2015). It is intended that these reviews will be informed by the assessment of sub-Basin water resource plans, monitoring and evaluation of environmental flow outcomes, and investigations of the impacts of climate change. The efficacy of such measures has yet to be tested with concerns about overly narrow or mal-adapted perspectives that could reduce the resilience of the freshwater ecosystems (e.g., Pittock and Finlayson, 2013). A major concern is the reliance on engineering works and measures rather than the adoption of ecosystem-based measures to maintain a more diverse range of ecological processes that would spread the risks and conserve a more diverse range of biota.

While the connected nature of freshwater systems poses challenges to conservation through PAs, it also presents opportunities to reimagine and reframe the role of PAs in society. Ecologically intact ecosystems such as wetlands and rivers provide many essential services with benefit pathways extending well beyond PA boundaries (Palomo et al., 2014). Water-based ecosystem services include water provision and increased water security to downstream users (Harrison et al., 2016), fisheries and other food resources, regulating services such as filtering pollutants from water, regulating flows and contributing to erosion control, and cultural services such as new knowledge generation through research, opportunities for education, sense of place, recreation opportunities and tourism (Figure 8.1). In a PA context, an ecosystem services approach can help to link conservation objectives with social, economic and cultural values of ecosystems (Martín-López et al., 2014; Boerema et al., 2016) and to contextualise the contribution of PAs to human well-being within their larger social-ecological landscapes (Palomo et al., 2014; García-Llorente et al., 2016). This broader justification of PAs is increasingly deemed necessary (Watson et al., 2014; Cumming, 2016).

Notwithstanding that wetlands and rivers are among the most threatened ecosystems in the world (Millennium Ecosystem Assessment, 2005; Dudgeon et al., 2006; Vörösmarty et al., 2010), reporting on the performance of PAs in achieving global conservation targets commonly concentrates on terrestrial and marine ecosystems only (Watson et al., 2014; Butchart et al., 2015). The absence of freshwater indicators in these performance assessments is at least in part due to the complex nature of freshwater conservation, as outlined above and in the ecosystem sections below. While the sum of protected hectares may serve as a proxy for the achievement of terrestrial and marine conservation

Figure 8.1 Many freshwater ecosystems are important for the provision of ecosystem services, in particular a) the supply for fresh water and b) for fisheries (photographs © C.M. Finlayson).

targets, rivers that flow through PAs are typically not in an ecologically intact condition due to disturbance of upstream and downstream connectivity pathways and threats beyond PA boundaries (Arthington et al., Chapter 3, this volume). Indicators to measure the ecological condition of freshwater ecosystems in PAs need to reflect the true protection status of these systems (Nel et al., 2007). The EU Water Framework Directive targets for "good ecological status" in all surface waters provide one example of attempts to define rigorous condition assessments at a continental scale (Finlayson et al., Chapter 12, this volume). Indicators of protection status and ecological condition would also link directly to the capacity of freshwater ecosystems to provide ecosystem services (Brauman et al., 2007), which in turn relate to the achievement of both Aichi conservation targets and the UN Sustainable Development Goals (Osborn et al., 2015).

Rivers, environmental flows and wetland water regimes

Rivers and flow regimes

Rivers and lakes hold 100,000 km^3 of fresh water and groundwater holds about 15,000 km^3, much of it stored in deep aquifers not in active exchange with the earth's surface (Jackson et al., 2001). The world's largest river systems comprise 60 per cent of the world's river runoff. Flowing or 'lotic' surface waters are prominent features of most landscapes, even deserts, and they exert a significant influence on landscape form and function. Rivers are the major agents of erosion, transport and deposition of materials from the mountains to valleys, to inland wetlands and lakes, and to estuaries and the oceans. Their biotic communities reflect a long evolutionary history of adaptations to dynamic, heterogeneous environments and the biological processes of competition, predation, colonization, succession and extinction. Fresh waters in rivers and their floodplain wetlands support ecosystems with diverse life forms and biogeochemical processes that provide ecological and cultural goods and services of critical importance to human societies (MEA, 2005). These goods and services are increasingly threatened by human activities in rivers and their catchments through over-exploitation of water and biological resources (e.g., fish), water pollution, fragmentation and destruction of habitat, invasion by alien species, and climate change (Dudgeon et al., 2006). All of these threatening processes are linked and exacerbated through the modification of river flows and wetland inundation regimes (Figure 8.2). Land-use change, river impoundment, surface and groundwater abstraction and artificial inter/ intra-basin transfers of water are the major drivers of change in water regimes. However, dams have the most impact on rivers and aquatic biodiversity by altering river flows, sediment dynamics, water quality and water temperatures, degrading habitats and blocking species movement. As discussed further

Figure 8.2 Dams have been built across many rivers, increasing the water storage and modifying the downstream flow patterns (photograph © C.M. Finlayson).

in the section on lakes, they also exhibit a buffering function in regard to pollutant and other inputs, therefore potentially masking upstream degradation in a drainage basin assessed at its downstream end. On a global scale, there are 945,000 dams above 15 m high, capable of holding back 96,500 km³ of water or about 15 per cent of the total annual river runoff globally. Thousands more are proposed, particularly in developing countries wishing to enhance 'green' energy supplies by means of hydropower developments (Winemiller et al., 2016).

Threats to rivers and freshwater biodiversity are increasing in many regions as human populations expand and seek higher living standards and improved quality of life (Vörösmarty et al., 2010). Paradoxically, the increasing modification of rivers and their flow regimes will rarely bring benefits to millions of people who depend on river resources for livelihoods, health and well-being (e.g., in the Mekong river system). PA offer a powerful means to conserve species-rich habitats and vital resources, important species radiations, aquatic biodiversity and ecosystem services, cultural values and belief systems. To be most effective, freshwater PAs should have control over the upstream drainage network, the surrounding land, the riparian zone, and downstream reaches (Dudgeon et al., 2006), and maintain both connectivity pathways and habitat patchiness. At the very least, within PAs, the goal should be to protect the natural water regimes of rivers, wetlands and floodplains from over-use, diversion and impoundment.

Environmental flows

Freshwater management has been integrated into the broader scope of ecological sustainability through the provision of environmental flows, defined as (Brisbane Declaration, 2007): "the quantity, timing and quality of water flows required to sustain freshwater and estuarine ecosystems and the human livelihoods and well-being that depend upon these ecosystems".

There is now wide recognition that a dynamic variable water regime is required to maintain species phenology (seasonal timing of events in the life cycle) and the native biodiversity and ecological processes characteristic of every river and wetland ecosystem (Poff et al., 1997). The natural flow regime paradigm and diverse eco-hydrological principles (e.g., Bunn and Arthington, 2002) flesh out the influence of flow volume, seasonal timing and variability on aquatic habitats, biodiversity, population recruitment and ecosystem productivity. These eco-hydrological principles inform assessment of the environmental flow requirements of aquatic and riparian plants, invertebrates, fish, waterbirds and other water-dependent animals (e.g., frogs, snakes and lizards, beavers and platypus). Other principles elaborate water-related ecosystem characteristics and processes such as foodweb structure, energy flow and biological interactions.

The key challenge for management where PAs receive water from unreserved upstream catchments is to engage water managers and users in a collaborative process for envisioning, assessing, implementing and monitoring environmental flows.

Over 250 practical methods, models and frameworks have been developed to link water volumes and patterns of flow to biodiversity and ecological processes, and thereby to define the environmental flow requirements of rivers and streams (Dyson et al., 2003; Tharme, 2003). While environmental flow assessment may seem complex, a simple guide to the main technical options available for protected area managers to assess what is required is given in Table 8.1. These methods focus largely on rivers and streams. However, they are applicable in concept and practice to lakes, ponds and wetlands that rarely flow but nevertheless experience natural spatial and seasonal patterns of water level fluctuation, wetting and drying, and hydrological links to rivers and groundwater. Estuaries also need to receive freshwater inflows to maintain their salinity regime, biodiversity, fisheries productivity and amenity values (see estuaries section below). Water regime assessment methods and applications for rivers, wetlands, groundwater-dependent ecosystems and estuaries can be found in Arthington (2012) and references therein.

Table 8.1 Environmental flow methods. Comparison of the four main types of methods used worldwide to estimate environmental flows = environmental water allocations – EWA (adapted from Tharme, 2003, for examples see Arthington, 2012). Resource intensity is represented in terms of time, cost and technical capacity.

Type	River ecosystem components	Data requirements and resource intensity	Resolution of output (EWA)	Appropriate levels of application
Hydrological	Whole ecosystem, non-specific, or only some biota (e.g., Montana method, Tennant, 1976)	Low Primarily desktop Use virgin/naturalised historical flow records Some use historical ecological data	Low Expressed as % of monthly or annual flow (median or mean); or as limits to change in flow parameters (e.g., Range of Variability Approach = RVA, Richter et al., 1996).	Reconnaissance level of water resource developments, or as a tool within habitat simulation or holistic (ecosystem) methodologies **Used widely**
Hydraulic rating	In-stream habitat for target biota	Low–medium Desktop or limited field work Historical flow records Discharge linked to hydraulic variables – typically at single river cross-section	Low–medium. Hydraulic variables (e.g., wetted perimeter) used as surrogate for habitat-flow needs of target species or assemblages	Water resource developments where little negotiation is involved, or as a tool within habitat simulation or holistic (ecosystem) methodologies **Used widely**

Habitat simulation	Primarily in-stream habitat for target biota. Some consider channel form, sediment transport, water quality, riparian vegetation, wildlife, recreation and aesthetics (e.g., PHABSIM, Bovee, 1982)	Medium–high Desktop and field Historical flow records. Many hydraulic variables are modelled at range of discharge at multiple stream cross-sections Physical habitat suitability or preference data needed for target species	Medium–high Output in form of Weighted Useable Area (WUA) of habitat for target species (fish, inverts, plants). May involve time-series of habitat availability	Water resource developments, often large-scale, involving rivers of high strategic importance, often with complex, negotiated trade-offs among users, or as method within holistic (ecosystem) approaches ***Primarily used in developed countries***
Holistic (ecosystem) frameworks	Whole ecosystem, all or several ecological components Most consider in-stream and riparian biota, some also consider: groundwater, wetlands, floodplains, estuary and coastal waters May assess social and economic dependence on species/ecosystem (e.g., DRIFT, King et al., 2003)	Medium–high Desktop and field Use virgin/naturalised flow records or rainfall records cf. current gauge records Many hydraulic variables at multiple cross-sections and river sites Biological data on flow-related and habitat-related requirements of biota and some/all ecological components	Medium–high Advanced fish methods use data on movement and migration, spawning, larval/juvenile requirements, water quality tolerances; exotic species included (e.g., DRIFT (Arthington et al., 2003) ELOHA (Ecological Limits of Hydrologic Alteration) quantifies e-flow 'rules' for rivers of contrasting hydrological type at user-defined regional scale (Poff et al., 2010)	Water resource developments, typically large-scale, involving rivers of high conservation value or strategic importance, and/or with complex user trade-offs Expert panels often used where flow-ecology knowledge is limited, or there are limited trade-offs among users, and/or time constraints. ***Used in developing and developed countries***

Setting limits to hydrological alteration

Protecting a river's flow regime and other drivers should be a major goal of river conservation and PA management. Legitimate activities that can interfere with the natural flow regimes of rivers within a protected area may include water storage to support park ranger facilities and tourism infrastructure, the barriers formed by roads and bridges, and water extraction or diversion to maintain habitats and park flora and fauna. Furthermore, in numerous cases, the PA may not incorporate the entire catchment supplying its water resources. For example, the upper tributaries of the Kruger National Park in South Africa flow through developed catchments (Roux et al., 2008).

Setting environmental flow targets for biodiversity and ecosystem conservation in PAs and other high-value aquatic systems has challenged scientists and water managers to define how much change in vital attributes of natural flow regimes is acceptable. In spite of tremendous advances in methods, setting a limit on hydrologic alteration remains the most challenging aspect of environmental flow science and sustainable water management. Simple methods set this limit as a percentage of the mean natural annual or seasonal flow, whereas methods such as PHABSIM estimate the river discharge that maintains fish habitat and connectivity though a valued river reach or the channel network (Table 8.1). Holistic environmental flow frameworks have been particularly effective in formulating scenarios of possible ecological impact and risk arising from alterations to river flow regimes. In the holistic Downstream Response to Imposed Flow Transformations (DRIFT) and Ecological Limits of Hydrologic Alteration (ELOHA) frameworks (Table 8.1), and several restoration protocols (e.g., Poff et al., 2003; Richter et al., 2006), scientists, stakeholders and managers consider a suite of relationships between flow regime alterations and ecological changes for the rivers under study. An important concept is the idea of a critical threshold of hydrological change above or below which ecological functions or elements of the ecosystem are impaired or lost (Arthington et al., 2006). These threshold discharges may include sufficient high flow to connect the river with its riparian corridor or floodplain, or to move sediment, or sufficient low flow to maintain within-channel connectivity between shallow (e.g., riffle) areas, and promote oxygenation of stream habitats. Some of the greatest successes in environmental flow management have seen close collaboration of scientists, water managers, community groups and farmers, for example, in preparing the Murray-Darling Basin Plan (Commonwealth of Australia, 2012).

Adapting to climate change

The natural environmental regimes that govern aquatic ecosystems, especially water regimes, have been replaced by altered regimes in many areas of the world under increasing human pressure for fresh water and in response to shifting climates (see IPCC, 2007). The combination of climate change and flow regulation is now driving structurally novel ecosystems that may require new thinking and a range of approaches to water management to cope with increasingly uncertain futures (Palmer et al., 2008). Research that identifies flow regime characteristics and associated ecological responses to variability by quantitative methods and models is one of the best options for preparedness. The study of ecological responses along contemporary gradients of flow variability (wet to dry tropics, coastal to arid-zone regions) may provide analogues for future climatic shifts (Arthington et al., 2006). Another approach is the application of models that make use of empirical data and expert judgements to compare and contrast scenarios of water use or climate change (e.g., Chan et al., 2012). An important way to advance understanding of the ecological roles of flow, and to improve water use for ecosystem and human benefit, is through well-designed monitoring of ecological outcomes over time in river systems receiving an environmental flow (Arthington, 2012; Davies et al., 2014).

Guiding principles for environmental flows

As implementation of environmental flows becomes accepted international best practice, protected area managers can take the lead in applying these concepts and methods to the diverse PAs and conservation systems they manage. Common guiding steps in the different environmental flow methods outlined above include:

- consult stakeholders to identify the different, flow-related elements of the environment that are valued (such as water quality, the riparian corridor and fish migrations) and collaboratively set environmental flow objectives;
- describe flow-ecology relationships quantitatively, as the thresholds for water quality, volume and timing of flows needed to sustain the identified values. These flow-ecology relationships must inform the stakeholder values and the agreed environmental flow objectives;
- identify the spatial and temporal scale of the environmental flows assessment and decide which specific methods and tools (Table 8.1) can best be applied (e.g., river reach, tributary, main large river channel, several rivers or basins);
- consider the natural seasonal variability of river flows and wetland water regimes, and seek opportunities to mimic these patterns as much as possible, for instance, with restrictions on water diversions or managed water releases from dams;
- negotiate agreements with water agencies and other stakeholders to implement the environmental flows, including water departments and utilities; and
- monitor and evaluate the ecological and social outcomes, and adjust the environmental flows to achieve the desired environmental and social objectives.

Environmental flows and water regimes are relevant to the conservation and management of lakes, peatlands, groundwater-dependent ecosystems and estuaries, as described in the next sections (see also Arthington, 2012).

Lake conservation and management

Lakes

Lakes are among the most dramatic, sometimes even mystical, features of our global landscape. There are an estimated 27 million natural lakes greater than one hectare in area. Indeed, the 17 largest lakes on the planet cover an area of approximately one million square kilometres. An estimated additional half a million artificial lakes (reservoirs) with surface areas greater than one hectare add to this total. These water systems collectively contain more than 90 per cent of the liquid freshwater on the surface of our planet at any given instant. It is estimated, for example, that the North American Great Lakes and Lake Baikal in the Russian Federation together contain approximately 38 per cent of all this freshwater. In fact, lakes and reservoirs collectively provide a wider range of ecosystem goods and services than other aquatic ecosystems. The Millennium Ecosystem Assessment (MEA, 2005) define ecosystem goods and services as benefits humans derive from ecosystems, including domestic, agricultural and industrial water supply, fisheries, recreation, tourism, hydropower generation, and transportation. In addition to providing habitat for a diverse plethora of aquatic species, lakes provide fundamental regulating services to humanity, including waste assimilation, climate, water and erosion regulation, water purification and refugia. They also have significant religious or cultural meanings for some societies. Against this background, lakes (and all surface freshwater systems) can be characterized as being finite, in that the vast majority of our planet's fixed quantity of liquid freshwater resides in lakes and reservoirs; sensitive, in that lakes and reservoirs are easily polluted and degraded; and irreplaceable, in that there is no substitute for freshwater in its many uses (Illueca and Rast, 1996). In spite of their obvious importance, however, the unfortunate reality is that many lakes and reservoirs are being degraded at an accelerating rate, due mainly to a growing global population and associated agricultural and economic development activities, as well as the hydrologic impacts of a changing global climate.

The term 'lakes' is henceforth used to refer to both natural lakes and artificial reservoirs, noting that the biodiversity values of artificial lakes is generally lower than for natural ones. Noting that much of humanity's readily accessible liquid freshwater exists in lakes, managing these water bodies for their conservation is a complex undertaking involving a range of scientific, socioeconomic and governance elements. Further, rather than being isolated water bodies, lakes are hydrologically linked to upstream rivers or tributaries flowing into them, to downstream water systems into which they discharge, and sometimes also to subsurface groundwater aquifers. Lakes and reservoirs, as well as ponds, wetlands and estuaries, are 'lentic' water systems that pool or store freshwater (Figure 8.3). As such, they represent an expression of the ecological and anthropogenic state of water, with evolutionary and historic memories of human–nature interaction. This is certainly the case for many ancient lakes, examples being Biwa,

Figure 8.3 Lakes are widespread, vary in size and depth, and contain vast quantities of freshwater (photographs © C.M. Finlayson).

Baikal, Issyk-Kul, Lanao, Malawi, Maracaibo, Prespa, Tanganyika and Titicaca. This lentic characteristic is in contrast to 'lotic' water systems such as rivers and streams, characterized by flowing waters, thereby being the expression of the physical state and dynamics of water. The hydrologic linkages between these lentic and lotic water systems have important scientific and management implications. Downstream water needs, for example, can sometimes significantly dictate the management requirements of upstream lakes that supply water to them, an example being the Lake Biwa-Yodo River complex in Japan (Nakamura et al., 2012).

Lake ecology

The ecology and limnology of lakes is a major branch of aquatic science. In addition to providing the above-noted ecosystem goods and services, lakes also provide habitats for multiple interacting aquatic organisms and communities, ranging from microscopic free-floating phytoplankton and zooplankton, to fish and aquatic mammals. The topic of aquatic science, however, cannot be comprehensively discussed within the context of this short overview. Useful sources for further discussion of lake and reservoir limnology include the works of Rast and Straskraba (2000), Wetzel (2001), Kalff (2002), O'Sullivan and Reynolds (2004), Dodson (2005) and Jorgensen et al. (2005).

In contrast to the strictly scientific concerns, lake conservation management essentially represents managing lakes for sustainable use. This translates into managing lakes, their basins and their resources for sustainable ecosystem services, whether these services are provisioning, regulating, cultural or supporting in nature (MEA, 2005). Managing lakes for conservation purposes involves a range of scientific/technical and socioeconomic/governance elements. The former include consideration of the quantity and quality of surface and groundwater sources, drainage basin characteristics, flora and fauna, soils, topography and land use, climate, etc., all of which collectively define the physical presence and condition of lake waters, including quality, quantity, condition, trends, stresses, etc. The latter include the legal and institutional framework within a lake drainage basin, economic considerations, demography, cultural and social customs, stakeholder participation possibilities, political realities, etc. The latter arguably comprise the most important elements in this regard, in that they fundamentally define the factors controlling how humans use their water resources, including lakes, their basins and their resources (GWP, 2000; Solanes and Gonzales-Villarreal, 1999).

Effectively managing lakes for conservation and sustainability also requires recognition of three unique features characterising these and other lentic water systems. As previously noted, these include an integrating nature; long water retention time; and complex response dynamics (ILEC, 2005). Because of their location at the hub of a drainage basin, lakes comprise the flow-regime integrators within the entire lake-river basin complex. The integrating nature of a lake refers to its function essentially as a 'mixing pot' for everything entering it from its surrounding drainage basin, and sometimes even from beyond its basin via the long-range transport of airborne pollutants. This integrating nature transcends the entire lake and riparian land interfaces, thereby causing both the lake resources and the problems associated with them to form a complex web of cause–effect relationships propagating throughout the lake. Simply stated, lake stresses come from many sources, with everything coming together in the lake, making the issues mostly inseparable. The long water retention time refers to the average time water spends in a given lake. Long water retention times typically result from large water volumes, giving lakes a buffering capacity against materials entering them via inflowing waters, often without exhibiting immediate negative impacts. Thus, lake problems develop gradually, and may not become evident until they have become serious lake-wide problems that can significantly impact human water uses and ecosystem integrity. This same buffering trait also can produce a 'lag' phenomenon in response to remedial programs implemented to restore them, sometimes for decades or longer. In other words, while problems may take a long time to become evident, solutions to these problems also may take a long time to become evident. The complex response dynamics means lakes do not respond to perturbations or pollution in a linear manner, but rather exhibit hysteresis in response to these disturbances. Thus, problems are often unpredictable and uncontrollable. In fact, all lake problems are essentially lake-wide problems, with lakes experiencing serious degradation, including to the aquatic communities for which they provide habitats, typically not returning to the condition they exhibited prior to the degradation (Nakamura and Rast, 2011).

Integrated lake basin management

Against this backdrop, the underlying cause of nearly all lake and other aquatic ecosystem degradation or overexploitation is inadequate governance. In focusing on aquatic ecosystem provisioning and cultural services, governance inadequacies ensure that individuals, organizations and communities can readily overwhelm the ecosystem regulating and support services upon which these former services are directly dependent. The result can be a feedback that can ultimately degrade all these ecosystem services. Based on examining lake management experiences around the world, the International Lake Environment Committee (2005) has identified six major lake governance pillars requiring recognition and consideration. These include: 1) policies, which essentially represent the "rules of the game"; 2) institutions, representing the entities responsible for carrying out the rules of the game; 3) stakeholder participation, meaning the meaningful involvement of all relevant stakeholders in implementing effective management plans; 4) technology, involving selection of hard (constructions) versus soft (behavioural changes) management approaches; 5) knowledge and information, focusing on obtaining the most accurate information and data, which can comprise both scientific studies and indigenous knowledge; and 6) finances, including identifying and ensuring sustainable sources of adequate financial support. These six pillars comprise the essential governance elements that collectively form the management regime for an integrated approach to managing lakes and their basins, and which are directed to ensuring sustainable lake ecosystem services, as discussed in detail by Nakamura and Rast (2011). Lake Chilwa in Malawi (see Box 8.1) highlights these challenges.

A practical lake management approach that considers both the scientific and governance elements is encompassed within the concept of Integrated Lake Basin Management (ILBM), as exemplified in the ILBM Platform Process developed by the International Lake Environment Committee. The platform represents a virtual stage for collective stakeholder actions for improving lake basin governance through ILBM, as a strategic means of facilitating its gradual and continuous improvement over time. This approach, which focuses on lakes and other lentic water systems upon which we depend for most of our provisioning ecosystem services, is being increasingly utilized to address lake management issues. Having been applied to lakes in many countries throughout the world, whether as a basic or cyclic process (Nakamura and Rast, 2011) it provides a means to address the above-noted management elements within an integrated approach focusing on sustainable ecosystem services. It also allows the consideration not only of lakes, but also the other water systems to which they are hydrologically connected in their basins, enhancing its utility for addressing conservation issues for other types of water systems as well. In fact, the majority of the accessible freshwater resources on our planet are linked in varying degrees to the lentic–lotic environment, and this linked lentic–lotic environment requires special care in its assessment and management. This aspect is typically overlooked, in spite of its important scientific and management implications, and experience to date indicates that the ILBM Platform Process can play an important complementary role in managing not only lakes and reservoirs, but also the range of other freshwater systems within their basins and beyond.

Box 8.1   Lake Chilwa, Malawi

Lake Chilwa covers approximately 2,284 km^2 and is located in the south of Malawi and along the border with Mozambique. It is a shallow enclosed endorheic saline lake with an average depth of 1–2 m, and is surrounded by an area of marsh dominated by dense stands of *Typha* with a seasonally inundated floodplain (Rebelo et al., 2011). The catchment covers 8,784 km^2, with 5,724 km^2 in Malawi and 3,060 km^2 in Mozambique. The annual rainfall across the catchment is approximately 1,362 mm, with water levels in the lake fluctuating three to four metres seasonally. These seasonal variations are superimposed on longer wetting and drying cycles, and occasionally it dries completely, most recently in 1996/1997.

The lake and surrounding marsh and floodplain were designated as a Ramsar site by Malawi in 1996 on the basis of regularly supporting one per cent or more of the individuals in a population of one species or subspecies of a waterbird.

Although Lake Chilwa is one of the most productive lakes in Africa, it is susceptible to major changes such as those seen for fishery catches (Jamu et al., 2006; Rebelo et al., 2011). During productive years, the annual fishery catch was more than 20,000 tonnes, whereas the average over the period 1962–2003 was a much lower 9,000 tonnes.

The lake is expected to be influenced by changes in the hydrology as a consequence of changes in the inflowing streams, with climate change projections indicating a decrease in the available water. As it is a shallow lake, with high levels of evaporation, it is vulnerable to drying during low rainfall years. Future reduction in lake levels due to increased extraction in the catchment, and to climatic variability, may result in drying of the marsh and grassed floodplain, thereby reducing the productivity of the lake (Jamu et al., 2006). The lake is also under increasing pressure from changes in agriculture in its catchment, and in-lake fisheries, as the human population rises and food demands increase. The "integrated small scale economy of farming, fishing and cattle-rearing" described by Kalk et al. (1979) has changed with less livestock rearing and more cultivation, including irrigation, and fishing (Allison and Mvula, 2002). The Mozambican side of the lake has a much lower population density and relatively undisturbed forests and a low level of agricultural activity, compared with that in Malawi (Jamu et al., 2006).

Although Malawi has a management plan for the lake, an integrated transboundary plan that encompasses the entire catchment is needed. This is of particular importance due to the seasonal and periodic water-level fluctuations and frequent drying of the lake, which result in shifts in livelihood activities within the wetland and the catchment. The Integrated Lake Basin Management (ILBM) Platform Process (Nakamura and Rast, 2011; Nakamura et al., 2012) provides a positive contribution to such efforts.

Peatland conservation and management

Peatlands

Peatlands are widespread and cover around 4 million km², although in many parts of the world there is a large degree of uncertainty about their extent

(Joosten, 2009). There are several definitions of peatlands, but they are generally considered to be areas of land with a naturally accumulated layer of peat, formed from carbon-rich dead and decaying plant material under waterlogged conditions, and generally seen as comprising at least 30 per cent dry mass of dead organic material and greater than 30 cm deep. They can develop under a wide range of vegetation types including lowland or upland fens, reed beds, wet woodland, bogs, and under tidal conditions and in mangroves. At the ecosystem level, the size, shape and biological features of peatlands are determined by climate and geomorphology as well as the quality and quantity of associated water. In Eurasia in particular the term 'mire' is used to refer to wetlands that are actively accumulating peat to differentiate them from the more general term 'peatland' that also includes ecosystems where peat is not actively being accumulated (Joosten and Clarke, 2002). For systems where the accumulation of peat has ceased, the term is not applicable.

Peatlands occur in many countries and could represent more than a third of global wetlands. Joosten (2009) provides an overview of the distribution of freshwater peatlands with peat layers greater than 30 cm depth, hence excluding shallower peats, drawing on multiple data sources. The largest areas are found in the northern hemisphere, especially in the boreal zone with 1,375,690 km^2 in Russia and 1,133,926 km^2 in Canada. Estimates of peatlands in tropical regions range from 275,424 to 570,609 km^2, although most of the data are from pre-1990 sources and there has been extensive destruction in recent years (Hooijer et al., 2010). These data do not extend to analyses of the distribution of peatlands in freshwater protected areas. The Ramsar site database (www.ramsar.org, accessed 13 November 2016) lists 584 wetlands of international importance that contain peatland, 26 per cent of the total number of listed sites, covering 53.79 m ha.

Importance of peatlands

The importance of peatlands is well known (Figure 8.4), although this has not prevented their degradation. They contain 10 per cent of the global freshwater volume and are significant in maintaining freshwater quality and hydrological integrity of many river valleys. They play an important role in maintaining permafrost and preventing desertification. In recent years their importance as global carbon stores and sinks has come to the fore as the global community addresses measures to mitigate climate change (Lindsay, 2010; Hooijer et al., 2010; Joosten, 2009; Joosten et al., 2012). Peatlands support important biological diversity and refugia for some of the rarest and most unusual species of wetland-dependent flora and fauna (Joosten and Clarke, 2002). Under waterlogged conditions they preserve a unique palaeo-ecological record, including valuable archaeological remains and records of environmental contamination. They support human needs for food, fresh water, shelter, heating, warmth, and employment (Joosten and Clarke, 2002).

Figure 8.4 Peatlands are widely distributed with many being used for grazing (photograph © C.M. Finlayson).

Management issues

Human pressures on peatlands are both direct, through drainage, land conversion (e.g., for oil palms and oil sands), excavation and inundation, and indirect, as a result of air pollution, water contamination, water removal and infrastructure development. When they are destroyed peats release large amounts of carbon and are not easily restored. In response to the degradation of peatlands the Ramsar Convention has adopted detailed *Guidelines for Global Action on Peatlands* (Ramsar Convention, 2002) including:

- establishing a global database of peatlands;
- detecting changes and trends in the quantity and quality of peatlands;
- developing and promoting education, training, and public awareness programmes;
- reviewing national networks of peatland PAs;
- developing and implementing peatland management guidelines and actions plans;
- establishing regional centres of expertise and research networks; and
- stimulating international cooperation on research and technology transfer.

Recent initiatives have stimulated actions to limit the loss of carbon from peatlands and to encourage their retention and restoration as part of climate change mitigation measures (Joosten et al., 2012; Biancalani and Avagyan, 2014). This is particularly important given the past loss of peatlands globally and the more recent degradation of tropical peatlands (Hooijer et al., 2010; Joosten et al., 2012). Approaches for restoring northern peatlands are well established (Quinty and Rochefort, 2003) and are being developed for tropical systems (Page et al., 2009).

Groundwater-dependent ecosystems

What are GDEs?

The notion of groundwater-dependent ecosystems (GDEs) implies groundwater is an important contributor to the maintenance of the hydrological regime supporting the ecosystem. Furthermore, a change in the quantity or quality of groundwater, often associated with human activity, will impact on the state and condition of the ecosystem (Eamus and Froend, 2006).

Richardson et al. (2011a) recognised three types of groundwater dependent ecosystems:

1   aquifer and cave ecosystems which provide unique habitats for living organisms (e.g., stygofauna and troglofauna, the animals that live underground), including karst aquifer systems, fractured rock and saturated sediments;
2   ecosystems fully or partly dependent on the surface expression of groundwater including wetlands, lakes, seeps, springs, river baseflow, coastal areas, estuaries and marine ecosystems;
3   ecosystems dependent on subsurface presence of groundwater (via the capillary fringe) including terrestrial vegetation that depends on groundwater fully or on an irregular basis to meet water requirements.

The degree of dependence on groundwater relative to other sources of water is important in differentiating these ecosystems and their response to changes in groundwater availability (Eamus et al., 2006). Of particular significance are the spatial and temporal variability in water tables and the nature of groundwater discharge into flowing or still surface water bodies. According to these interactions, different physicochemical properties and species assemblages will develop and become characteristic of the predominant hydrological regime (Horwitz et al., 2008).

Management issues

Interest in GDEs has largely developed from a need to understand the consequences of direct use or pollution of aquifers that maintain ecosystems. Management issues arise when the conditions under which groundwater contributes to the maintenance of GDEs are disrupted by human activity. Both the quantity and quality of groundwater is important as well as the spatial and temporary variability. These relationships can be disrupted by changes to the groundwater through abstraction, pollution and reduction in rainfall recharge. Effective management of GDEs therefore requires integration of associated surface and groundwater resources and necessitates an understanding of the origins, pathways and storages of water. For example, some GDEs are entirely maintained by continuous groundwater discharge whilst others are maintained by minor but critical groundwater inflows restricted to particular seasons or interannual episodes.

In general, processes that threaten GDEs are no different to those that threaten other ecosystems. Changes in groundwater can arise from reduced rainfall recharge, land clearing, forestry and agriculture, urbanization and direct groundwater abstraction for water supply. The ecological changes brought about by these activities will vary between types of GDEs, depending on their hydrological requirements (Richardson et al., 2011a). Any significant change to the quantity and quality of groundwater can be expected to have an impact on the associated biota and ecological processes. Hatton and Evans (1998) identified types of GDEs that differed in their 'degree of dependence' on groundwater, and the impact of altered groundwater regimes was considered proportional to inferred ecosystem dependence.

Identifying the importance of groundwater in ecosystems prior to development of groundwater resources (or other activities in a catchment) will inform resource planning and potential trade-offs against increasing demand for drinking water. The array of current approaches to identifying groundwater requirements of GDEs are summarised by Richardson et al. (2011b) and range from measurement of groundwater transpiration by individual trees to hydrological water balances and remote sensing at the landscape scale. In most cases an integration of different approaches and associated disciplines and knowledge is required to adequately understand the potential responses of GDEs to altered water regimes.

Management of GDEs can also be informed by improved understanding of the potential for ecosystems to adapt to change in groundwater availability. For example, some GDEs of the Swan Coastal Plain in Western Australia may have shifted to an alternative state (defined by biota and ecological processes that correspond to a particular groundwater regime) in accordance with changes in the extant hydrological support conditions (Froend and Sommer, 2010; Sommer and Froend, 2014). However, the potential of GDEs to adapt can be limited under catastrophic (and largely irreversible) changes in the availability of groundwater, such as the exacerbation of drought-induced drawdown by groundwater abstraction, which has resulted in widespread mortality of groundwater dependent (phreatophytic) vegetation (Sommer and Froend, 2011). In response, management agencies have assessed the threats to phreatophytic vegetation (Barron et al., 2013) and restricted groundwater pumping near vulnerable wetland ecosystems to enhance the potential for GDEs to adapt (McFarlane et al., 2012). In order to avoid such scenarios, adaptive management actions aimed at applying learned outcomes, integrating catchment management and balancing water demands with conservation, are required.

Estuary management and conservation

Estuaries

Estuaries are among the most complex of all the types of freshwater or freshwater-dependent ecosystems discussed in this chapter. Their position at the interface of the terrestrial and marine environment makes them vulnerable to the impacts of just about all human activities, whether land-based, freshwater or marine, including the impacts of climate change. To add to these pressures, estuaries tend to be a magnet for human activity, irrespective of their protection status. Thus managing estuaries as PAs can be particularly challenging, and its effectiveness often depends on managing external influences even more than on managing *in situ* activities. Thus the successful management of estuarine PAs hinges on co-operative governance between a range of government agencies.

Managing hydrodynamics and water quality

Estuarine functioning is primarily driven by the quantity and quality of freshwater inputs and their temporal distribution, in conjunction with inputs from the marine environment, with the former having a major influence on the latter (Allanson and Baird, 1999; Borja et al., 2011; Whitfield et al., 2012). Mediated by freshwater inflows and tides, fresh and salt waters mix in a nutrient-rich environment that supports a diversity of aquatic species. Run-of-river freshwater abstraction, large impoundments and small farm dams are mainly responsible for the decrease in the overall quantity of freshwater entering estuaries. On the other hand, inter-basin transfer schemes, waste water treatment works and increased runoff from 'hardened' catchments (e.g., road networks) are responsible for increased freshwater inflow that historically received a lower inflow (Nirupama and Simonovic, 2007).

Ideally, the freshwater flow into an estuary should be maintained in all its variability, as the components of a system's natural flow regime support its overall habitat structure and dynamics (van Niekerk and Turpie, 2012). Base flows are generally responsible for maintaining the salinity regime, and in the case of temporarily open systems, their connectivity to the sea (mouth state). In contrast, floods shape the geomorphological aspects such as size and shape of an estuary and its characteristic sediment structure. These processes also help to maintain the linkages between estuaries and their surrounding terrestrial, freshwater and marine systems. The life-history strategies of many species involve movement between these systems, for which the maintenance of open-mouth conditions at the right time of year is essential. This includes many marine species of conservation and commercial value whose populations and viability depend on well-functioning networks of estuaries. Thus, although they are discrete ecosystems, estuaries should not be managed as isolated systems (van Niekerk and Turpie, 2012).

In addition to the quantity of water entering estuaries, catchment activities and infrastructure also affect the quality of this water, in terms of the loads of sediments, nutrients and other pollutants entering the system (Turner et al., 2004). This can result in smothering of habitats, increased turbidity and eutrophication, all of which can result in significant changes in biotic communities, and local extinctions. While some of the pollution entering estuaries arises from estuary users and adjacent settlements, these are largely problems that arise from the entire catchment area, and cannot be effectively dealt with by PA managers operating within the system.

The protection of an estuary therefore entails ensuring that the quantity and quality of freshwater inflows are maintained as close to natural as possible, in order to maintain ecological functioning and biodiversity in a relatively natural state. In reality, however, estuary managers have to deal with many changes that are difficult to reverse to the extent desired, if at all. Where this is the case, protection of estuaries can involve imposing artificial means such as flood-flow releases from dams, and breaching the estuary artificially. These interventions are far more complex than trying to maintain natural processes, and require considerable investment in research and monitoring in order to devise strategies that can achieve conservation goals. The Chilika Lagoon (see Box 8.2) is such an example.

Managing in-situ pressures on habitats and biota

The main pressures that have to be managed within estuary systems are developments that encroach on estuary habitats, harvesting of resources such as fish and mangroves, aquaculture, human disturbance and the eradication or control of invasive alien species (Perissinotto et al., 2013).

While population and development tends to be concentrated along the coast generally, estuaries are particularly attractive, due to their productivity and their suitability for harbours, aquaculture and recreation. Estuarine PAs can limit this development, especially when they are part of terrestrial PAs, but often have dense development on their boundaries. With this comes a high level of industrial or recreational use. It is rare that estuaries are included entirely within large PAs that buffer them from intense use and, in many instances, PAs are either limited to part of an estuary, or incorporate multiple uses through zonation.

Managing the use of an estuary involves making trade-offs between the different types of value that it can generate (Turpie et al., 2007). For example, allowing subsistence fishing or angling will impact on the provision of ecosystem services such as their functioning as nursery areas to support marine fisheries, and allowing excessive development and access will impact on the biodiversity of the system and its value as an ecotourism destination.

Key interventions

In order for protection of estuaries to be successful, all of the following interventions at the local to national scale will be necessary.

- Integrated conservation planning that takes landscape processes and socio-economic trade-offs into account (see Turpie and Clark, 2007).
- Catchment management and the setting of environmental flow requirements to assure provision of adequate quantity and quality of inflows to maintain the protected estuaries in a desired state of health (Adams, 2013).
- The development of management plans to control competing uses within estuaries.
- Effective restriction of consumptive use to prioritize conservation of biodiversity and the supply of regulating services such as nursery areas for crustaceans and fish, carbon sequestration and coastal protection.
- Delineation of development setback lines to protect landscape value as well as to accommodate mouth migration, and water levels associated with changes in mouth state and sea level rise.
- Variability in abiotic drivers is a critical natural feature of estuaries, this manifests itself as variability in flow, water levels, salinity regimes, oxygen, turbidity, morphology and ultimately biota. A natural outcome of water resource development is the regulation of flow and the inability to provide the natural resetting processes required to ensure long-term productivity. Loss of regular stresses ultimately reduces an estuary's natural resilience to events such as floods and droughts in the future.
- Ensuring connectivity to both catchment and sea is key to estuary health and productivity. This speaks to the need for 'free flowing rivers' and the mitigation of impoundment barriers (e.g., fish ladders) as well as less clearly defined 'obstructions', e.g., poor water quality (low oxygen) can act as a migration barrier, while predatory alien invertebrate/fish species can similarly disrupt/prevent the natural flow of species and genes.

Box 8.2   Restoration of Lake Chilika, India

Chilika is a brackish coastal lagoon situated in Orissa State, India. It covers an area of 906 km^2 to 1,165 km^2 in the dry and wet seasons respectively, being flanked by an ephemeral floodplain of 400 km^2. It comprises shallow to very shallow marine, brackish and freshwater ecosystems with estuarine characters, and is a hotspot of biodiversity with more than one million overwintering migratory birds (Kumar and Pattnaik, 2012). It is also one of only two lagoons that support the Irrawaddy Dolphin (*Orcaella brevirostris*). The diverse and dynamic assemblage of fish, invertebrates and crustaceans supports a rich fishery that generates more than USD $17.3 million of annual revenue. Chilika was designated as a Wetland of International Importance (Ramsar Site) in 1981.

Degradation

Sustained provision of the wide range of ecosystem services of Chilika and maintenance of the livelihoods of dependent communities, including some 0.2 million fishers and 0.4 million farmers, is associated with the coastal and freshwater hydrology of the lagoon. From 1950–2000, increased sediment from a degrading catchment reduced the connectivity of the lagoon to the sea, causing a rapid decline in fisheries (Mohapatra et al., 2007). The introduction of shrimp culture, as well as the decline in fisheries, led to the breakdown of traditional resource management systems, with a loss of traditional occupations, and resentment between traditional fishers and immigrants (Dujovny, 2009). Due to changes in ecological characteristics Chilika was included in the Montreux Record by Ramsar in 1993.

Restoration

The Government of Odisha created the Chilika Development Authority (CDA) in 1991 as a node for the ecological restoration of the lake, chaired by the Chief Minister and comprising senior representatives of all concerned departments, as well as representatives of the fishing communities. It has established programs for catchment restoration, hydro-biological monitoring, sustainable development of fisheries, wildlife conservation, community participation and development and capacity building. In 2000, it was agreed to create a channel to reconnect the lagoon to the sea – a major intervention based on modelling and stakeholder consultations. An intensive awareness campaign about the values and functions of the lake was undertaken, and a hydro-ecological monitoring program put in place. These programs are coordinated through the Wetland Research and Training Center constructed on the shoreline of Chilika in 2002.

Restoration of the hydrological regime and re-establishment of the salinity regime (Ghosh et al., 2006) led to the recovery of the fisheries and biodiversity. The average fish landing increased from 1,747 mt in 2000 to 14,228 mt in 2012. The Irrawaddy Dolphin population increased from 89 to 142 individuals between 2003 and 2012. The sea grass meadows expanded from 20 km^2 in 2000 to 80 km^2. There has also been a decline in the area covered by invasive weeds. These ecological improvements, in particular the increase in dolphins, led to a resurgence of community-based

(continued)

(continued)

ecotourism. The success of these interventions was recognized with the Ramsar Wetland Conservation Award and Evian Special Prize for "wetland conservation and management initiatives." Chilika Lake was also removed from the Montreux Record by Ramsar Convention in 2002 due to improvements in its ecosystem.

An integrated management planning process involving key stakeholders was initiated in 2008 to guide the ongoing conservation and wise use of Chilika. A management planning framework was developed (Kumar and Pattnaik, 2012), with a plan being released by the Chief Minister in October 2012. The CDA also initiated a program for the revitalization of a community-based fisheries institution with the State Government establishing the Chilika Fishermen Central Cooperative Society (CFCCS) Ltd in July 2010.

References

Abell, R., Allan, J.D. and Lehner, B. (2007) 'Unlocking the potential of protected areas for freshwaters', *Biological Conservation*, vol 134, pp48–63.

Adams, J.B. (2013) 'A review of methods and frameworks used to determine the environmental water requirements of estuaries', *Hydrological Sciences Journal*, vol 59, pp451–465.

Allan, D., Esselman, P., Abell, R., McIntyre, P., Tubbs, N., Biggs, H., Castello, L., Jenkins, A. and Kingsford, R. (2010) 'Protected areas for freshwater ecosystems: Essential but underrepresented'. In Mittermeier, R.A., Farrell, T., Harrison, I.J., Upgren, A.J. and Brooks, T. (eds) *Fresh Water: The Essence of Life*, CEMEX and ILCP, Arlington.

Allanson, B. and Baird, D. (eds) (1999) 'Estuaries of South Africa', Cambridge University Press, New York.

Allison, E.H. and Mvula, P.M. (2002) *Fishing Livelihoods and Fisheries Management in Malawi. LADDER Working Paper No 22*, Department for International Development, London.

Arthington, A.H. (2012) *Environmental Flows. Saving Rivers in the Third Millennium*, University of California Press, Berkeley.

Arthington, A.H., Bunn, S.E., Poff, N.L. and Naiman, R.J. (2006) 'The challenge of providing environmental flow rules to sustain river ecosystems', *Ecological Applications*, vol 16, pp1311–1318.

Arthington, A.H., Rall, J.L., Kennard, M.J. and Pusey, B.J. (2003) 'Environmental flow requirements of fish in Lesotho Rivers using the DRIFT methodology', *River Research and Applications*, vol 19, pp641–666.

Barron, O., Froend, R.H., Hodgson, G., Ali, R., Dawes, W., Davies, P. and McFarlane, D. (2013) 'Projected risks to groundwater-dependent terrestrial vegetation caused by changing climate and groundwater abstraction in the Central Perth Basin, Western Australia', *Hydrological Processes*, doi: 10.1002/hyp.10014.

Biancalani, R. and Avagyan A. (eds) (2014) *Towards Climate-Responsible Peatlands Management.* Rome: Food and Agriculture Organization of the United Nations (FAO), p. 117.

Boerema, A., Rebelo, A.J., Bodi, M.B., Esler, K.J. and Meire, P. (2016) 'Are ecosystem services adequately quantified?', *Journal of Applied Ecology*, doi:10.1111/1365-2664. 12696.

Borja, A., Basset, A., Bricker, S., Dauvin, J.-C., Elliott, M., Harrison, T.D., Marques, J.-C., Weisberg, S.B. and West, R. (2011) 'Classifying ecological quality and integrity of estuaries'. In E. Wolanski and D. McLusky (eds), *Treatise on Estuarine and Coastal Science*, pp125–62. Academic Press, Waltham, MA.

Bovee, K.D. (1982) *A Guide to Stream Habitat Analysis using the IFIM*, US Fish and Wildlife Service Report FWS/OBS-82/26, Fort Collins.

Brauman, K.A., Daily, G.C., Duarte, T.K.E. and Mooney, H.A. (2007) 'The nature and value of ecosystem services: an overview highlighting hydrologic services', *Annual Review of Environment and Resources*, vol 32, pp67–98.

Brisbane Declaration (2007). *Environmental Flows are Essential for Freshwater Ecosystem Health and Human Well-Being*. Declaration of the 10th International River Symposium and International Environmental Flows Conference, 3–6 September: Brisbane, Australia. (www.eflownet.org.)

Bunn, S.E. and Arthington, A.H. (2002) 'Basic principles and ecological consequences of altered flow regimes for aquatic biodiversity', *Environmental Management*, vol 30, pp492–507.

Butchart, S.H., Clarke, M., Smith, R.J., Sykes, R.E., Scharlemann, J.P., Harfoot, M., Buchanan, G.M., Angulo, A., Balmford, A., Bertzky, B. and Brooks, T.M. (2015) 'Shortfalls and solutions for meeting national and global conservation area targets', *Conservation Letters*, vol 8(5), pp329–337.

Chan, T.U., Hart, B.T., Kennard, M.J., Pusey, B.J., Shenton, W., Douglas, M.M., Valentine, E. and Patel, S. (2012) 'Bayesian network models for environmental flow decision making in the Daly River, Northern Territory, Australia', *River Research and Applications*, vol 28, pp283–301.

Commonwealth of Australia (2012) *Basin Plan*, Commonwealth of Australia, Canberra.

Cumming, G.S. (2016) 'The relevance and resilience of protected areas in the Anthropocene', *Anthropocene*, vol 13, pp46–56.

Davies, P.M., Naiman, R.J., Warfe, D.M., Pettit, N.E., Arthington, A.H. and Bunn, S.E. (2014) 'Flow–ecology relationships: closing the loop on effective environmental flows', *Marine and Freshwater Research*, vol 65, pp133–141.

Dodson, S. (2005) *Introduction to Limnology*, McGraw-Hill, New York.

Dudgeon, D., Arthington, A.H., Gessner, M.O, Kawabata, Z., Knowler, D., Lévêque, C., Naiman, R.J., Prieur-Richard, A.-H., Soto, D., Stiassny, M.L.J. & Sullivan C.A. (2006). Freshwater biodiversity: importance, threats, status, and conservation challenges. *Biological Reviews* 81 (2): 163–182.

Dujovny, E. (2009) 'The deepest cut: Political ecology in the dredging of a new sea mouth in Chilika Lake, Orissa, India', *Conservation and Society*, vol 7, pp 192–204.

Dyson, M., Bergkamp, G. and Scanlon, J. (eds) (2003) *Flow: The Essentials of Environmental Flows*, IUCN, Gland.

Eamus, D. and Froend, R. (2006) 'Groundwater-dependent ecosystems: the where, what and why of GDEs', *Australian Journal of Botany*, vol 54, pp91–96.

Eamus, D., Froend, R., Loomes, R., Hose, G. and Murray, B. (2006) 'A functional methodology for determining the groundwater regime needed to maintain the health of groundwater-dependent vegetation', *Australian Journal of Botany*, vol 54, pp97–114.

Froend, R. and Sommer, B. (2010) 'Phreatophytic vegetation response to climatic and abstraction-induced groundwater drawdown: Examples of long-term spatial and temporal variability in community response', *Ecological Engineering*, vol 36, pp1191–1200.

García-Llorente, M., Harrison, P.A., Berry, P., Palomo, I., Gómez-Baggethun, E., Iniesta-Arandia, I., Montes, C., del Amo, D.G. and Martín-López, B. (2016) 'What can conservation strategies learn from the ecosystem services approach? Insights from ecosystem assessments in two Spanish protected areas', *Biodiversity and Conservation*, doi:10.1007/s10531-016-1152-4.

Ghosh, A.K., Pattnaik, A.K. and Ballatore, T.J. (2006) 'Chilika Lagoon: restoring ecological balance and livelihoods through re-salinization', *Lakes and Reservoirs: Research and Management*, 11: pp239–55.

GWP (2000) *Integrated Water Resources Management*, Global Water Partnership, Stockholm.

Harrison, I.J., Green, P.A., Farrell, T.A., Juffe-Bignoli, D., Sáenz, L. and Vörösmarty, C.J. (2016) 'Protected areas and freshwater provisioning: a global assessment of freshwater provision, threats and management strategies to support human water security', *Aquatic Conservation: Marine and Freshwater Ecosystems*, vol 26(S1), pp103–120.

Hatton, T. and Evans, R. (1998) *Dependence of Ecosystems on Groundwater and its Significance to Australia*. Land and Water Research and Development Corporation, Canberra.

Herbert, M.E., Mcintyre, P.B., Doran, P.J., Allan, J.D. and Abell, R. (2010) 'Terrestrial reserve networks do not adequately represent aquatic ecosystems', *Conservation Biology*, vol 24, pp1002–1011.

Hooijer, A., Page, S., Canadell, J.G., Silvius, M., Kwadijk, J., Wösten, H. and Jauhiainen, J. (2010) 'Current and future CO_2 emissions from drained peatlands in Southeast Asia', *Biogeosciences*, vol 7, pp1505–1514.

Horwitz, P., Bradshaw, D., Hopper, S., Davies, P., Froend, R. and Bradshaw, F. (2008) 'Hydrological change escalates risk of ecosystem stress in Australia's threatened biodiversity hotspot', *Journal of the Royal Society of Western Australia*, vol 91, pp1–11.

Illueca, J. and Rast, W. (1996) 'Precious, finite and irreplaceable', *Our Planet*, vol 8, www.ourplanet.com/imgversn/83/rast.html.

ILEC (International Lake Environment Committee) (2005) *Managing Lakes and Their Basins for Sustainable Use. A Report for Lake Basin and Stakeholders*. International Lake Environment Committee, Kusatsu, Japan, p146. (www.ilec.or.jp/en/pubs/p2/lbmi.)

IPCC (2007) Impacts, Adaptation and Vulnerability. Intergovernmental Panel on Climate Change 2007: Working Group II. Contribution to the Intergovernmental Panel on Climate Change Fourth Assessment Report, International Panel on Climate Change, Geneva.

Jackson, R.B., Carpenter, S.R., Dahm, C.N., McKnight, D.M., Naiman, R.J., Postel, S.L. and Running, S.W. (2001) 'Water in a changing world', *Ecological Applications*, vol 11, pp1027–1045.

Jamu, D., Delaney, L.M. and Campbell, C.E. (2006) *Transboundary Management Plan for the Lake Chilwa Catchment Area*, Association of Universities and Colleges of Canada, Ottawa.

Joosten, H. (2009) *The Global Peatland CO2 Picture: Peatland Status and Drainage Related Emissions in all Countries of the World*, Greifswald University & Wetlands International, Waganingen.

Joosten, H. and Clarke, D. (2002) *Wise Use of Mires and Peatlands – Background and Principles Including a Framework for Decision-making*, International Mire Conservation Group and International Peat Society, Saarijärvi.

Joosten, H., Tapio-Biström, M.L. and Tol, S. (eds) (2012) *Peatlands – Guidance for Climate Change Mitigation Through Conservation, Rehabilitation and Sustainable Use*, FAO and Wetlands International, Rome and Ede.

Jorgensen, S., Loffler, H., Rast, W. and Straskraba, M. (2005) *Lake and Reservoir Management*, Elsevier, Amsterdam.

Kalff, J. (2002) *Limnology*, Prentice Hall, Upper Saddle River.

Kalk, M., McLachlan, A. and Howard Williams, C. (1979) *Lake Chilwa: Studies of Change in a Tropical Ecosystem*, Dr W. Junk, The Hague.

King, J., Brown, C. and Sabet, H. (2003) 'A scenario-based holistic approach to environmental flow assessments for rivers', *River Research and Applications*, vol 19, pp619–639.

Kingsford, R.T., Biggs, H.C. and Pollard, S.R. (2011a) 'Strategic adaptive management in freshwater protected areas and their rivers', *Biological Conservation*, vol 144, pp1194–1203.

Kumar, R. and Pattnaik, A.K. (2012) *Chilika – An Integrated Management Planning Framework for Conservation and Wise Use*, Wetlands International – South Asia and Chilika Development Authority, New Delhi and Bhubaneswar.

Lindsay, R. (2010) *Peatlands and Carbon: A Critical Synthesis to Inform Policy Development in Oceanic Peat Bog Conservation and Restoration in the Context of Climate Change*. RSPB, Scotland.

Martín-López, B., Gómez-Baggethun, E., García-Llorente, M. and Montes, C. (2014) 'Trade-offs across value-domains in ecosystem services assessment', *Ecological Indicators*, vol 37, pp220–228.

McFarlane, D., Strawbridge, M., Stone, R. and Paton, A. (2012) 'Managing groundwater levels in the face of uncertainty and change: a case study from Gnangara', *Water Science and Technology: Water Supply*, vol 12, pp321–328.

McLoughlin, C.A. and Thoms, M.C. (2015) 'Integrative learning for practicing adaptive resource management', *Ecology and Society*, vol 20(1), dx.doi.org/10.5751/ES-07303-200134.

MEA (Millennium Ecosystem Assessment) (2005) *Ecosystems and Human Well-being: Wetlands and Water Synthesis*, World Resources Institute, Washington DC.

Mohapatra, A., Mohanty, R.K., Mohanty, S.K., Bhatta, K.S. and Das, N.R. (2007) 'Fisheries enhancement and biodiversity assessment of fish, prawn and mud crab in Chilika lagoon through hydrological intervention', *Wetlands Ecology and Management*, vol 15, pp229–251.

Nakamura, M. and Rast, W. (2011) *Development of ILBM Platform Process. Evolving Guidelines through Participatory Improvement*. Research Center for Sustainability and Environment, Shiga University, and International Lake Environment Committee, Kusatsu, Japan, p76. (www.ilec.or.jp/en/pubs/p2/ilbm-platform-process.)

Nakamura, M., Rast, W., Kagatsume, T. and Tomohiro, S. (2012) *Primer: Development of ILBM Platform Process. Evolving Guidelines through Participatory Improvement*. Research Center for Sustainability and Environment, Shiga University, and International Lake Environment Committee, Kusatsu, Japan, p26. (www.ilec.or.jp/en/pubs/p2/primer-ilbm-platform-process.)

Neave, I., McLeod, A., Raisin, G. and Swirepik, J. (2015) 'Managing water in the Murray–Darling Basin under a variable and changing climate', *AWA Water Journal*, vol 42, pp102–107.

Nel, J.L., Reyers, B., Roux, D.J. and Cowling, R.M. (2009) 'Expanding protected areas beyond their terrestrial comfort zone: Identifying spatial options for river conservation', *Biological Conservation*, vol 142, pp1605–1616.

Nel, J.L., Roux, D.J., Maree, G., Kleynhans, C.J., Moolman, J., Reyers, B., Rouget, M. and Cowling, R.M. (2007) 'Rivers in peril inside and outside protected areas: a systematic approach to conservation assessment of river ecosystems', *Diversity and Distributions*, vol 13, pp341–352.

Nirupama, N. and Simonovic, S.P. (2007) 'Increase of flood risk due to urbanisation: a Canadian example', *Natural Hazards*, 40: pp25–41.

Osborn, D., Cutter, A. and Ullah, F. (2015) 'Universal sustainable development goals. Understanding the transformational challenge for developed countries', Report of the Stakeholder Forum, IUCN Global Gender Office, Washington DC.

O'Sullivan, P.E. and Reynolds, C.S. (2004) *The Lakes Handbook: Limnology and Limnetic Ecology*, Blackwell Publishing, Oxford.

Page, S., Hoscilo, A., Wösten, H., Jauhiainen, J., Silvius, M., Rieley, J., Ritzema, H., Tansey, K., Graham, L., Vasander, H. and Limin, S. (2009) 'Restoration ecology of lowland tropical peatlands in Southeast Asia: current knowledge and future research directions', *Ecosystems*, vol 12, pp888–905.

Pahl-Wostl, C., Sendzimir, J., Jeffrey, P., Aerts, J., Berkamp, G. and Cross, K. (2007) 'Managing change toward adaptive water management through social learning', *Ecology and Society*, vol 12.

Palmer, M.A., Reidy Liermann, C.A., Nilsson, C., Flörke, M., Alcamo, J., Lake, P.S. and Bond, N. (2008) 'Climate change and the world's river basins: anticipating management options', *Frontiers in Ecology and the Environment*, vol 6, pp81–89.

Palomo, I., Montes, C., Martín-López, B., González, J.A., García-Llorente, M., Alcorlo, P. and Mora, M.R.G. (2014) 'Incorporating the social–ecological approach in Protected Areas in the Anthropocene', *BioScience*, vol 64, pp181–191.

Perissinotto, R., Stretch, D.D. and Taylor, R.H. (eds) (2013) *Ecology and Conservation of Estuarine Ecosystems: Lake St Lucia as a Global Model*. Cambridge University Press, New York.

Pittock, J. and Finlayson, C.M. (2013) 'Climate change adaptation in the Murray-Darling Basin: reducing resilience of wetlands with engineering', *Australian Journal of Water Resources*, vol 17, pp161–168.

Poff, N.L., Allan, J.D., Bain, M.B., Karr, J.R., Prestegaard, K.L., Richter, B.D., Sparks, R.E. and Stromberg, J.C. (1997) 'The natural flow regime: a paradigm for river conservation and restoration', *Bioscience*, vol 47, pp769–784.

Poff, N.L., Allan, J.D., Palmer, M.A., Hart, D.D., Richter, B.D., Arthington, A.H., Rogers, K.H., Meyer, J.L. and Stanford, J.A. (2003) 'River flows and water wars: emerging science for environmental decision making', *Frontiers in Ecology and the Environment*, vol 1, pp298–306.

Poff, N.L., Richter, B.D., Arthington, A.H., Bunn, S.E., Naiman, R.J., Kendy, E., Acreman, M., Apse, C., Bledsoe, B.P., Freeman, M.C., Henriksen, J., Jacobson, R.B., Kennen, J.G., Merritt, D.M., O'Keeffe, J.H., Olden, J.D., Rogers, K., Tharme, R.E. and Warner, A. (2010) 'The ecological limits of hydrologic alteration (ELOHA): a new framework for developing regional environmental flow standards', *Freshwater Biology*, vol 55, pp147–170.

Pollard, S. and Du Toit, D. (2008) 'Integrated water resource management in complex systems: How the catchment management strategies seek to achieve sustainability and equity in water resources in South Africa', *Water SA*, vol 34, pp671–679.

Pollard, S. and Du Toit, D. (2011) 'Towards adaptive integrated water resources management in southern Africa: the role of self-organisation and multi-scale feedbacks for learning and responsiveness in the Letaba and Crocodile catchments', *Water Resources Management*, 25: pp4019–35.

Pollard, S., Shackleton, C. and Carruthers, J. (2003) 'Beyond the fence: people and the Lowveld Landscape'. In Du Toit, J.T., Rogers, K.H. and Biggs, H.C. (eds), *The Kruger Experience: Ecology and Management of Savanna Heterogeneity*. Island Press, Washington, DC, pp422–446.

Quinty, F. and Rochefort L. (2003) *Peatland Restoration Guide*, 2nd ed. Québec: Canadian Sphagnum Peat Moss Association. Joint publication with New Brunswick Department of Natural Resources and Energy.

Ramsar Convention (2002) *Guidelines for Global Action on Peatlands*. Resolution VIII.17, 8th Meeting of the Conference of the Contracting Parties to the Convention on Wetlands (Ramsar, Iran, 1971), Valencia, Spain, 18–26 Nov. Available online at: www.ramsar.org/sites/default/files/documents/pdf/res/key_res_viii_17_e.pdf, accessed 23 Jan. 2016.

Rast, W. and Straskraba, M. (2000) *Lakes and Reservoirs: Similarities, Differences and Importance*, International Environment Technology Centre, UNEP, and International Lake Environment Committee, Kusatsu.

Rebelo, L.-M., McCartney, M.P. and Finlayson, C.M. (2011) 'The application of geospatial analyses to support an integrated study into the ecological character and sustainable use of Lake Chilwa', *Journal of Great Lakes Research*, vol 37, supplement 1, pp83–92.

Richardson, S., Irvine, E., Froend, R., Boon, P., Barber, S. and Bonneville, B. (2011a) *Australian Groundwater-dependent Ecosystems Toolbox Part 1: Assessment Framework*. Waterlines Report Series 69, National Water Commission, Canberra.

Richardson, S., Irvine, E., Froend, R., Boon, P., Barber, S. and Bonneville, B. (2011b) *Australian Groundwater-dependent Ecosystems Toolbox Part 2: Assessment Tools*. Waterlines Report Series 70, National Water Commission, Canberra.

Richter, B.D., Baumgartner, J.V., Powell, J. and Braun, D.P. (1996) 'A method for assessing hydrologic alteration within ecosystems', *Conservation Biology*, vol 10, pp1163–1174.

Richter, B.D., Warner, A.T., Meyer, J.L. and Lutz, K. (2006) 'A collaborative and adaptive process for developing environmental flow recommendations', *River Research and Applications*, vol 22, pp297–318.

Roux, D.J., Nel, J.L., Ashton, P.J., Deacon, A.R., de Moor, F.C., Hardwick, D., Hill, L., Kleynhans, C.J., Maree, G.A., Moolman, J. and Scholes, R.J. (2008) 'Designing protected areas to conserve riverine biodiversity: Lessons from a hypothetical redesign of the Kruger National Park', *Biological Conservation*, vol 141, pp100–117.

Solanes, M. and Gonzales-Villarreal, F. (1999) *The Dublin Principles for Water as Reflected in a Comparative Assessment of Institutional and Legal Arrangements for IWRM*, Global Water Partnership, Stockholm.

Sommer, B. and Froend, R. (2011) 'Resilience of phreatophytic vegetation to groundwater drawdown: is recovery possible under a drying climate?', *Ecohydrology*, 4: pp67–82.

Sommer, B. and Froend R.H. (2014) 'Alternative states of phreatophytic vegetation in a drying Mediterranean-type landscape', *Journal of Vegetation Science*, vol 25, pp1045–1055.

Swirepik, J.L., Burns, I.C., Dyer, F.J., Neave, I.A., O'Brien, M.G., Pryde, G.M. and Thompson, T.M. (2015) 'Establishing environmental water requirements for the Murray-Darling Basin, Australia's largest developed water system', *River Research and Applications*, vol 32, pp1153–1165.

Tennant, D.L. (1976) 'Instream flow regimens for fish, wildlife, recreation and related environmental resources', *Fisheries*, vol 1, pp6–10.

Tharme, R.E. (2003) 'A global perspective on environmental flow assessment: Emerging trends in the development and application of environmental flow methodologies for rivers', *River Research and Applications*, vol 19, pp397–441.

Turner, L., Tracey, D., Tilden, J. and Dennison, W.C. (2004) *Where River Meets Sea: Exploring Australia's Estuaries*. Cooperative Research Centre for Coastal Zone, Estuary and Waterway Management, Brisbane.

Turpie, J.K. and Clark, B.M. (2007) *The Health Status, Conservation Importance, and Economic Value of Temperate South African Estuaries and Development of a Regional Conservation Plan*, Anchor Environmental Consultants AEC/07/01.

Turpie, J.K., Sihlope, N., Carter, A., Maswime, T. and Hosking, S. (2007) 'Maximising the socio-economic benefits of estuaries through integrated planning and management: a rationale and protocol for incorporating and enhancing estuary values in planning and management', Appendix 1 in *Profiling Estuary Management in Integrated Development Planning in South Africa with Particular Reference to the Eastern Cape Province*, Water Research Commission publication 1485/1/07, Water Research Commission, Pretoria.

van Niekerk, L. and Turpie, J.K. (eds) (2012) *South African National Biodiversity Assessment 2011: Technical Report. Volume 3: Estuary Component*, Council for Scientific and Industrial Research, Stellenbosch.

Vörösmarty, C.J., McIntyre, P.B., Gessner, M.O., Dudgeon, D., Prusevich, A., Green, P., Glidden, S., Bunn, S.E., Sullivan, C.A., Liermann, C.R. and Davies, P.M. (2010) 'Global threats to human water security and river biodiversity', *Nature*, vol 467, pp555–561.

Watson, J.E., Dudley, N., Segan, D.B. and Hockings, M. (2014) 'The performance and potential of protected areas', *Nature*, vol 515, pp67–73.

Wetzel, R. (2001) *Limnology: Lake and River Ecosystems*, Elsevier/Academic Press, San Diego.

Whitfield, A.K., Bate, G.C., Adams, J.B., Cowley, P.D., Froneman, P.W., Gama, P.T., Strydom, N.A., Taljaard, S., Theron, A.K., Turpie, J.K., van Niekerk, L. and Wooldridge, T.H. (2012) 'A review of the ecology and management of temporarily open/closed estuaries in South Africa, with particular emphasis on river flow and mouth state as primary drivers of these systems', *African Journal of Marine Science*, 34: pp163–80.

Winemiller, K.O., McIntyre, P.B., Castello, L., Fluet-Chouinard E, Giarrizzo, T., Nam, S., Baird, I.G., Darwall, W., Lujan, N.K. and Harrison, I. (2016) 'Balancing hydropower and biodiversity in the Amazon, Congo and Mekong'. *Science*, vol 351, pp128–129.

Chapter 9

Freshwater protected area corridors

J. Pittock, M. Thieme, E. Blom and D. Willems

Key messages

- Riparian and floodplain corridors are particularly biodiverse and often form key habitat for animals in the terrestrial landscape, and in most parts of the world they support more species of plants and animals than any other landscape unit.
- The maintenance and restoration of riparian and floodplain corridors is a conservation priority for both freshwater and terrestrial ecosystems with considerable benefits to be gained from restoring riparian and floodplain forests. These forests play key roles in providing organic matter that drives elements of the aquatic food chain, forming physical habitat, filtering out pollutants and providing shade for maintaining appropriate light environments and water temperatures. Floodplains provide habitat, food and recruitment opportunities for fishes and other fauna.
- A key question for managers restoring riparian corridors is 'how wide is wide enough'? The minimal answer could be wide enough to enable full development of the vegetation canopy to maximize shade across the relevant water body and form an adequate mesic (moist, humid) microclimate. A more informed answer is that the full width of the regularly inundated riparian and floodplain land should be restored.
- Systems of river corridor Protected Areas (PAs) have been established in many jurisdictions around the world based on criteria such as biological importance, maintaining free-flowing ecological processes and supporting cultural values. Increasingly dams and levee banks are being removed from rivers and their floodplains to restore ecosystem functions and services. Restoration of functional riparian and floodplain systems may aid flood management and enhance other climate change adaptation measures.

Riparian corridors

Rivers are nature's natural corridors. The flow of water, nutrients and sediments, and the movement of species along streams, generates diverse habitat in riparian and floodplain corridors across terrestrial landscapes. These riparian and floodplain corridors are particularly biodiverse and often form key habitat for animals in the terrestrial landscape (Naiman, et al. 1993; Naiman, et al. 2005). Tockner et al. (2008) conclude that: "far more species of plants and animals occur on floodplains than in any other landscape unit in most regions of the world."

Consequently the maintenance and restoration of riparian corridors and floodplains is a conservation priority for both freshwater and terrestrial ecosystems.

There are considerable benefits to be gained from restoring riparian forests as outlined in the preceding section on borders and the upcoming section on climate change (Lukasiewicz et al., 2013; Finlayson and Pittock, Chapter 13, this volume). Riparian and floodplain forests play key roles in providing organic matter that drives elements of the aquatic food chain, forming physical habitat, filtering out pollutants and maintaining appropriate illumination, shade and water temperatures (Naiman, et al. 2005). Many river fishes have evolved feeding and life-history strategies to exploit food resources and recruitment opportunities in floodplain habitats inundated during the high-flow season (Lobon-Cervia et al. 2015; Godfrey et al. 2016). Substantial changes in the range of freshwater species are projected under climate change, but these may be constrained by natural geomorphic and anthropogenic barriers (Bond et al., 2011). As a result of their geomorphic evolution, lowland rivers provide some of the most gentle elevation gradients in the landscape and thus the ideal corridors for changes in distribution of many species under climate change, as shown with changes in fish communities in large rivers in France (Daufresne and Boet, 2007).

Riparian and floodplain management

A key question for managers restoring riparian corridors in areas where land use is contested is 'how wide is wide enough'? The simple answer for optimal biodiversity conservation is as wide as possible (Keller et al., 1993). Specific assessment is required in each case (Spackman and Hughes, 1995). The minimal answer where trade-offs between land use and effective conservation are required could be a corridor wide enough to enable full development of the vegetation canopy to maximize shade across the relevant water body and form an adequate mesic (moist, humid) micro-climate. Riparian vegetation is often thick and forms extensive shade and reduces air movement forming a mesic micro-climate that supports particular species and resists fire. A more informed answer from the perspective of maintaining connectivity and ecological processes is that restoration of the full width of the regularly inundated riparian land should be restored, that is, the floodplain as distinguished by wetland vegetation and soils (Kotze et al., 1996; DWAF, 2008).

In recent years landscape-scale linking projects (connectivity conservation projects) have commenced in many regions of the world, including Australia, New Zealand, the United States and Europe (Wyborn, 2011; Fitzsimons et al., 2013). Surprisingly, very few of these initiatives are centred on river corridors, which are natural pathways for movement of biota, unlike many linking projects that are replete with biophysical barriers. An exception is the 'room for rivers' floodplain restoration programs along major rivers, such as in Europe along the Danube (Ebert et al., 2009) and Rhine (see the Millingerwaard case study, Box 9.1) that combine habitat restoration, corridor establishment, and ecosystem-based adaptation to climate change and reducing flood risk. Similarly, in the central Yangtze River floodplain in China, programs since the late 1990s have focussed on removing levees to restore large areas of floodplain lakes, and opening sluice gates to restore connectivity between the floodplain and river channels (Yu, Jiang et al., 2009, Pittock and Xu 2011). These initiatives have restored populations of fish and other aquatic wildlife, as well as helping to better manage flood peaks, improve water quality and enhance the livelihoods of local people.

Protecting river corridors

Connectivity is one of the fundamental attributes necessary for ecological integrity of aquatic ecosystems and particularly for rivers (Karr, 1991; Arthington, et al., Chapter 8, this volume). Longitudinal, lateral and vertical connectivity allow for natural ecosystem processes and functions to occur such as the natural periodicity and magnitude of water flows, associated thermal regimes and sediment and nutrient transport, and movement of species through the riverscape (Poff et al., 1997; Bunn and Arthington, 2002; Fausch et al., 2002). Unfortunately, many rivers have experienced alterations in their natural connectivity due to water infrastructure such as dams, irrigation structures, channelization, inter-basin transfers and other human interventions. Examining the effects of dams globally, Grill et al. (2015) found that 48 per cent of river volume is moderately to severely impacted by either flow regulation, fragmentation, or both and that this number would nearly double to 93 per cent with the construction of planned dams. Nilsson et al. (2005) similarly found that over half (172 out of 292) of large river catchments are affected by dams in terms of fragmentation and flow regulation.

In certain parts of the world there are policies and efforts intended to protect stretches of river that maintain a high level of connectivity and other aspects of ecological integrity. For example, in 1968, the United States passed The National Wild and Scenic Rivers Act, which enabled the identification and designation of wild free-flowing rivers and their immediate environments with outstandingly remarkable values such as scenic, recreational, geologic, fish and wildlife, historic, cultural, and/or other values worthy of conservation. An inventory of candidate rivers was completed and there are currently 12,709 miles of 208 rivers or river segments protected under this designation (Figure 9.1) (United States National

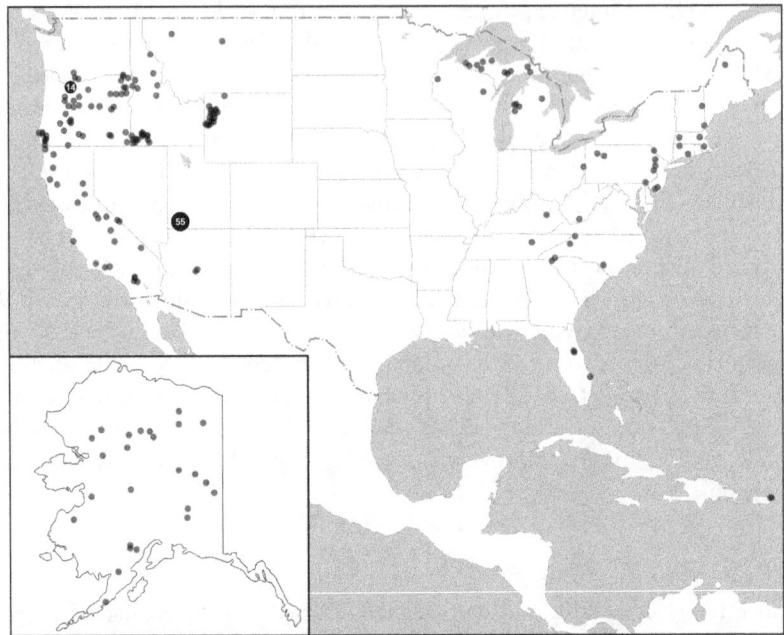

Figure 9.1 Rivers in the United States designated under the National Wild and Scenic Rivers Act. Source: Clive Hilliker © The Australian National University, adapted from US Fish and Wildlife Service (2016).

Park Service, 2011; United States Fish and Wildlife Service, 2016). However, less than 1 per cent of all rivers nationally, only about 0.35 per cent, are designated as wild and scenic rivers. Similarly the Swedish Government protected the Kalix, Pite, Torne and Vindel as National Rivers in the 1990s (Figure 9.2) (Anon, 2016).

Other countries and sub-national jurisdictions have similar designations or volunteer programs in place including Australia, New Zealand, Canada, Finland, Norway and Spain, and there is high variation in how widespread and successful these efforts have been across countries (Moir et al., 2016).

Three jurisdictions illustrate different ways of conserving ecologically significant river corridors. The Australian state of Victoria protected in law two classes of rivers in its 1992 Heritage Rivers Act (Victorian Government, 2007). This law reserves from further development 26 small "natural catchment areas" that have not been subject to significant disturbance by post-colonial society. Further, 18 "heritage river areas" are reserved that comprise corridors on major rivers selected on a combination of biologically significant and cultural value criteria (Figure 9.3). A further set of ecologically representative rivers were identified as priorities for non-legislative conservation programs.

Figure 9.2 Designated Swedish national rivers. Source: Clive Hilliker © The Australian National University.

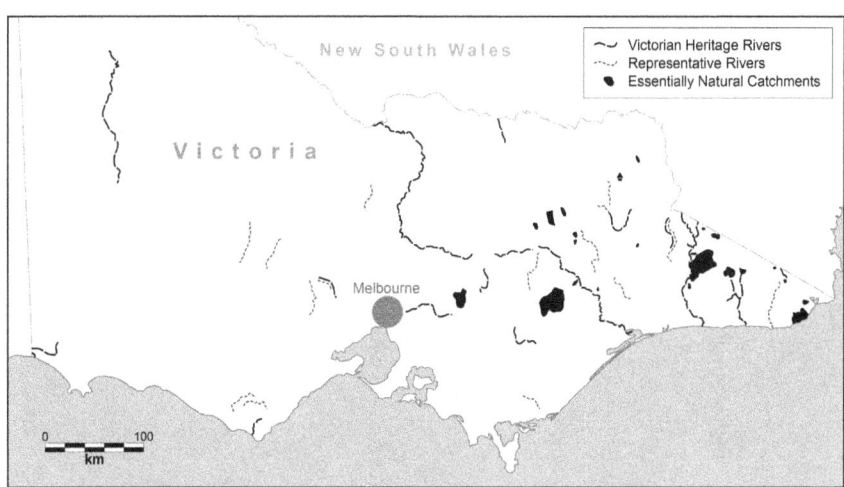

Figure 9.3 Rivers in the Australian state of Victoria designated under the Heritage Rivers Act as 'Heritage Rivers' comprising heritage and representative rivers, as well as 'Essentially Natural Catchments.' Source: Clive Hilliker © The Australian National University, adapted from Victorian Government (2007).

By contrast with this systematic approach to reservation under the Victorian law, the non-legislative Canadian Heritage Rivers System relies on voluntary nominations of rivers by sub-national governments and other stakeholders for conservation. Currently there are 42 Heritage Rivers (39 designated and three nominated) across Canada that cover nearly 12,000 river kilometres (Figure 9.4) (CHRS, 2017). Although not protected in law, the greater status of the designated rivers and consensus on values achieved among federal and local governments, and other stakeholders, has engendered active conservation of these rivers.

In another example, South Africa's National Freshwater Ecosystem Priority Areas project "aimed to identify a national network of freshwater conservation areas and to explore institutional mechanisms for their implementation" (SANBI, 2011: online). While the resulting detailed mapping and prioritization of the freshwater systems completed in 2011 has not yet resulted in greater legal protection for significant places, it does inform water allocation decisions (Figure 9.5).

China is another example of a developing country beginning to conserve rivers. The Chinese State Council designated the middle and downstream sections of the Chishui River as a "Natural Conservation Area for Rare, Treasure and Special Fishes of Upstream Yangtze River" in 2005 (Figure 9.6). This tributary of the Yangtze River was conserved in large part to offset the impacts of hydropower projects elsewhere in the Yangtze Basin (WWF, 2006; Moir et al., 2016).

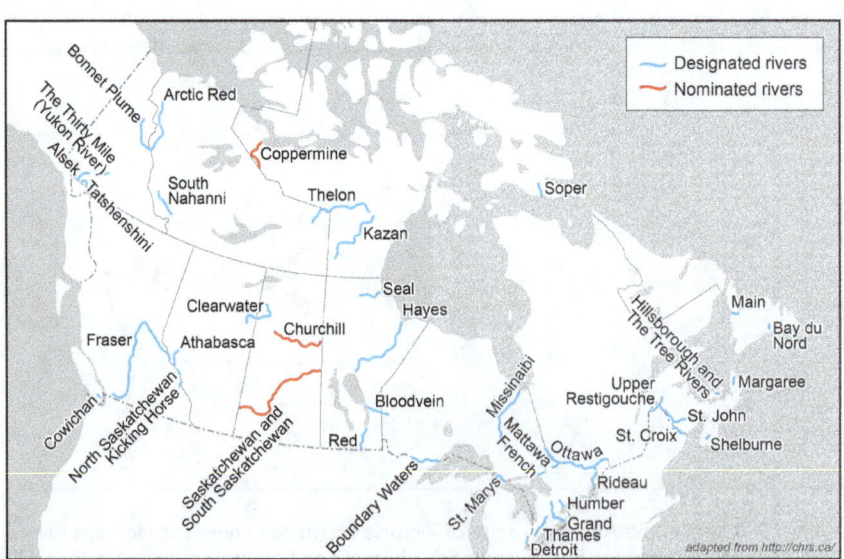

Figure 9.4 Rivers in Canada designated under the Heritage Rivers System. Source: Clive Hilliker © The Australian National University, adapted from CHRS (2017).

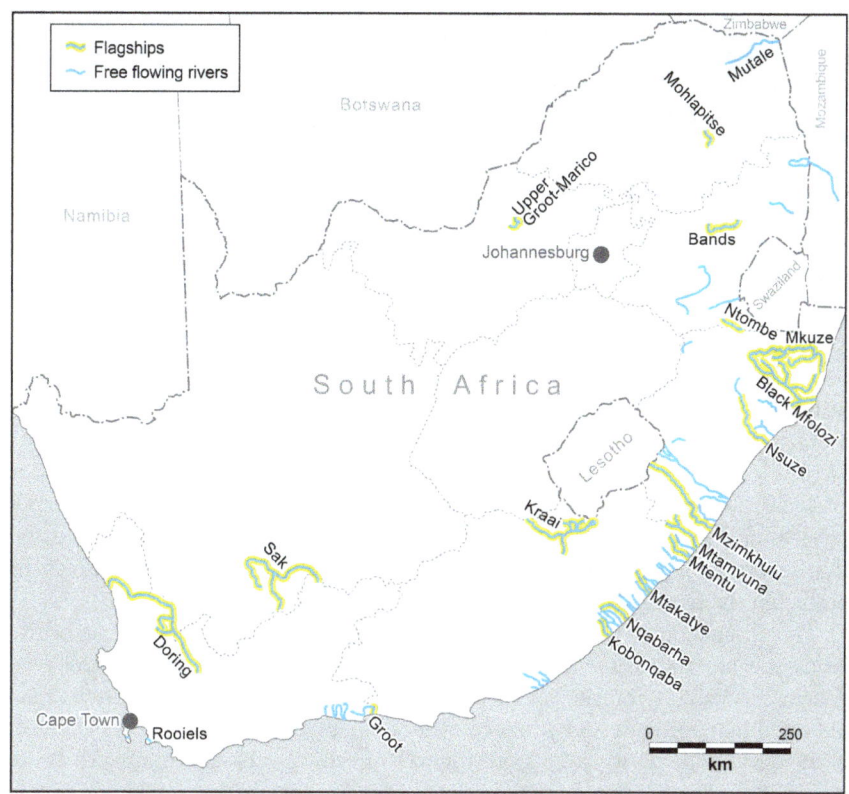

Figure 9.5 Rivers in South Africa identified by the National Biodiversity Institute as free-flowing and 'flagship' rivers that are priorities for conservation. Source: SANBI (2011).

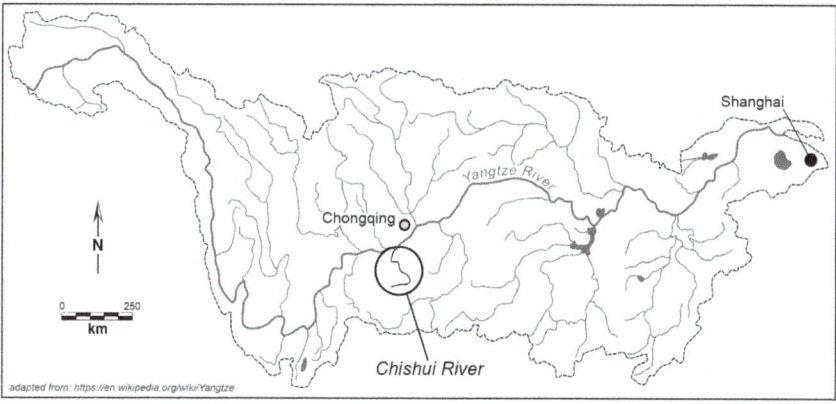

Figure 9.6 Location of the Chishui River Natural Conservation Area in the Yangtze River basin, China. Source: Clive Hilliker © The Australian National University.

Recently there have been calls for more proactive and system-level planning for the placement of water infrastructure, e.g., through hydropower master planning, to maintain a high level of riverine connectivity and biodiversity in certain parts of a basin (Opperman et al., 2015; Winemiller et al., 2016). In particular, planned large hydropower developments are expected to significantly fragment the Congo, Amazon and Mekong and it has been recommended that early-stage basin planning is an approach that can mitigate some of these effects (Winemiller et al., 2016). Norway provides an example of proactive planning. It has a complex legal framework regarding river protection, hydropower licensing and water management, which includes a Master Plan that excludes certain rivers from eligibility for hydropower licensing (Moir et al., 2016).

Additionally, dam removal efforts have increased in the last decade as the utility of some dams has waned and their negative consequences have become more apparent (Arthington et al., Chapter 3, this volume). This trend is greatest in North America where over 1,000 dams have been removed (O'Connor et al., 2015). The costs associated with dam removal and loss of critical ecosystem services and functions of connected rivers argue for better long-term planning that incorporates a system-level perspective and designation of certain rivers as protected riverscapes (Auerbach et al., 2014).

This discussion has emphasised that: river corridors are a priority part of the landscape for management to maximize conservation of biodiversity and ecosystem services; restoration of functional riparian and floodplain systems may aid flood management and enhance other climate change adaptation measures; barriers to connectivity are a great threat to freshwater biota but are also being modified and removed in some countries; and in many developed and a few developing countries there are programs to assess and protect key rivers from further development.

Case study   Millingerwaard, The Netherlands

The Millingerwaard is an area of former farm land in the floodplain along the River Rhine (Figure 9.7). Here, alluvial forests, marshlands, natural grasslands, surface waters and river dunes have been restored for nature conservation, recreation and flood management (Bekhuis et al., 2005). The 800 hectares are a Natura2000 site and IUCN protected area Category II 'National Park' managed by the State Forestry Commission.

In 2013 the Millingerwaard celebrated its 20th anniversary. Within two decades, the agricultural area along the River Rhine was turned into a mature nature reserve with spectacular results. Commercial clay and sand extraction companies have proven to be allies of nature organisations and Rijkswaterstaat (Directorate

General for Public Works and Water Management). By extracting clay deposits, following the underlying geographic relief, the natural pattern and structure of the historic riverine landscape is uncovered. In this way, mineral companies earn their incomes but also nature is developed, river safety is improved by giving room for the river, and a highly valued recreational area is created. Species like beaver, badger, black stork and the white-tailed eagle have returned to the floodplains of the Millingerwaard. Clay and sand lay extraction continues today, financing the nature development, following the pace of the demands in the mineral market (Bekhuis et al., 2005). In this way the Millingerwaard has become a pilot project for the 'Living Rivers' vision developed by WWF-Netherlands in the '90s (Helmer et al., 1992). In many other parts of the River Rhine the approach has been replicated contributing to reduced flood risk, recreation and biodiversity conservation and restoration along the river and its distributaries.

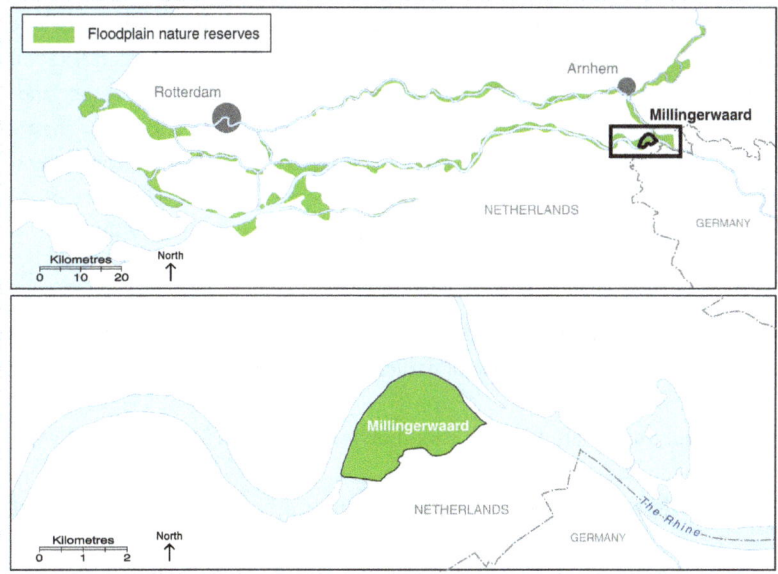

Figure 9.7 Map of Millingerwaard, the Netherlands. Source: Clive Hilliker © The Australian National University.

After the initial design and the establishment of the Millingerwaard, daily management has been taken over by the State Forestry Commission. An important characteristic of the management is the use of large herbivores. Old breeds of cattle and horses (mimicking extinct herbivores) roam the area, and

(continued)

(continued)

together with beavers, deer and geese, they perform vegetation control and improve spatial variety, creating various (micro) habitats for many other species. The herds live in naturally functioning social groups that function without human interference and daily care.

As an additional management tool, from time to time natural disasters are imitated, which would otherwise not take place in the highly regulated Dutch rivers. Then the managers remove part of the alluvial forest area. The wood is sold as biomass for green energy. These 'disturbances' help to regenerate river-bound pioneer communities and lead to a greater biodiversity. Moreover, in this way flood safety is assured, leaving sufficient space for the river to pass by during flood peaks.

The Millingerwaard has become a very popular recreational area, where visitors are allowed to roam freely. Maps with hikes and information are provided, but there are no restrictions on where to explore. The experienced freedom and nature perception is unlike any other nature reserve in The Netherlands. There are large financial and social spin-offs, providing jobs and increasing the values of surrounding real estate. It is estimated that there has been an increase of six million euro per year in the regional economy since The start of the Millingerwaard, due to the recreational activities (Bekhuis et al., 2005).

Success factors include:

- Financial: successful alliance between commercial parties and nature development; the maximum use of natural processes in design and maintenance lower the costs;
- Institutional: fruitful cooperation between nature management and water management;
- Social: free access, no restrictions for visitors, school education programmes.

Challenges include:

- The combination of developing high natural values and maintaining flood safety is one of the key management challenges of the area. For example, inundation of free refuge areas for the self-sustainable herds of wild herbivores (including cattle and horses) obstruct the river flows.
- Climate change: when more room for the river is needed due to increasing discharges of the River Rhine, spatial quality and natural values may be compromised.

References

Anon. (2016). 'Five parties in the bloc deal on energy policy' (in Swedish). *State Track.* Stockholm, Statskoll, statskoll.se/nyheter/fem-partier-i-blockoverskridande-uppgorelse-om-energipolitiken/.

Auerbach, D. A., Deisenroth, D. B., McShane, R. R., McCluney, K. E. and LeRoy Poff, N. (2014). 'Beyond the concrete: Accounting for ecosystem services from free-flowing rivers', *Ecosystem Services* 10:1–5.

Bekhuis, J., Litjens, G. and Braakhekke, W. (2005). *A Policy Field Guide to the Gelderse Poort: A New, Sustainable Economy under Construction*, WWF Netherlands, Zeist.

Bond, N., Thomson, J., Reich, P. and Stein, J. (2011). 'Using species distribution models to infer potential climate change-induced range shifts of freshwater fish in south-eastern Australia', *Marine and Freshwater Research* 62: 1043–1061.

Bunn, S. E. and Arthington, A. H. (2002). 'Basic principles and ecological consequences of altered flow regimes for aquatic biodiversity', *Environmental Management* 30: 492–507.

CHRS (2017). *Canadian Heritage Rivers System. Canada's National River Conservation Program*, Canadian Heritage Rivers System, Gatineau.

Daufresne, M. and Boet, P. (2007). 'Climate change impacts on structure and diversity of fish communities in rivers', *Global Change Biology* 13: 2467–2478.

DWAF (2008). *Updated Manual for the Identification and Delineation of Wetlands and Riparian Areas*, Department of Water Affairs and Forestry, Pretoria.

Ebert, S., Hulea, O. and Strobel, D. (2009). 'Floodplain restoration along the Lower Danube: a climate change adaptation case study', *Climate and Development* 1: 212–219.

Fausch, K. D., Torgersen, C. E., Baxter, C. V. and Li, H. W. (2002). 'Landscapes to river-scapes: bridging the gap between research and conservation of stream fishes', *BioScience* 52: 483–498.

Fitzsimons, J., Pulsford, I. and Wescott, G. (2013). 'Lessons from large-scale conservation networks in Australia', *Parks* 19: 115–125.

Godfrey, P.C., Arthington, A.H., Pearson, R.G., Karim, F. and Wallace, J. (2016) 'Fish larvae and recruitment patterns in floodplain lagoons of the Australian Wet Tropics'. *Marine and Freshwater Research*, 68 (5), pp964–979.

Grill, G., Lehner, B., Lumsdon, A. E., MacDonald, G. K., Zarfl, C. and Liermann, C. R. (2015). 'An index-based framework for assessing patterns and trends in river fragmentation and flow regulation by global dams at multiple scales', *Environmental Research Letters* 10(1): 015001.

Helmer, W., Litjens, G., Overmars, W., Barneveld, H., Kink, A., Sterenburg, H. and Janssen, B. (1992). *Living Rivers*, WWF Netherlands, Zeist.

Karr, J. (1991). 'Biological integrity: a long neglected aspect of water resource manage-ment', *Ecological Applications* 1: 66–84.

Keller, C. E., Robbins, C. and Hatfield, J. (1993). 'Avian communities in riparian forests of different widths in Maryland and Delaware', *Wetlands* 13: 137–144.

Kotze, D., Klug, J., Hughes, J. and Breen, C. (1996). 'Improved criteria for classifying hydric soils in South Africa', *South African Journal of Plant and Soil* 13: 67–73.

Lobon-Cervia, J., Hess, L. L., Melack, J. M. and Araujo-Lima, C. A. (2015). 'The impor-tance of forest cover for fish richness and abundance on the Amazon floodplain', *Hydrobiologia* 750:245–255.

Lukasiewicz, A., Finlayson, C. M. and Pittock, J. (2013). *Identifying Low Risk Climate Change Adaptation in Catchment Management while Avoiding Unintended Consequences*, National Climate Change Adaptation Research Facility, Gold Coast.

Moir, K., Thieme, M. L. and Opperman, J. (2016). *Securing a Future that Flows: Case Studies of Protection Mechanisms for Rivers*, World Wildlife Fund and The Nature Conservancy, Washington, DC.

Naiman, R. J., Décamps, H. and McClain, M. C. (2005). *Riparia*, Academic Press, San Diego.

Naiman, R. J., Décamps, H. and Pollock, M. (1993). 'The role of riparian corridors in maintaining regional biodiversity', *Ecological Applications* 3: 209–212.

Nel, J. L., Driver, A., Strydom, W., Maherry, A., Petersen, C., Hill, L., Roux, D. J., Nienaber, S., van Deventer, H., Swartz, E. and Smith-Adao, L. B. (2011). *Atlas of Freshwater Ecosystem Priority Areas in South Africa: Maps to Support Sustainable Development of Water Resources*. WRC Report No. TT 500/11, Water Research Commission, Pretoria.

Nilsson, C., Reidy, C. A., Dynesius, M. and Revenga, C. (2005). 'Fragmentation and flow regulation of the world's large river systems', *Science* 308: 405–408.

O'Connor, J. E., Duda, J. J. and Grant, G. E. (2015). '1000 dams down and counting', *Science* 348: 496–497.

Opperman, J., Grill, G. and Hartmann, J. (2015). *The Power of Rivers: Finding Balance between Energy and Conservation in Hydropower Development*, The Nature Conservancy, Washington, DC.

Pittock, J. and Xu, M. (2011). *World Resources Report Case Study. Controlling Yangtze River Floods: A New Approach*, World Resources Report 2010: Decision making in a changing climate, World Resources Institute, Washington, DC.

Poff, N. L., Allan, J. D., Bain, M. B., Karr, J. R., Prestegaard, K. L., Richter, B. D., Sparks, R. E. and Stromberg, J. C. (1997). 'The natural flow regime: a paradigm for river conservation and restoration', *BioScience* 47: 769–784.

SANBI (2011). *2011 National Freshwater Ecosystem Priority Areas (NFEPA)*, South African National Biodiversity Institute, Claremont, bgis.sanbi.org/nfepa/NFEPAmap.asp

Spackman, S. C. and Hughes, J. W. (1995). 'Assessment of minimum stream corridor width for biological conservation: Species richness and distribution along mid-order streams in Vermont, USA', *Biological Conservation* 71: 325–332.

Tockner, K., Bunn, S. E., Gordon, C., Naiman, R. J., Quinn, G. P., Standord, J. and Polunin, N. (2008). 'Flood plains: critically threatened ecosystems' in N. Polunin (ed.) *Aquatic Ecosystems: Trends and Global Prospects*, pp. 45–61, Cambridge University Press, Cambridge.

United States Fish and Wildlife Service (2016). *National Wild and Scenic Rivers System: About the WSR Act*, www.rivers.gov/wsr-act.php

United States National Park Service (2011). *Nationwide Rivers Inventory*, www.nps.gov/ncrc/programs/rtca/nri/hist.html#creation

Victorian Government (2007). Version No. 014, Heritage Rivers Act 1992, No. 36 of 1992, Version incorporating amendments as at 7 December 2007, Victorian Government, Melbourne.

Winemiller, K., McIntyre, P., Castello, L., Fluet-Chouinard, E., Giarrizzo, T., Nam, S., Baird, I. G., Darwall, W., Lujan, N. K., Harrison, I., Stiassny, M. L. J., Silvano, R. A. M., Fitzgerald, D. B., Pelicice, F. M., Agostinho, A. A., Gomes, L. C., Albert, J. S., Baran, E., Petrere, M., Zarfl, C., Mulligan, M., Sullivan, J. P., Arantes, C. C., Sousa, L. M., Koning, A. A., Hoeinghaus, D. J., Sabaj, M., Lundberg, J. G., Armbruster, J., Thieme, M. L., Petry, P., Zuanon, J., Vilara, G. T., Snoeks, J., Ou, C., Rainboth, W., Pavanelli, C. S., Akama, A., Soesbergen, A. v. and Sáenz, L. (2016). 'Balancing hydropower and biodiversity in the Amazon, Congo, and Mekong', *Science* 351: 128–129.

WWF (2006). *Free-flowing Rivers – Economic Luxury or Ecological Necessity?* WWF, Gland.

Wyborn, C. (2011). 'Landscape scale ecological connectivity: Australian survey and rehearsals', *Pacific Conservation Biology* 17:121–131.

Yu, X., Jiang, L., Wang, J., Wang, L., Lei, G. and Pittock, J. (2009). 'Freshwater management and climate change adaptation: experiences from the central Yangtze in China', *Climate and Development* 1: 241–248.

Chapter 10

Planning ecologically

The importance of management at catchment scales

R. Flitcroft, C. Little, J. Cabrera and I. Arismendi

Key messages

- Management of rivers and other wetlands within Protected Areas (PAs) requires integration with landscape management. Integrated management has the potential to address issues regarding both scientific questions (such as multi-scale ecosystem functions, processes and interactions), and applied challenges (such as management of land uses in diverse multi-user landscapes).
- Rather than approaching rivers as disconnected reaches where organisms could be analysed without a catchment or network context, broader conceptual frameworks link ecological communities to underlying geophysical systems. These include the river continuum concept, hierarchical organization of instream habitat, the natural flow regime, process domains, the network dynamics hypothesis, and the riverscape concept.
- Catchment-management plans are a means of integrating the diverse land uses and owners who, in combination, may directly or indirectly influence the quality of a shared river system. There is no clear template for catchment management that works well everywhere. Rather, there are a variety of examples of successful work, and different tools. Here, we present novel catchment management plans using ecosystem services as currency in the small Mechaico and Quilahuillque catchments of Chile, broad-scale multi-state water management planning in the Murray-Darling Basin in Australia, and a massive restoration program intended to restore the Everglades in South Florida, USA. Each of these examples offers insights into different elements of integrated catchment management through innovative approaches to solving complex water management challenges.

Rivers are ecologically connected networks

Rivers are inextricably linked to the landscapes that they drain, and stream condition is directly or indirectly affected by landscape condition (Arthington et al., Chapter 3, this volume). Terrestrial landscapes are generally influenced by a variety of drivers including climate, geomorphic setting, vegetation, natural disturbance regimes, and anthropogenic land use. Throughout much of the world, anthropogenic land use is a critical driver of terrestrial conditions that directly affects the structure, function, and resilience of aquatic ecosystems (Dudgeon et al., 2006), including systems in protected areas. Management of rivers and other wetlands within protected areas, then, requires integration with landscape management. Integrated management has the best potential to address issues regarding both scientific questions (such as multi-scale ecosystem processes and interactions), and applied challenges (such as management of land uses in diverse multi-user landscapes).

Conceptual frameworks are used by aquatic ecologists to structure enquiry about rivers as ecosystems. Rather than approaching rivers as disconnected reaches in which organisms can be analysed without a catchment or network context, broader conceptual frameworks link ecological communities to underlying geophysical systems. Some common examples include: the river continuum concept (Vannote et al., 1980), hierarchical organization of instream habitat (Frissell et al., 1986), the natural flow regime (Poff et al., 1997), process domains (Montgomery, 1999), the network dynamics hypothesis (Benda et al., 2004), and the riverscape concept (Fausch et al., 2002). Further, the geomorphic setting of streams—elevation profile, gradient, upstream catchment size—can be used to predictably define the intrinsic potential of streams to support different habitats (Burnett et al., 2007). The capacity of streams to support habitat is mediated by flow regime, disturbance history and land use. These drivers of instream habitat condition manifest differently over varying periods of time (Reeves et al., 1995) and at different spatial scales and extents.

The quantity and quality of instream habitat will change over time, and may vary throughout a catchment. This variation reflects the directional movement of water and the predictable changes in river size and flow associated with the accumulation of tributary water. Rivers are, by nature, corridors of movement for both biotic and abiotic elements. Abiotic components of the environment such as sediment and water are delivered to rivers by precipitation or disturbance events, and are then transported downstream based on the physics of waterflow and underlying topography (Wipfli et al., 2007). The expression "we all live downstream" is certainly apt, as pathways for movement by sediments, pollution, and water chemistry follow downstream gradients. For many biotic elements, such as macroinvertebrates and fishes, the concept of movement within the stream channel is complicated by the ability of individuals to move upstream as well as downstream, and both into and out of tributaries that form a network of stream habitats. Hence, PAs located in the middle of a catchment need to influence management both upstream and downstream to best conserve instream biodiversity.

Different locations within a catchment will support varied movement pathways for biotic and abiotic elements that, in turn, drive different aquatic processes. For example, the effect of a disturbance such as a road failure with associated sediment flow in the headwaters of a catchment will directly affect the biota in the intersecting stream section. Sediment from such an event will not remain in the affected stream section, but will wash downstream (Junk et al., 1989). However, the magnitude of habitat impairment will diminish as sediments settle, are captured, or spread out, and as flow increases. The portions of the river network not connected by flow to the disturbance (upstream of the failure, or in tributaries that join the stream below the failure) will be unaffected.

The river network drains a specific portion of the landscape, a catchment, based on upslope topography (Lotspeich, 1980). The catchment of a large river system will also include multiple smaller catchment sub-basins nested within it. River catchments, at most scales of organization, generally do not coincide with lines of human ownership, including protected-area boundaries (Figure 10.1). Although catchments are sometimes used as the basis for management plans, such plans are

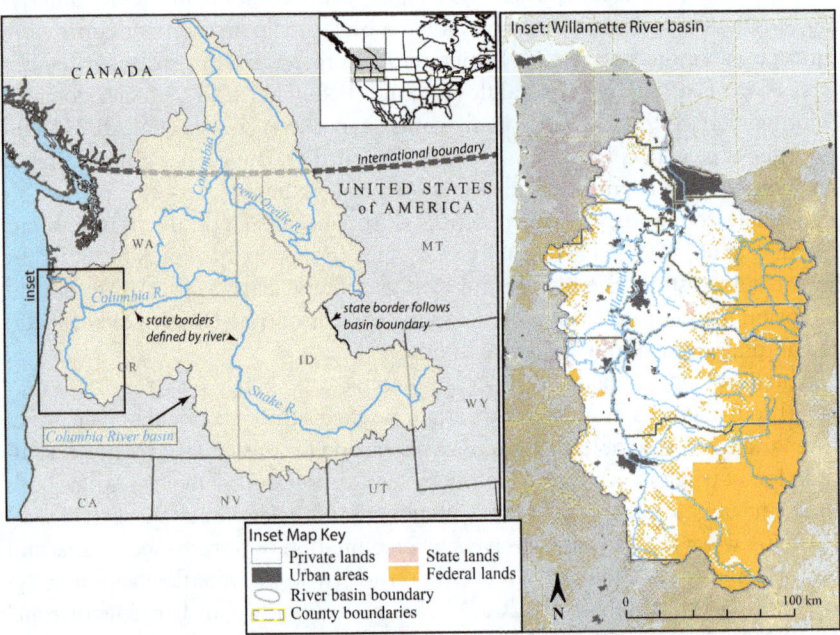

Figure 10.1 Catchments and jurisdictional boundaries. At any scale of organization, river catchments will most likely cross boundaries of human ownership or management jurisdiction. At the scale of the Columbia River, the entire catchment crosses international borders as well as state boundaries. The smaller Willamette River catchment (inset) crosses multiple county jurisdictions with landownerships divided among the US federal government, state of Oregon, and private holdings.

often challenging to implement because lines of human ownership and jurisdictions do not match catchment boundaries. The human landscape is an overlay upon the terrestrial setting. It is critical for protected area managers to become familiar with the concepts of rivers as ecosystems that have been developed by aquatic ecologists in order to understand how river reaches in their jurisdiction form part of the continuum of aquatic habitats. The spatial context of where stream habitats lie within the river network (i.e., headwaters compared with lowlands) must be combined with land-management history in order to understand the factors affecting water quality and habitat condition for any location within a catchment.

Some land-management activities have clear effects on aquatic habitat. These activities are often called point sources for pollution or some other form of water quality impairment. For example, hydroelectric or flood-control dams have a quantifiable influence on water flow and temperature throughout the year (Poole and Berman, 2002). Beyond flow, these structures may be impediments to movement for some native species, or may modify stream habitat by increasing temperature or removing flow (impounding lakes), thereby altering water quality and enhancing conditions for invasive species (Bunn and Arthington, 2002). These point-source effects may coincide with other land uses whose effect on aquatic habitat is more difficult to quantify. For example, run-off from agricultural fields or roads is more difficult to quantify and pinpoint, and is therefore referred to as a non-point source of pollution. The cumulative effects of non-point-source pollution may be considerable, and may also modify the type and quality of available aquatic habitats.

Catchment management

Catchment-management plans are a means of integrating diverse land uses with owners who, in combination, may directly or indirectly influence the quality of a shared river system. New approaches that encourage collaboration among the diverse set of stakeholders present in a river catchment are required for management that integrates terrestrial and aquatic planning (Abell et al., 2011; Allan et al., 2010; Russi et al., 2013). Catchment-management plans can offer opportunities for PA managers to favourably influence stakeholders and neighbouring land use, as well as to promote resilient ecosystems and ecosystem services. This type of management also requires the scope of planning to move beyond jurisdictions to the natural boundary of the catchment. Broad-scale cross-jurisdiction management has been approached from a variety of angles, including community-based organizations, government-sponsored programs, and programs led by non-profit organizations. Some of the most successful examples of catchment management and planning may be found in places where all three of these broad categories of stakeholders have invested in the work. Examples of successful catchment-scale planning, often spearheaded by grass-roots efforts and supported by government entities and non-profit organizations, have been documented in the USA (Flitcroft et al., 2009; Margerum, 2012), Australia (Curtis and Lockwood, 2000) and Europe (Warner et al., 2012).

Globally, catchment management goes by many names. The water sector often calls it Integrated Water Resources Management (IWRM), describing management across water-using sectors and stakeholders with the goal of sustainable development (Schoeman et al., 2014). A related approach to IWRM are Ecosystem-based Approaches (EBAs) that also seek to find balance between sectors, but with a strong scientific basis to guide decision-making. This is similar to Adaptive Management (AM) that is strongly driven by science to iteratively refine management goals over time. Both EBAs and AMs have been applied in terrestrial settings, but are being used more often in the water sector (Schoeman et al., 2014). To focus on ecological units, many governments and stakeholders have focused on Integrated River Basin Management (IRBM) and Integrated Lake Basin Management (ILBM; see Arthington et al., Chapter 8, this volume). In North America, catchments are usually called watersheds. The concept is also applied to groundwater aquifer management. Regardless, good catchment management engages multiple stakeholders in applying a common vision for sustainably managing a shared basin. Defining and managing for sustainable levels of water withdrawal and water quality are common elements of most catchment-management institutions (Baron et al., 2002).

Management of catchments is inherently difficult. Catchment-scale management lies at the intersection of ecological interests and human uses, including social and economic issues. Catchments extend across jurisdictional boundaries and force multi-agency, or sometimes international, involvement. There is no clear template for catchment management that works well everywhere. Rather, there is a variety of examples of successful work, and different tools are available that can inform the organic development of catchment-management programs. As catchment management becomes more mainstream, more examples of what works and what does not will be available as models for the future (Sadoff et al., 2008).

Examples of integrated catchment management

To explore applied elements of integrated catchment management, we will review three case studies. In Chile, a novel cost-payment system implemented by a single government entity to manage water resources has been tried in two small catchments (Mechaico and Quilahuilque River catchments). Results of this program have the potential to be more broadly used to better regulate water use and distribution. At a larger scale, work in the Murray-Darling Basin of southeast Australia requires the coordinated cooperation of five states/territories. An innovative basin plan that seeks to balance anthropogenic and environmental uses has been implemented, but there are indications that elements of this large program may require modification to be fully successful (Bond et al., 2014; Pittock et al., 2015; Pittock, 2016). In another example of broad-scale complex basin planning, the Florida Everglades presents an opportunity to explore management that integrates extensive freshwater-protected areas and diverse water uses and needs.

Figure 10.2 Catchment maps for case study locations: (a) Mechaico and Quilahuilque catchments, Chile; (b) Murray-Darling Basin, Australia; (c) South Florida—Everglades, Florida, USA.

Figure 10.2 (continued)

These three examples offer opportunities to learn about different approaches to catchment-scale management, integration among different governing bodies, and challenges to effective implementation of complex planning goals.

Mechaico and Quilahuilque catchments, Chile

A sustainable society will not be achieved without water security. In Chile, problems related to the increase in detrimental water quality and scarcity have been associated with shifts in precipitation regimes from climate change (Boisier et al., 2016); from land-use changes in which native forest was replaced by industrial, exotic fast-growing tree plantations (Lara et al., 2009; Little et al., 2009, 2015); and by a lack of best practices in catchment management (Little et al., 2016). Because the Chilean central government is primarily responsible for providing public access to drinking water, during the past decade the Forestry Institute (Instituto Forestal de Chile [INFOR] part of the Ministry of Agriculture) has being promoting forest-management practices to maintain and improve water quality. Two small forested catchments (Mechaico and Quilahuilque; Figure 10.2a) have been proposed as a case study for sustainable catchment management through a model of payment for ecosystem services (PES). These two catchments (1,670 ha) provide drinking water to the city of Ancud (population of 20,000) on Chiloe Island. The Mechaico and Quilahuilque catchments are shaped by small mountains (30–300 m above sea level) with gradual slopes covered in native forests, meadows, and shrublands. The regional climate is temperate and rainy with ocean influence (annual precipitation of 2,500 mm, average annual temperature of 11 °C). The dominant soils are of volcanic origin, are nutrient poor, and are of variable depths (<30 cm to over 1 m).

In the Mechaico and Quilahuilque catchments there are about 30 landowners with properties of varying extent (range of 0.5–157 ha). The main economic activities in these catchments support family subsistence and include small-scale tourism, production of livestock and cheese, and management for forestry and rangeland. A lack of land-use regulation coupled with non-existent public–private incentives or subsidies has impeded the promotion of best management practices, leading to deforestation, soil overuse, and unprotected riparian areas. Because these forested catchments have low productivity overall, there is an urgent need to develop alternative mechanisms to achieve sustainability of human activities as well as the ecosystem services that these catchments provide (Little et al., 2016; Penaluna et al., 2016). A pilot program is currently being implemented by INFOR based on a concept of whole-ecosystem management, using water as currency referred to as "water funds". The criteria being evaluated under this pilot program are: (a) identification of the ecosystem service to be provided (i.e., water quality and quantity); (b) identification of land uses that best provide the ecosystem service (i.e., native forests); (c) identification of landowners able to offer the ecosystem service; (d) identification of the benefactors of the ecosystem service (users of drinking water and the public works company); and (e) management activities that meet the criteria to receive the PES (clearly quantified and identified). Cooperators include national and international non-governmental agencies (Agrupación de Ingenieros Foretales por el Bosque Nativo AIFBN, and the Forest Stewardship Council FSC); the public works company (Empresa de Servicios Sanitarios de Los Lagos ESSAL S.A.) and government agencies (Forest Service CONAF and the Instituto de Desarrollo Agropecuario INDAP). Currently, 11 landowners have voluntarily adopted some of the recommended best practices of forest management to improve drinking water. The INFOR is creating a third-party institution to manage these "water funds", and is looking into ways to improve public incentives or subsidies to support all proposed management practices. This initiative of public–private cooperation is expected to serve as a model of catchment management that can be replicated elsewhere in Chile.

Murray-Darling Basin, Australia

Complex and competing water needs and uses often characterize large freshwater river systems. The vast Murray-Darling Basin (>1 million km^2) in south-eastern Australia is an example of diverse and independent governing bodies that act largely independently across an extensive river system (Figure 10.2b). In fact, the Murray-Darling Basin is distributed across five states or territories (Queensland, New South Wales, Victoria, South Australia, and the Australian Capital Territory), each with different governance structures. However, the river basin is a shared water resource critical to Australia's national agriculture and domestic food production (Swirepik et al., 2015) and biodiversity. Approximately 77,000 km of rivers and 30,000 wetlands (covering 5.7 million ha of the basin, nearly 5 per cent of total area, Kingsford et al., 2004) are encompassed within this catchment, which receives little rainfall, leading to modest annual flows of 31,600 gigaliters (GL) (variable between 6,700 and 117,900 GL) (Water Act 2007 – Basin Plan, 2012). In the past 200 years, extensive development of water resources has occurred in all five states/territories that manage the Murray-Darling Basin. Development has included dams, reservoirs, land clearance, and agricultural irrigation which has affected extensive areas of the basin and have resulted in current problems that include limited water for natural ecosystems, drying of wetlands, salinity, acidification, cyanobacteria blooms, and loss of biodiversity (Pittock and Finlayson, 2011; Leblanc et al., 2012). The need for comprehensive management that crosses political boundaries became apparent with the recent Millennium Drought (from the mid-1990s to 2009), which highlighted the compounding effect of water demand and environmental degradation in a changing climate (Leblanc et al., 2012).

Australia has been a leader in developing novel management plans based on evolving paradigms of large-scale natural resource management (Pittock et al., 2012). However, implementation of plans under times of environmental stress (such as the Millennium Drought) has been a challenge (Neave et al., 2015). The Murray-Darling Basin Authority (MDBA) was formed in 2008 in recognition that centrally coordinated water management had a greater chance to balance competing water needs for environmental and economic interests than an independent management approach (www.mdba.gov.au/). Coordinated decision-making reached by concensus was one goal of the MDBA. However, finding common ground among disparate stakeholders for the development of comprehensive water-management plans that balance environmental and anthropogenic water needs has been challenging (Arthington, 2012; Pittock et al., 2012). In particular, developing and implementing mechanisms to quantify the benefit of healthy ecosystems (as with ecosystem services) requires consistent institutional support, which has not always characterized management under the MDBA (Pittock et al., 2012).

Recovering hydrologic function in heavily managed systems has been recognized as a driver of resilience for aquatic ecosystems in the face of future climate changes (Morrongiello et al., 2011; Pittock and Finlayson, 2011). Although this may be true, in a system as altered as the Murray-Darling River system, ongoing water management may represent a more intense stress on native fishes than climate change (Balcombe et al., 2011). Large-scale and ambitious work to enhance environmental conditions has been started with the formation of the MDBA. However, necessary actions to achieve long-term resilience may require considerable changes in water management, possibly threating current alliances in the Murray-Darling Basin (Pittock et al., 2015). Successful applications and work to date may provide a foundation for future innovations should involvement and commitment by partners continue.

South Florida – Everglades, USA

Natural flow regimes include flooding (Poff et al., 1997). Public outcry at the economic costs of flooding often leads to extensive water-control projects that may have enormous environmental costs. Such was the case in South Florida, where water-control structures built in the early 20th century effectively modified flooding, but at great cost to the world-renowned Everglades ecosystem (designated as a Wetland of International Importance under the Ramsar Convention in 1987, as a United Nations Education, Scientific and Cultural Organization (UNESCO) International Biosphere Reserve in 1976, and as a UNESCO World Heritage Site in 1979) (Aumen and Havens, 2015). The Everglades ecosystem stretches from the headwaters of the Kissimmee River to Florida Bay, covering ~ 47,000 km², including Lake Okeechobee, a shallow, freshwater system (Figure 10.2c). The ecological damage from flood control was first noticed in the 1950s with declines in fish populations. In ensuing decades, 68 animals and plants were listed as threatened or endangered under the US Endangered Species Act, and seagrass beds in Florida Bay began to perish (USACE and SFWMD, 2000). Over time, the total area of Everglades habitat declined from approximately 10,000 to 5,000 km². In response to the ecological disaster that was unfolding, and as a result of strong public support for the Everglades ecosystem, the US Congress approved the Comprehensive Everglades Restoration Plan (CERP) in 2000. With an anticipated cost of between US$8 and US$11 billion over 30+ years, the CERP became the most expensive and comprehensive restoration plan in US history (Best, 2000). The CERP also offered an opportunity to explore how socially supported restoration efforts can become institutionalized in administration through government agencies and mandates. Restoration driven by public support may offer insights for implementation and ongoing stakeholder participation in long-term efforts (Kiker et al., 2001).

Enhancing and restoring the remaining Everglades ecosystem is the primary goal of CERP. The overland sheet flow of water that naturally characterized the Everglades ecosystem was effectively interrupted by water diversions and flood-control projects. Restoring hydrologic function in the lower Everglades (i.e., Everglades National Park) while also maintaining landholdings and human uses of the landscape is a significant challenge. The 68 projects proposed in CERP were intended to be implemented over time. However, in the first decade and a half of CERP, limited projects were completed (Kiker et al., 2001; Auman and Havens, 2015). Concern regarding the pace of implementation led to increased efforts by the US Army Corps of Engineers to begin new projects. However, it is possible that the longer timeframe of restoration (perhaps 50 years rather than the anticipated 30) may allow for the integration of projects designed to respond to climate change (Auman and Havens, 2015), which was not a strong consideration at the outset of the program.

Large-scale, long-term restoration projects like CERP may offer avenues for adaptive management of natural resources and responsiveness to emerging environmental stressors (such as climate change). As with work in the Murray-Darling Basin, the ongoing commitment and involvement of stakeholders will be critical for the success of CERP in coming decades.

Conclusions

New, innovative and ingenious tools are being used to solve water management problems around the world. The best examples are informed by the complex and interconnected nature of rivers themselves. Such approaches are shifting management toward a more holistic direction that integrates instream and upslope conditions throughout the continuum of habitats laid out along the river corridor. Broad-scale and interdisciplinary approaches to integrated catchment management require collaboration and cooperation among stakeholders, governments, and sometimes nations. The complex network of connections inherent in river courses is the template for understanding the physical processes that drive water and abiotic elements of the river. Through this frame, we can better conceptualize and provide for ecological water requirements while simultaneously fulfilling human needs and uses. There is still much room for improvement, but through creative experiments at varying spatial scales, we can inform the broader discussion about sustainable water management.

References

Abell, R., Thieme, M., Ricketts, T.H., Olwero, Ng. R., Petry, P., Dinerstein, E., Revenga, C. and Hoekstra, J. (2011) 'Concordance of freshwater and terrestrial biodiversity', *Conservation Letters*, vol 4, no 2, pp 127–136.

Allan, D., Esselman, P., Abell, R., McIntyre, P., Tubbs, N., Biggs, H., Castello, L., Jenkins, A. and Kingsford, R. (2010) 'Protected areas for freshwater ecosystems: Essential but underrepresented' in Mittermeier, R. A., Farrell, T. A., Harrison, I. J., Upgren, A. J. and Brooks T. M. (eds) *Fresh Water: The Essence of Life*, CEMEX & ILCP, Arlington, Virginia, USA.

Arthington, A.H. (2012) *Environmental Flows. Saving Rivers in the Third Millennium*, University of California Press, Berkeley, CA.

Auman, N.G. and Havens, K.E. (2015) 'Predicting ecological responses of the Florida Everglades to possible future climate scenarios: Introduction', *Environmental Management*, vol 55, pp 741–748.

Balcombe, S.R., Sheldon, F., Capon, S.J., Bond, N.R., Hadwen, W.L., Marsh, N. and Bernays, S.J. (2011) 'Climate-change threats to native fish in degraded rivers and floodplains of the Murray-Darling Basin, Australia', *Marine and Freshwater Research*, vol 62, pp 1099–1114.

Baron, J.S., Poff, N.L., Angermeier, P.L., Dahm, C.N., Gleick, P.H., Hairston, N.G., Jackson, R.B., Johnston, C.A., Richter, B.D. and Steinman, A.D. (2002) 'Meeting ecological and societal needs for freshwater', *Ecological Applications*, vol 12, pp 1247–1260.

Benda, L.E., Poff, N.L., Miller, D., Dunne, T., Reeves, G.H., Pess, G. and Pollock, M. (2004) 'The network dynamics hypothesis: how channel networks structure riverine habitats', *BioScience*, vol 54, pp 413–427.

Best, G.R. (Ed.) (2000) 'Greater Everglades Ecosystem Restoration (GEER) science conference: Defining success proceedings', *South Florida Ecosystem Restoration Task Force and Working Group and US Geological Survey*, Miami, Florida, USA.

Boisier, J.P., Rondanelli, R., Garreaud, R.D. and Muñoz, F. (2016) 'Anthropogenic and natural contributions to the Southeast Pacific precipitation decline and recent megadrought in central Chile', *Geophysical Research Letters*, vol 43, pp 413–421, doi:10.1002/2015GL067265.

Bond, N.R., Costello, J., King, A., Warfe, D., Reich, P., Balcombe, S. (2014) 'Risks and opportunities from artificial watering as a means of achieving environmental flow objectives', *Frontiers in Ecology and Environment*, vol 12, pp 386–394.

Bunn, S.E. and Arthington, A.H. (2002) 'Basic principles and ecological consequences of altered flow regimes for aquatic biodiversity', *Environmental Management*, vol 30 no 4, pp 492–507.

Burnett, K.M., Reeves, G.H., Miller, D.J., Clarke, S., Vance-Borland, K. and Christiansen, K. (2007) 'Distribution of salmon-habitat potential relative to landscape characteristics and implications for conservation', *Ecological Applications*, vol 17 no 1, pp 66–80.

Curtis, A. and Lockwood, M. (2000) 'Landcare and catchment management in Australia: lessons for state-sponsored community participation', *Society and Natural Resources: An International Journal*, vol 13, no 1, pp 61–73.

Dudgeon, D., Arthington, A.H., Gessner, M.O, Kawabata, Z., Knowler, D., Lévêque, C., Naiman, R.J., Prieur-Richard, A.-H., Soto, D., Stiassny, M.L.J. & Sullivan C.A. (2006). Freshwater biodiversity: importance, threats, status, and conservation challenges. *Biological Reviews* 81 (2): 163–182.

Fausch, K.D., Torgersen, C.E., Baxter, C.V. and Li, H.W. (2002) 'Landscapes to riverscapes: Bridging the gap between research and conservation of stream fishes', *Bioscience*, vol 52, no 6, pp 483–498.

Flitcroft, R.L., Dedrick, D.C., Smith, C.L., Thieman, C.A. and Bolte, J.P. (2009) 'Social infrastructure to integrate science and practice: the experience of the Long Tom Watershed Council', *Ecology and Society*, vol 14, no 2, 36, www.ecologyandsociety.org/vol14/iss2/art36/.

Frissell, C.A., Liss, W.J., Warren, C.E. and Hurley, M.D. (1986) 'A hierarchical framework for stream habitat classification: viewing streams in a watershed context', *Environmental Management*, vol 10, no 2, pp 199–214.

Junk, W.J., Bayley, P. B. and Sparks, R. E. (1989) 'The flood pulse concept in river-floodplain systems', *Canadian Journal of Fisheries and Aquatic Sciences*, vol 106, pp 110–127.

Kiker, C.F., Milon, J.F. and Hodges, A.W. (2001) 'South Florida: the reality of change and the prospects for sustainability: Adaptive learning for science-based policy: the Everglades restoration', *Ecological Economics*, vol 37, no 3, pp 403–416.

Kingsford, R.T., Brandis, K., Thomas, R.F., Crighton, P., Knowles, E. and Gale, E. (2004) 'Classifying landform at broad spatial scales: the distribution and conservation of wetlands in New South Wales, Australia', *Marine and Freshwater Research*, vol 55, pp 17–31.

Lara, A., Little, C., Nahuelhual, L., Urrutia, R. and Díaz, I. (2011) 'Lessons, challenges and policy recommendations for the management, conservation and restoration of native forests in Chile'. In Figueroa, E. (ed), *Successful and Failed Experiences in Biodiversity Conservation: Lessons and Policy Recommendations from the American Continent*, Universidad de Chile, Programa de Investigación Domeyco, Santiago, Chile, pp 281–327.

Lara, A., Little, C., Urrutia, R., McPhee, J., Álvarez-Garretón, C., Oyarzún, C., Soto, D., Donoso, P., Nahuelhual, L., Pino, M. and Arismendi, I. (2009) 'Assessment of ecosystem services as an opportunity for the conservation and management of native forest in Chile', *Forest Ecology and Management*, vol 258, pp 415–424.

Leblanc, M., Tweed, S., Van Dijk, A. and Timbal, B. (2012) 'A review of historic and future hydrological changes in the Murray-Darling Basin', *Global and Planetary Change*, vol 80–81, pp 226–246.

Little, C., Cuevas, J.G., Lara, A., Pino, M. and Schoenholtz, S. (2015) 'Buffer effects of streamside native forests on water provision in watersheds dominated by exotic forest plantations', *Ecohydrology*, vol 8, pp 1205–1217.

Little, C., Lara, A., McPhee, J. and Urrutia, R. (2009) 'Revealing the impact of forest exotic plantations on water yield in large scale watersheds in south-central Chile', *Journal of Hydrology*, vol 374, pp 162–170.

Little, C., Zambrano, M., Benitez, S. and Rivera, A. (2016) 'Aguas continentales'. In *Informe País: Estado del Medio Ambiente en Chile, Comparación 1999–2015*, Universidad de Chile, Instituto de Asuntos Públicos, Centro de Análisis de Políticas Públicas, Santiago de Chile, pp 115–164.

Lotspeich, F.B. (1980) 'Watersheds as the basic ecosystem: this conceptual framework provides for a natural classification system', *Water Resources Bulletin*, vol 16, pp 581–586.

Margerum, R. (2012) 'Integrated water resources management in the United States: The Rogue and Willamette River cases'. In Warner, J.F., Van Buuren, A. and Edelenbos J. (eds) *Making Space for the River: Governance Experiences with Multifunctional River Flood Management in the US and Europe*, IWA Publishing, London, UK.

Montgomery, D.R. (1999) 'Process domains and the river continuum', *Journal of the American Water Resources Association*, vol 35, pp 397–410.

Morrongiello, J.R., Beatty, S.J., Bennett, J.C., Crook, D.A., Ikedife, D.N.E.N., Kennard M.J., Kerezsy, A., Lintermans, M., McNeil, D.G., Pusey, B.J. and Rayner, T. (2011) 'Climate change and its implications for Australia's freshwater fish', *Marine and Freshwater Research*, vol 62, pp 1082–1098.

Neave, I., McLeod, A., Raisin, G. and Swirepik, J. (2015) 'Managing water in the Murray-Darling basin under a variable and changing climate', *Water –Technical Papers April 2015*, pp 102–107.

Penaluna, B.E., Olson, D.H., Flitcroft, R.L., Weber, M., Bellmore, J.R., Wondzell, S.M., Dunham, J.B., Johnson, S.L. and Reeves, G.H. (2016) 'Aquatic biodiversity in forests: A weak link in ecological and ecosystem resilience', *Biodiversity and Conservation*, doi:10.1007/s10531-016-1148-0.

Pittock, J. (2016) 'Murray-Darling Basin: Conservation and Law'. In Finlayson, C.M, Everard, M., Irvine, K., McInnes, R., Middleton, B., van Dam, A. and Davidson, N.C. (eds), *The Wetland Book I: Structure and Function, Management and Methods*. Dordrecht, Springer Netherlands, vol 1–9.

Pittock, J., Cork, S. and Maynard, S. (2012) 'The state of the application of ecosystem services in Australia', *Ecosystem Services*, vol 1, pp 111–120.

Pittock J. and Finlayson C.M. (2011) 'Australia's Murray-Darling Basin: Freshwater ecosystem conservation options in an era of climate change', *Marine and Freshwater Research*, vol 62, pp 232–243.

Pittock, J., Grafton, R.Q. and Williams, J. (2015) 'The Murray-Darling Basin Plan fails to deal adequately with climate change', *Water. Journal of the Australian Water Association*, vol 42, no 6, pp 8–34.

Poff, N.L., Allan, J.D., Bain, M.B., Karr, J.R., Prestegaard, K.L., Richter, B.D., Sparks, R.E. and Stromberg, J.C. (1997) 'The natural flow regime a paradigm for river conservation and restoration', *BioScience*, vol 47, no 11, pp 769–784.

Poole, G.C. and Berman, C. H. (2002) 'An ecological perspective on in-stream temperature: natural heat dynamics and mechanisms of human-caused thermal degradation', *Environmental Management*, vol 27, no 6, pp 787–802.

Reeves, G.H., Benda, L.E., Burnett, K.M., Bisson, P.A. and Sedell, J.R. (1995) 'A disturbance-based ecosystem approach to maintaining and restoring freshwater habitats of evolutionarily significant units of anadromous salmonids in the Pacific Northwest'. In Nielson, J.L. and Powers, D.A. (eds) *Evolution and the Aquatic Ecosystem: Defining Unique Units in Population Conservation, American Fisheries Society Symposium 17*. American Fisheries Society, Bethesda, Maryland, USA.

Russi, D., ten Brink, P., Farmer, A., Badura, T., Coates, D., Förster, J., Kumar, R. and Davidson, N. (2013) *The Economics of Ecosystems and Biodiversity for Water and Wetlands*, IEEP, London, Brussels.

Sadoff, C., Greiber, T., Smith, M. and Bergkamp, G. (2008) *Share: Managing Water Across Boundaries*, International Union for Conservation of Nature, Gland, Switzerland.

Schoeman, J., Allan, C. and Finlayson C.M. (2014) 'A new paradigm for water? A comparative review of integrated, adaptive and ecosystem-based water management in the Anthropocene', *International Journal of Water Resources Development*, doi: 10.1080/07900627.2014.907087.

Swirepik, J.L., Burns, I.C., Dyer, F.J., Neave, I.A., O'Brien, M.G., Pryde, G.M. and Thompson, R.M. (2015) 'Establishing environmental water requirements for the Murray-Darling Basin, Australia's largest developed river system', *River Research and Applications*, doi: 10.1002/rra.2975.

US Army Corps of Engineers and South Florida Water Management District (USACE and SFWMD) (2000) *Master Program Management Plan, Comprehensive Everglades Restoration Plan*. South Florida Water Management District, West Palm Beach, Florida, USA.

Vannote, R.L., Minshall, G.W., Cummins, K.W., Sedell, J.R. and Cushing, C.E. (1980) 'The river continuum concept', *Canadian Journal of Fisheries and Aquatic Sciences*, vol 37, pp 130–137.

Warner, J.F., Van Buuren, A. and Edelenbos, J. (Eds) (2012) *Making Space for the River: Governance Experiences with Multifunctional River Flood Management in the US and Europe*, IWA Publishing, London, UK.

Water Act 2007 – Basin Plan (2012). Prepared by the Murray-Darling Basin Authority, www.comlaw.gov.au/Details/F2012L02240 [accessed 28 November 2016].

Wipfli, M.S., Richardson, J.S. and Naiman, R.J. (2007) 'Ecological linkages between headwaters and downstream ecosystems: transport of organic matter, invertebrates, and wood down headwater channels', *Journal of the American Water Resources Association*, vol 43, no 1, pp 72–85.

Chapter 11

Planning for the protection and management of freshwater ecosystems inside and outside protected areas

J.L. Nel and D.J. Roux

Key messages

- The inherent connectivity of most freshwater ecosystems necessitates that Protected Area (PA) authorities engage in water planning processes both within their protected area boundary and beyond in associated catchments. For planning to be effective and achieve its desired outcomes it needs to be embedded in a management approach that acknowledges the inherent complexity and uncertainty not only of ecosystems, but of linked social-ecological systems.

- Three water management approaches – Integrated Water Resource Management, Ecosystem-Based Management and Adaptive Management – have evolved in recent years that move beyond traditional 'command-and-control' approaches to incorporate a much better understanding of the interconnections between people and ecosystems, the factors that shape these interactions, and options to achieve sustainable futures and improved human wellbeing. These approaches, and their underlying principles, provide useful guidance for water planning, whether PA authorities are planning inside or outside PAs.

- Planning within PAs should ensure that freshwater ecosystems are managed explicitly as part of the PA landscape. PA strategies and management plans should include freshwater conservation objectives, such as freshwater-friendly design and management of PA boundaries, visitor facilities, road-river crossings, and weirs and dams inside protected areas. Management plans should also allow for monitoring of freshwater indicators to inform the management of freshwater ecosystems.

- Planning beyond the PA boundary should ideally be embedded within a regional water management framework that provides a platform for the cross-sectoral cooperative governance of water resources at multiple scales of governance. This will enable the PA authority to engage with the social-ecological linkages that may impact on freshwater ecosystems within their PAs. It also offers the opportunity to identify new societal roles that PAs may contribute to in the future, such as the provision of ecosystem services to downstream communities.

- In engaging with regional water planning, PA authorities should insist that scientifically credible outputs from freshwater ecosystem assessments, such as environmental flow assessment, systematic conservation planning and ecosystem service assessment, are used to assess the needs of freshwater ecosystems when negotiating trade-offs between different water development options. By engaging in water planning processes within and beyond their PA the managing authorities can act as powerful stakeholders in the negotiation for freshwater ecosystem protection within regional water planning processes.

Introduction

The Earth's seemingly vast natural resources are not infinite. Planning is considered an important step in making best use of these limited resources and achieving desired outcomes. However, planning alone will not achieve its desired outcomes unless embedded within an effective implementation process that is geared for today's modern world. As the world population continues to grow, and the environmental impacts of economic activity increase, humanity's effect on the biosphere threatens to exceed vital planetary boundaries and destabilise critical biophysical systems (Rockström et al., 2009). There is mounting evidence that we have entered a new geological era, the Anthropocene, where humans have become a global force of change at the planetary scale (Williams et al., 2015). The Anthropocene is characterised by a world of increasing complexity and turbulence, in which humans and ecosystems are inextricably linked and nested across scales, from individual to global. Effective water planning and implementation within this new era needs to pay careful attention to the social, economic and ecological drivers, and the interconnections, trade-offs and synergies between them across multiple sectors, scales and regions – or face unintended consequences. Being aware of these interactions and feedbacks is even more critical for freshwater ecosystems that are inherently connected across space and time (Pringle, 2003; Arthington et al., Chapter 3, this volume) as well as to human use.

In response to the challenges presented by the Anthropocene, several pragmatic operational models to achieving sustainable management of the Earth's resources have emerged, such as Integrated Water Resource Management, Ecosystem-Based Management and Adaptive Management (Schoeman et al., 2014). These can serve as over-arching implementation frameworks in which to embed water planning. Although the management approaches have different origins and objectives, they share some fundamental principles on managing the resilience of linked social–ecological systems to ensure a sustainable supply of ecosystem services on which humanity depends (Biggs et al., 2012). These 'resilience principles' serve as working principles for planning and management efforts, and provide PA authorities with some helpful guidance when planning for the protection of freshwater ecosystems.

In most countries water and biodiversity are managed through separate legislation (Gilman et al., 2004). This has led to different sectors developing for water and biodiversity, each with its own distinct policies, terminologies and tools (Roux et al., 2008). Water sector planning has largely focussed on meeting human water supply needs through engineering solutions; while biodiversity sector planning has focused largely on protecting terrestrial ecosystems. Until recently, neither sector has given sufficient attention to the protection and management of freshwater ecosystems. To rectify this situation, PA authorities should work at closing the gap between the water and biodiversity sectors in two ways: (1) by making sure that freshwater ecosystem considerations are incorporated into biodiversity sector plans, such as PA strategies and management plans; and (2) by engaging in regional water planning processes beyond the protected area boundary, which is often viewed as a water sector mandate and not part of their biodiversity responsibilities.

This chapter focuses on water planning in a changing world in which humans and nature are inextricably linked. The chapter aims to provide PA authorities with some of the main considerations when planning for the protection of freshwater ecosystems and associated biodiversity. It targets PA authorities at the operational level of individual PA, as well as PA systems managers with a more national or international focus. We begin by outlining some emerging water management approaches in which water planning can be embedded, explaining some of the 'resilience principles' that they have in common. The second part of this chapter describes key freshwater considerations that should be incorporated into planning within PAs. We then turn our attention to regional water planning beyond the PA, where we describe three assessment approaches that have been developed to inform catchment-scale water planning: environmental flow assessment, systematic conservation planning and ecosystem service assessment. We discuss the role that PA authorities can play in ensuring that the outputs from these assessments inform and shape water planning processes, and the opportunities these perspectives provide for re-imagining the future role of PAs in the Anthropocene.

Implementation frameworks for water planning in the Anthropocene

Conventional water planning and management has focused largely on engineering solutions to control the natural variability of freshwater ecosystems and optimise water supply. Such 'command-and-control' approaches usually have limited stakeholder involvement and typically disregard interactions across scales and sectors. They are less useful in the Anthropocene, which demands approaches that foster capacity to adapt to unforeseen changes that may emerge from the complex interconnections and feedbacks between societies, economies and the environment (Folke et al., 2011). More recent operational models for sustainable management of resources seek to incorporate a much better understanding of the interconnections between people and ecosystems, the factors that shape these interactions, and options to achieve sustainable futures and improved human wellbeing (Schoeman et al., 2014). These operational frameworks are briefly described below.

Integrated Water Resource Management

This management approach promotes 'coordinated development and management of water, land and related resources, in order to maximise economic and social welfare in an equitable manner without compromising the sustainability of vital systems' (Global Water Partnership, 2000). The approach provides a governance platform for actors to negotiate integrated land and water management at a catchment scale. Key elements include: catchment-scale planning and management; stakeholder participation and decentralised management; coordination across sectors to plan and manage water resource development; long-term planning; and consideration of multiple interests, scenarios and trade-offs. The approach has become a widely-used water management framework in many countries. Despite significant uptake of the approach, it has come under recent criticism as being vague, with no clear guidance on what needs to be integrated, how this should be achieved, who leads the approach and decisions, and who pays (Biswas, 2004; Crase and Cooper, 2015). Much of this criticism has focussed on the difficulty of achieving integration in decentralised management, however, Pahl-Wostl et al. (2007) make a strong case for a more dynamic actor landscape in which integration is achieved not by centralised bureaucratic hierarchies but rather by processes of network governance (*sensu* 'polycentric governance' in Box 11.1 below). Balancing power differentials among stakeholders and promoting social learning among them are essential ingredients for promoting the effectiveness of such a governance network (Mostert et al., 2007), and hence the effectiveness of integrated water resource management.

Ecosystem-Based Management

This management approach promotes integrated management of land, water and societies and seeks to balance conservation and sustainable use in an equitable way (Finlayson et al., 2011). The approach emphasises holistic ecosystem conservation and seeks to create an incentive for this through making explicit the trade-offs in ecosystem services under different resource-use options. Methodologies are varied but the most common principles include: considering ecosystem connections; assessing multiple spatial and temporal scales; conserving ecosystem structure and functioning; internalising conservation costs and benefits; adopting a long-term perspective;

applying adaptive management that recognises that change is inevitable; applying a range of protection mechanisms (strict protection to sustainable use); incorporating multiple social values and knowledge systems; and promoting decentralised management. The concept of Ecosystem-Based Adaptation to climate change embodies many of these principles and is frequently focussed on freshwater systems (Doswald et al., 2014). Like integrated water resource management approaches, ecosystem-based management faces criticism concerning ambiguity of definitions and protocols (Finlayson et al., 2011). Substantial criticism has also been levelled at the aim to internalise conservation costs and benefits by reducing them to a monetary valuation of ecosystem services, as this may undermine biodiversity, cultural and ethical values (Dudgeon, 2014). Tallis et al. (2010) provide suggestions for addressing some of the more frequently-occurring problems of ecosystem-based management, which include eliciting goals from stakeholders, identifying indicators of ecosystem condition, understanding the uncertainties associated with each indicator, providing a range of management scenarios for negotiating trade-offs, establishing monitoring systems and evaluating the effectiveness of the management strategies.

Adaptive Management

This management approach acknowledges that interacting systems of people and ecosystems – or social-ecological systems – are inherently complex, unpredictable and difficult or impossible to control (Holling, 2001; Rogers, 2003). In the face of this complexity and uncertainty, ongoing learning is considered key to enhancing the adaptive capacity that enables the social–ecological system to respond to change (Fabricius and Cundill, 2014; McLoughlin and Thoms, 2015). Adaptive management represents a significant departure from the conventional 'command-and-control' water management approaches (Kingsford et al., 2011). The approach emphasises experimentation, monitoring, learning and negotiation among individuals from diverse sectors and institutions in a continuous, never-ending cycle. Individuals, their social relations and networks serve as a web that ties together the adaptive management system, and the approach pays careful attention to complex social dynamics such as trust building and power relationships (Arnold et al., 2012). Despite some evidence of success (McLoughlin et al., 2011), adaptive management remains elusive in practice. Key challenges to implementing adaptive management include: insufficient resources committed to monitoring the biophysical and social environment, deeply rooted norms of action without reflection, box-ticking instead of learning, and lack of cooperation across institutions (McLain and Lee, 1996; Stirzaker et al., 2011; Walters, 1997).

Principles for water planning and management in the Anthropocene

The management approaches described in the previous section have emerged from different disciplines in response to the goal of achieving sustainable and equitable development in the face of change. They all acknowledge the importance of linked social–ecological systems, the inherent uncertainty of predicting future trajectories of change in these complex and adaptive systems, and hence the importance of building adaptive capacity – or 'resilience' – to respond to unforeseen changes. Adaptive capacity refers not only to the ability to persist through continuous development in the face of change but also to be able to innovate and transform into new and more desirable configurations if opportunities arise.

Resilience has been proposed as being critical for the sustainable delivery of ecosystem services in social–ecological systems in the Anthropocene (Folke, 2003). Seven principles for building social–ecological resilience have recently been distilled from a wide variety of case studies around the world, which included those based on integrated water resource management, ecosystem-based management and adaptive management (Box 11.1; Biggs et al., 2012). These principles provide guidance on: (1) how to manage generic social–ecological system properties (Principles 1–3 of Box 11.1); and (2) how to manage key properties of social–ecological governance (Principles 4–7 of Box 11.1). In managing the social–ecological system, the principles advise users to maintain diversity and redundancy, manage connectivity, and manage the slow variables of change and associated feedbacks. In managing the governance system, the principles encourage complex adaptive systems thinking, learning, broad stakeholder participation, and polycentric governance where multiple interacting governing bodies have autonomy to make and enforce rules within a specific policy arena and geography (Pahl-Wostl et al., 2013).

While the stakeholders and issues may differ depending on the context, these principles provide useful guidance for water planning, whether PA authorities are planning inside or outside PAs. They are also generic principles that should be incorporated into all levels of PA governance, whether planning at a local level as a ranger or PA manager, or planning at a regional level as a PA systems manager.

Box 11.1 Seven principles for building resilience in social-ecological systems

(After Biggs et al., 2012)

1 **Maintain diversity and redundancy:** Systems with many different components, be they species, actors or sources of knowledge, are generally more resilient than systems with few components. This leads to redundancy which provides 'insurance' by allowing some components to compensate for the loss or failure of others.

2 **Manage connectivity:** Well-connected systems can recover from disturbances more quickly, but overly connected systems may lead to

(continued)

(*continued*)

rapid spread of disturbances. In social systems, connectivity may affect the governance of ecosystem services, for example, by influencing the flow of information between actors.

3   **Manage slow variables and feedbacks:** Slow variables determine the underlying structure of social–ecological systems, whereas the dynamics of the system typically arise from interactions and feedbacks between fast variables that respond to the conditions created by the slow variables. Slow ecological variables are often linked to regulating ecosystem services. For example, phosphorus concentration in a lake may increase slowly with no noticeable difference to water quality regulation, but once it passes a certain threshold a regime shift is experienced and the lake becomes eutrophic, after which it is very costly and difficult to return to a non-eutrophied state. Slow social variables are typically associated with legal and institutional systems, values and traditions.

4   **Foster complex adaptive systems thinking:** Complex adaptive systems thinking acknowledges that interacting social–ecological systems are connected on many different levels, with a multitude of perspectives, and are unpredictable and difficult to control. This requires a fundamental shift from command-and-control management approaches to ones such as adaptive management.

5   **Encourage learning:** Social–ecological systems are always in development, knowledge is always incomplete, and uncertainty, change and surprise are inevitable. Therefore, there is a constant need to revise existing knowledge and stimulate learning. More collaborative processes can also help.

6   **Broaden participation:** Participation is considered fundamental to promoting the collective action required to respond to disturbance and changes in social–ecological systems. There are a range of advantages to a broad and well-functioning participation. The participation of a diversity of stakeholders improves legitimacy, facilitates monitoring and enforcement, promotes understanding of system dynamics, improves the capacity to detect and interpret shocks and disturbances, and builds trust and a shared understanding for cooperation.

7   **Promote polycentric governance:** In polycentric governance, multiple governing bodies interact and have the power to make and enforce rules within a specific policy arena and geography. This form of decentralised governance is believed to promote local self-organisation where more centralised formal procedures seem to fail. However, it is also vulnerable to tensions between actors and negative institutional interactions. Achieving stakeholder diversity means

striking a balance between openness and mandates for decision-making. It also means negotiating trade-offs between various competing demands for water resources. These trade-offs often lead to the third challenge ('scale-shopping') where dissatisfied groups at one scale simply frame their interests at a more favourable political venue at another scale. A key to successful polycentric governance is to maintain network relations and manage power differentials between stakeholders, which goes beyond information sharing and ad hoc collaboration.

Water planning processes within protected areas

Historically, planning and management processes in PAs have only incidentally conserved freshwater biodiversity and associated ecosystem services as part of a terrestrial-biased conservation strategy (Nel et al., 2007). This has contributed to freshwater ecosystems being overall the most threatened and least protected ecosystems compared with their terrestrial and marine counterparts (Vörösmarty et al., 2010).

Protected area authorities need to counter this bias in several ways. Protected area systems managers need to ensure that freshwater ecosystem considerations are addressed when planning for the expansion of protected area systems. These considerations include representation and persistence of freshwater biodiversity (Linke et al., 2011), as well as seeking to protect the ecosystem services that freshwater ecosystems supply (Brauman et al., 2007). In their interactions with stakeholders, all protected area authorities should be able to make a case for the contribution that freshwater biodiversity and ecosystem services make to human wellbeing and sustainable development. Freshwater ecosystems provide valuable ecosystem services, such as clean drinking water, nutrient sequestration, flood regulation, provision of exploitable plants and animals, and a recreational space (Millennium Ecosystem Assessment, 2005). Declaring freshwater protected areas for protecting specific freshwater ecosystem services is an increasingly applied protection strategy, and protected area authorities should seek to integrate ecosystem service considerations into future protected area expansion plans. These could include the identification and protection of important water source areas for provision of downstream water to communities, protection of riparian vegetation as a filtration system to enhance water quality, and the protection of nursery grounds for replenishment of recreational and commercial fishing stock (Cosman et al., 2012).

Protected area managers need to ensure that freshwater ecosystems are managed explicitly as part of the protected area landscape. Ideally, protected area managers should have some training in freshwater ecology, and as a minimum they should have an appreciation of hydrological connectivity and factors that negatively impact freshwater ecosystems (Arthington et al., Chapter 3, this volume).

They should ensure that all protected area strategies and management plans include explicit freshwater conservation objectives that will maintain freshwater biodiversity and ecosystem services. General freshwater conservation objectives should consider the freshwater-friendly design and management of protected area boundaries, visitor facilities, road-river crossings, and weirs and dams inside protected areas (Nel et al., 2009a). Key pressures and threats to freshwater conservation – stemming from sources both within and outside of the protected area – should be identified and managed.

To help identify key pressures and threats, management plans should include long-term seasonal monitoring of freshwater indicators: at a minimum at sites coming into and leaving the protected area, but preferably systematically within the protected area to inform management of freshwater ecosystems. Indicators are defined as 'measures, variables, or indices that represent or mimic either the structure or function of ecological processes and systems across a disturbance gradient' – over both space and time (Brooks et al., 1998). Indicators for freshwater monitoring range from physical and chemical properties of the water and aquatic habitats to attributes of biological assemblages including diatoms, invertebrates, fish, and aquatic plants. Examples of biological response monitoring programmes developed for rivers in various parts of the world are: River InVertebrate Prediction And Classification System (RIVPACS) in the United Kingdom (Wright et al. 1993); Environmental Monitoring and Assessment Program (EMAP) for Surface Waters in the United States (Lazorchak et al., 2000); AUStralian RIVer Assessment Scheme (AUSRIVAS) in Australia (Davies, 2000); and the River Health Programme (RHP) in South Africa (Dickens and Graham, 2002). While monitoring of water quality variables provides a relatively simple indication of potential stressors, biological or ecological indicators provide insights into the system's response to multiple physical and chemical conditions that it is exposed to both simultaneously and over time. Some indicators will reflect changes in the ecosystem almost instantly (e.g., water quality variables) while other indicators will respond to catchment changes over weeks or months (e.g., flow and invertebrate or fish community structure), and yet others will reflect slow changes over years or decades (e.g., macro channel form, riparian vegetation, water table depth). It is therefore important to make use of a spectrum of indicators to provide a variety of lenses though which to assess the state of aquatic ecosystems (Bunn et al., 2010).

Partial inclusion of rivers in protected areas is no guarantee for their protection since impacts outside protected area boundaries can still have negative consequences for freshwater biodiversity within them (Mancini et al., 2005). This means that protected area authorities need to acknowledge processes and threats external to their boundaries, and engage in regional water planning processes within the catchments associated with their protected areas. The next section provides guidance on the underlying principles that can be used to enhance the effectiveness of this engagement.

Regional water planning processes beyond the protected area boundary

Acknowledging the opportunities and challenges posed by the Anthropocene, and the inherently connected nature of freshwater ecosystems, protected area authorities should endeavour to embed protected area planning and management of freshwater ecosystems into a regional water management framework, such as integrated water resource management, ecosystem-based management or adaptive management. These regional water planning processes provide a platform for the cross-sectoral cooperative governance of water resources which will enable the protected area authority to engage with social–ecological linkages that may impact on the freshwater ecosystems within their protected areas.

In most countries, biodiversity and water resources are managed by different sectors and agencies (Gilman et al., 2004). A consequence of this sectoral disconnect has been that many protected area authorities do not engage in water planning processes in the catchment within which protected areas are embedded, as they regard such planning as outside of their mandate or responsibility. On the contrary, because of the connected nature of freshwater systems, protected area authorities need to strongly influence future water development plans within the whole catchment, to ensure that they do not negatively impact the freshwater ecosystems in their protected area (Flitcroft et al., Chapter 10, this volume). Thus, protected area authorities have a responsibility to engage in regional water planning to ensure that the needs of freshwater ecosystems are given due consideration when assessing the trade-offs among different development options. Where regional water planning is absent, protected area authorities should serve to catalyse such planning with the mandated agents responsible for water and the environment.

Three assessment/planning approaches for freshwater ecosystems have been developed to inform such regional water planning processes: environmental flow assessment (Poff et al., 2010), systematic conservation planning (Nel et al., 2009b) and ecosystem service assessment (Brauman et al., 2007). Each of these approaches is described briefly below. Protected area authorities should strongly encourage these assessments to be done as part of regional water planning, so that scientifically credible outputs can be used to assess the needs of freshwater ecosystems when negotiating trade-offs with different water development options.

Environmental flow assessment

Environmental flow assessments evaluate how much a freshwater ecosystem changes with alterations to its natural flow regime (Tharme, 2003). Resultant outputs describe the quality, quantity, and timing of water set aside to maintain the key ecological and societal values (Hirji and Davis, 2009). The main categories of environmental flow methods, their resource requirements, levels of resolution of water requirements, and utility in different settings are set out in Table 8.2 of Chapter 8 (Arthington et al., this volume), and described in more detail in Tharme (2003).

Where water development upstream of a protected area is necessary (e.g., building of dams and other water schemes), protected area authorities should insist on the establishment and enforcement of environmental flow requirements for sustaining the ecosystems that provide the water resource and sustain the values of the protected area. While environmental flow assessments have traditionally been undertaken on a river-by-river basis – usually at the reactive demand of intended dam development – they have more recently been expanded into holistic frameworks for several rivers or entire river basins. Holistic frameworks, such as DRIFT (King et al., 2003) and ELOHA (Poff et al., 2010) provide information on the likely ecological impact and risk associated with different water development options, and are discussed in greater detail in Chapter 8 (Arthington et al., this volume). Holistic assessments at the basin scale are now being applied in integrated water resources management as a proactive planning tool which explores the impact of different water development futures on the desired ecological condition of rivers and wetlands in entire basins (Dollar et al., 2010; King and Brown, 2010). Protected area authorities should encourage such basin-scale holistic assessments to be undertaken at an early stage of water resource planning as a means of seeking common vision and understanding for managing water resources and the underpinning ecological condition of freshwater ecosystems. It is imperative that they play an active role in the process of negotiating the desired condition of freshwater ecosystems in the catchment. Where water development is necessary, this will help to ensure that environmental flows are achieved in connected systems within and outside of protected areas. Moreover, if systematic conservation plans for freshwater ecosystems exist, then the protected area authority should also ensure that holistic assessments include scenarios that evaluate the impact of development on the ecological condition of priority areas identified in the conservation plan (Nel et al., 2011).

Systematic conservation planning

A systematic conservation plan identifies priority areas for conserving biodiversity in a planning region, which can be as large as a continent or a local landscape. The approach seeks to ensure representation of the full variety of biodiversity in the planning region, and its persistence in the long term (Margules and Pressey, 2000). Accordingly, representation and persistence of biodiversity are two fundamental principles of conservation planning. Although originally terrestrial in focus, systematic conservation planning has been adapted for use in aquatic ecosystems (Nel et al., 2009a; Linke et al., 2011). For example, freshwater conservation planning approaches have evolved to incorporate longitudinal, lateral and vertical connectivity which is critical to the persistence of freshwater ecosystems (Hermoso et al., 2012; Arthington et al., Chapter 3, this volume). Freshwater-specific approaches can also allow for the identification of priority areas with a range of protection levels, so that multiple-use zones within catchments can be allocated, thus preventing the need to 'lock away' entire catchments from human use (Abell et al., 2007).

Freshwater conservation plans provide guidance on which rivers and wetlands in a region need to be conserved to ensure representation and persistence of freshwater biodiversity (Roux et al., 2008). They can be regarded as strategic bio-diversity sector input into regional water planning and management (Nel et al., 2016). Where freshwater conservation plans exist, protected area authorities should insist that they are considered in these planning and decision-making pro-cesses. Where they are lacking, protected area authorities should encourage the development of a freshwater conservation plan, to understand which freshwater ecosystems are fundamental to the representation and persistence of freshwater biodiversity in the catchment and region.

Ecosystem service assessment

An ecosystem service assessment (sometimes also referred to as an 'ecosystem assessment') links the biophysical components of ecosystems to the benefits (tan-gible or intangible) that people derive from these (Haines-Young and Potchin, 2010; Martín-López et al., 2014). In a protected area context, ecosystem services thus provide a conceptual bridge between intrinsic conservation objectives and social, economic and cultural values. An ecosystem service assessment can be used to quantify the value derived from ecosystems and their services, not only by visi-tors to protected areas, but also by communities outside protected areas (Brauman et al., 2007). The approach can therefore help to contextualise the role of protected areas within their larger social–ecological landscapes (Palomo et al., 2014; García-Llorente et al., 2016). Provisioning ecosystem services would include quantity and quality of water for drinking as well as agriculture and industrial uses (Brauman et al., 2007). Regulating ecosystem services include flow regulation and dilution of waste (Fisher et al., 2009). Cultural ecosystem services are the non-material benefits that result from people–nature interactions such as spiritual enrichment, intellectual development, recreation and aesthetic experiences (Chan et al., 2012).

Water-related ecosystem services should be considered in decisions regard-ing the expansion of protected area systems. Protected area systems managers should seek to broaden the scope of protected areas to include areas important not only for the representation and persistence of freshwater biodiversity, but also for the delivery of water-related ecosystem services. These areas could include, for example, water source areas of strategic importance for downstream water provi-sion (Nel et al., 2011), or wetlands for flood attenuation (Bullock and Acreman, 2003). Protecting these areas can supply substantial benefit to people who often live far away from the protected area and conserves biodiversity at the same time (Egoh et al., 2009).

Protected area managers are also encouraged to assess the water-related eco-system services currently provided by individual protected areas, as well as the potential for future ecosystem service provision. Explicit identification of the ecosystem services provided by protected areas to current and potential ben-eficiaries offers an important opportunity to make the links between protected areas and people more obvious, which may uncover new support and revenue streams for managing protected areas. The ecosystem service assessment would also highlight current and potential beneficiaries as stakeholders that need to be engaged so that appropriate management options can be developed for the regional landscape within which the protected area is embedded.

Conclusions

PA authorities should seek to mend the sectoral disconnect that is often prevalent between those managing biodiversity and those managing water resources. This chapter has highlighted several ways in which this can be achieved, namely to:

1   ensure that all PA strategies and management plans consider the needs of freshwater ecosystems, and include monitoring of the status and trends of these ecosystems over time;
2   gain a basic understanding of hydrological connectivity and its ecological implications in the catchments associated with PAs, and how water developments can negatively impact downstream ecosystems in PAs;
3   make a strong case for protecting freshwater ecosystems, especially from an ecosystem services perspective;
4   engage in regional water planning processes in the broader catchment within which PAs are located; and
5   insist that outputs from freshwater ecosystem assessments, such as environmental flow assessment, systematic conservation planning and ecosystem service assessment, are used to inform regional water planning and decision-making.

By engaging in water planning processes within and beyond their PA, protected area authorities can serve as catalysts for sustainable water management in associated catchments. Indeed, they can act as powerful agents and stakeholders in the negotiation for freshwater ecosystem protection within regional water planning processes. The potential benefits go far beyond just managing the factors that could impact negatively on the freshwater ecosystems within PAs. The engagement also offers the opportunity to develop new collaborations, highlight benefits to the broader community, and provides an avenue for exploring new sources of revenue for managing PAs.

References

Abell, R., Allan, J.D. and Lehner, B. (2007) 'Unlocking the potential of protected areas for freshwaters', *Biological Conservation*, vol 134, pp48–63.
Arnold, J.S., Koro-Ljungberg, M. and W. Bartels (2012) 'Power and conflict in adaptive management: analyzing the discourse of riparian management on public lands', *Ecology and Society*, vol 17, p19, http://dx.doi.org/10.5751/ES-04636-170119.

Biggs, R., Schlüter, M., Biggs, D., Bohensky, E.L., BurnSilver, S., Cundill, G., Dakos, V., Daw, T.M., Evans, L.S., Kotschy, K., Leitch, A.M., Meek, C., Quinlan, A., Raudsepp-Hearne, C., Robards, M.D., Schoon, M.L., Schultz, L. and West, P.C. (2012) 'Toward principles for enhancing the resilience of ecosystem services', *Annual Review of Environment and Resources*, vol 37, pp421–448.

Biswas, A. K. (2004) 'Integrated water resources management: a reassessment', *Water International*, vol 29, pp248–256.

Brauman, K.A., Daily, G.C., Duarte, T.K.E. and Mooney, H.A. (2007) 'The nature and value of ecosystem services: an overview highlighting hydrologic services', *Annual Review of Environment and Resources*, vol 32, pp67–98.

Brooks, R.P., O'Connell, T.J., Wardrop, D.H. and Jackson, L.E. (1998) 'Towards a regional index of biological integrity: The example of forested riparian ecosystems', *Environmental Monitoring and Assessment*, vol 51(1–2), pp131–143.

Bullock, A. and Acreman, M. (2003) 'The role of wetlands in the hydrological cycle', *Hydrology and Earth System Sciences*, vol 7, pp358–389.

Bunn, S.E., Abal, E.G., Smith, M.J., Choy, S.C., Fellows, C.S., Harch, B.D., Kennard, M.J. and Sheldon, F. (2010) 'Integration of science and monitoring of river ecosystem health to guide investments in catchment protection and rehabilitation', *Freshwater Biology*, 55(s1), pp223–240.

Chan, K.M., Guerry, A.D., Balvanera, P., Klain, S., Satterfield, T., Basurto, X., Bostrom, A., Chuenpagdee, R., Gould, R., Halpern, B.S. and Hannahs, N. (2012) 'Where are cultural and social in ecosystem services? A framework for constructive engagement', *BioScience*, vol 62, pp744–756.

Cosman, D., Schmidt, R., Harrison-Cox, J. and Batker, D. (2012) 'How water utilities can spearhead natural capital accounting', *The Solutions Journal*, vol 2, pp28–31.

Crase, L. and Cooper, B. (2015) 'Politics, socio-economics and water allocations: a note on the limits of Integrated Water Resources Management', *Australasian Journal of Environmental Management*, doi:10.1080/14486563.2015.1041068.

Davies, P.E. (2000) 'Development of a national river bioassessment system (AUSRIVAS) in Australia'. In Wright, J.F., Sutcliffe, D.W. and Furse, M.T. (eds), *Assessing the Biological Quality of Fresh Waters: RIVPACS and Other Techniques*. Freshwater Biological Association, Ambleside.

Dickens, C.W. and Graham, P.M. (2002) 'The South African Scoring System (SASS) version 5 rapid bioassessment method for rivers', *African Journal of Aquatic Science*, 27(1), pp1–10.

Dollar, E.S.J., Nicolson, C.R., Brown, C.A., Turpie, J.K., Joubert, A.R., Turton, A.R., Grobler, D.F., Pienaar, H.H., Ewart-Smith, J. and Manyaka, S.M. (2010) 'The development of the South African Water Resource Classification System (WRCS): a tool towards the sustainable, equitable and efficient use of water resources in a developing country', *Water Policy*, vol 12, pp479–499.

Doswald, N., Munroe, R., Roe, D., Giuliani, A., Castelli, I., Stephens, J., Möller, I., Spencer, T., Vira, B. and Reid, H. (2014) 'Effectiveness of ecosystem-based approaches for adaptation: review of the evidence-base', *Climate and Development*, vol 6, pp185–201.

Dudgeon, D. (2014) 'Accept no substitute: biodiversity matters', *Aquatic Conservation: Marine and Freshwater Ecosystems*, vol 24, pp435–440.

Egoh, B., Reyers, B., Rouget, M., Bode, M. and Richardson, D.M. (2009) 'Spatial congruence between biodiversity and ecosystem services in South Africa', *Biological Conservation*, vol 142, pp553–562.

Fabricius, C. and Cundill, G. (2014) 'Learning in adaptive management: insights from published practice', *Ecology and Society*, vol 19, p29, doi:http://dx.doi.org/10.5751/ES-06263-190129.

Finlayson, M.C., Davidson, N., Pritchard, D., Milton, G.R. and MacKay, H. (2011) 'The Ramsar Convention and ecosystem-based approaches to the wise use and sustainable development of wetlands', *Journal of International Wildlife Law and Policy*, vol 14, pp176–198.

Fisher, B., Turner, R.K. and Morling, P. (2009) 'Defining and classifying ecosystem services for decision making', *Ecological Economics*, vol 68, pp643–653.

Folke, C., Jansson, A., Rockström, J., Olsson, P., Carpenter, S.R., Chapin III, F.S., Crépin, A., Daily, G., Danell, K., Ebbesson, J., Elmqvist, T., Galaz, V., Moberg, F., Nilsson, M., Osterblom, H., Ostrom, E., Persson, A., Peterson, G., Polasky, S., Steffen, W., Walker, B. and Westley, F. (2011) 'Reconnecting to the biosphere', *Ambio*, vol 40, pp719–738.

Folke, C. (2003) 'Freshwater for resilience: a shift in thinking', *Philosophical Transactions of the Royal Society of London Series B – Biological Sciences*, 358 ISS 1440, pp2027–2036.

García-Llorente, M., Castro, A.J., Quintas-Soriano, C., López, I., Castro, H., Montes, C. and Martín-López, B. (2016) 'The value of time in biological conservation and supplied ecosystem services: A willingness to give up time exercise', *Journal of Arid Environments*, vol 124, pp13–21.

Gilman, R.T., Abell, R.A. and Williams, C.E. (2004) 'How can conservation biology inform the practice of Integrated River Basin Management?', *International Journal of River Basin Management*, vol 2, pp1–14.

Global Water Partnership (2000) 'Integrated water resource development', GWP Technical Advisory Committee Background Paper 4, March. GWP, Stockholm, Sweden.

Haines-Young, R. and Potschin, M. (2010) 'The links between biodiversity, ecosystem services and human well-being', *Ecosystem Ecology: a new synthesis*, pp110–139.

Hermoso, V., Kennard, M.J. and Linke, S. (2012) 'Integrating multidirectional connectivity requirements in systematic conservation planning for freshwater systems', *Diversity and Distributions*, vol 18, pp448–458.

Hirji, R. and Davis, R. (2009) *Environmental Flows in Water Resources Policies, Plans, and Projects. Findings and Recommendations*, The International Bank for Reconstruction and Development/The World Bank, Washington, DC, 189pp.

Holling, C.S. (2001) 'Understanding the complexity of economic, ecological and social systems', *Ecosystems*, 4: 390–405.

King, J., Brown, C. and Sabet, H. (2003) 'A scenario-based holistic approach to environmental flow assessments for rivers', *River Research and Applications*, vol 19, pp619–639.

King, J.M. and Brown, C. (2010) 'Integrated basin flow assessments: concepts and method development in Africa and South-east Asia', *Freshwater Biology*, vol 55, pp127–146.

Kingsford, R.T., Biggs, H.C. and Pollard, S.R. (2011) 'Strategic adaptive management in freshwater protected areas and their rivers', *Biological Conservation*, 144, pp1194–1203, Koedoe, v53i2.1007.

Lazorchak, J.M., Hill, B.H., Averill, D.K., Peck, D.V. and Klemm D.J. (eds) (2000) 'Environmental monitoring and assessment program – surface waters: field operations impact of research in support of the river health program and methods for measuring the ecological condition of non-wadeable rivers and streams', US Environmental Protection Agency, Cincinnati OH.

Linke, S., Turak, E. and Nel, J. (2011) 'Freshwater conservation planning: the case for systematic approaches', *Freshwater Biology*, vol 56, pp6–20.

Mancini, L., Formichetti, P., Anselmo, A., Tancioni, L., Marchini, S. and Sorace, A. (2005) 'Biological quality of running waters in protected areas: the influence of size and land use', *Biodiversity and Conservation*, vol 14, pp351–364.

Margules, C.R. and Pressey, R.L. (2000) 'Systematic conservation planning', *Nature*, vol 405, pp243-253.

Martín-López, B., Gómez-Baggethun, E., García-Llorente, M. and Montes, C. (2014) 'Trade-offs across value-domains in ecosystem services assessment', *Ecological Indicators*, vol 37, pp220–228.

McLain, R.J. and Lee, R.G. (1996) 'Adaptive management: Promises and pitfalls', *Environmental Management*, vol 20, pp437–448, doi:10.1007/BF01474647, PMid:8661620.

McLoughlin, C.A. and Thoms, M.C. (2015) 'Integrative learning for practicing adaptive resource management', *Ecology and Society*, vol 20, p34.

McLoughlin, C.A., Deacon, A., Sithole, H. and Gyedu-Ababio, T. (2011) 'History, rationale, and lessons learned: Thresholds of potential concern in Kruger National Park river adaptive management', *Koedoe*, vol 53, pp69–95.

Millennium Ecosystem Assessment (2005) *'Ecosystems and Human Well-being: Synthesis. World Resources Institute'*. Island Press, Washington DC.

Mostert, E., Pahl-Wostl, C., Rees, Y., Searle, B., Tàbara, D. and Tippett J. (2007) 'Social learning in European river-basin management: barriers and fostering mechanisms from 10 river basins', *Ecology and Society*, 12(1): p19, www.ecologyandsociety.org/vol12/iss1/art19/.

Nel, J.L., Reyers, B., Roux, D.J. and Cowling, R.M. (2009a) 'Expanding protected areas beyond their terrestrial comfort zone: identifying spatial options for river conservation', *Biological Conservation*, vol 142, pp1605–1616.

Nel, J.L., Roux, D.J., Cowling, R.M., Abell, R., Ashton, P.J., Higgins, J.A., Thieme, M. and Viers, J.C. (2009b) 'Progress and challenges in freshwater conservation planning', *Aquatic Conservation: Marine and Freshwater Ecosystems*, vol 19, pp474–485.

Nel, J.L., Roux, D.J., Driver, A., Hill, L., Maherry, A., Snaddon, K., Petersen, C., Smith-Adao, L.B., Van Deventer, H. and Reyers, B. (2016) 'Knowledge co-production and boundary work to promote implementation of conservation plans', *Conservation Biology*, vol 30, pp76–188.

Nel, J.L., Roux, D.J., Maree, G., Kleynhans, C.J., Moolman, J., Reyers, B., Rouget, M. and Cowling, R.M. (2007) 'Rivers in peril inside and outside protected areas: a systematic approach to conservation assessment of river ecosystems', *Diversity and Distributions*, vol 13, pp341–352.

Nel, J.L., Turak, E., Linke, S. and Brown, C. (2011) 'Integration of environmental flow assessment and freshwater conservation planning: a new era in catchment management', *Marine and Freshwater Research*, vol 62, pp290–299.

Pahl-Wostl, C., Arthington, A.H., Bogardi, J., Bunn, S.E., Hoff, H., Lebel, L., Nikitina, E., Palmer, M.A., Poff, N.L., Richards, K. and Schlüter, M. (2013) 'Environmental flows and water governance: managing sustainable water uses', *Current Opinion in Environmental Sustainability*, vol 5, pp341–351.

Pahl-Wostl, C., Craps, M., Dewulf, A., Mostert, E., Tabara, D. and Taillieu, T. (2007) 'Social learning and water resources management', *Ecology and Society*, 12(2): p5, www.ecologyandsociety.org/vol12/iss2/art5/.

Palomo, I., Montes, C., Martín-López, B., González, J.A., García-Llorente, M., Alcorlo, P. and Mora, M.R.G. (2014) 'Incorporating the social–ecological approach in protected areas in the Anthropocene', *BioScience*, p.bit033.

Poff, N.L., Richter, B.D., Arthington, A.H., Bunn, S.E., Naiman, R.J., Kendy, E., Acreman, M., Apse, C., Bledsoe, B.P., Freeman, M.C., Henriksen, J., Jacobson, R.B., Kennen, J.G., Merritt, D.M., O'Keeffe, J.H., Olden, J.D., Rogers, K., Tharme, R.E. and Warner, A. (2010) 'The ecological limits of hydrologic alteration (ELOHA): a new framework for developing regional environmental flow standards', *Freshwater Biology*, vol 55, pp147–170.

Pringle, C. (2003) 'What is hydrologic connectivity and why is it ecologically important?', *Hydrological Processes*, vol 17, pp2685–2689.

Rockström, J., Steffen, W., Noone, K., Persson, A., Chapin III, F.S., Lambin, E., Lenton, T.M., Scheffer, M., Folke, C., Schellnhuber, H., Nykvist, B., De Wit, C.A., Hughes, T., van der Leeuw, S., Rodhe, H., Sorlin, S., Snyder, P.K., Costanza, R., Svedin, U., Falkenmark, M., Karlberg, L., Corell, R.W., Fabry, V.J., Hansen, J., Walker, B., Liverman, D., Richardson, K., Crutzen, P. and Foley J. (2009) 'Planetary boundaries: exploring the safe operating space for humanity', *Ecology and Society*, vol 14, p32, www.ecologyandsociety.org/vol14/iss2/art32/.

Rogers, K. (2003) 'Adopting a heterogeneity paradigm: Implications for management of protected savannas', in J.T. du Toit, K.H. Rogers and H.C. Biggs (eds), *The Kruger experience. Ecology and Management of Savanna Heterogeneity*, pp41–58, Island Press, Washington DC.

Roux D.J., Ashton P.J., Nel J.L. and MacKay H.M. (2008) 'Improving cross-sector policy integration and cooperation in support of freshwater conservation', *Conservation Biology*, vol 22, pp1382–1387.

Schoeman, J., Allan, C. and Finlayson, C.M. (2014) 'A new paradigm for water? A comparative review of integrated, adaptive and ecosystem-based water management in the Anthropocene', *International Journal of Water Resources Development*, vol 30, pp377–390.

Stirzaker, R.J., Roux, D.J. and Biggs, H.C. (2011) 'Learning to bridge the gap between adaptive management and organisational culture', *Koedoe* 53, Art #1007, 6 pages, doi:10.4102/systems', *Ecosystems* 4, pp390–405, doi:10.1007/s10021-001-0101-5.

Tallis, H., Levin, P.S., Ruckelshaus, M., Lester, S.E., McLeod, K.L., Fluharty, D.L. and Halpern, B.S. (2010) 'The many faces of ecosystem-based management: making the process work today in real places', *Marine Policy*, vol 34, pp340–348.

Tharme, R.E. (2003) 'A global perspective on environmental flow assessment: emerging trends in the development and application of environmental flow methodologies for rivers', *River Research and Applications*, vol 19, pp397–441.

Vörösmarty, C.J., McIntyre, P.B., Gessner, M.O., Dudgeon, D., Prusevich, A., Green, P., Glidden, S., Bunn, S.E., Sullivan, C.A., Reidy Liermann, C. and Davies, P.M. (2010) 'Global threats to human water security and river biodiversity', *Nature*, 467, pp555–561.

Walters, C. (1997) 'Challenges in adaptive management of riparian and coastal ecosystems', *Conservation Ecology*, vol 1, 1, www.consecol.org/vol1/iss2/art1/.

Williams, M., Zalasiewicz, J., Haff, P.K., Schwägerl, C., Barnosky, A.D. and Ellis, E.C. (2015) 'The Anthropocene biosphere', *The Anthropocene Review*, vol 2, pp196–219.

Wright, J.F., Furse, M.T. and Armitage, P.D. (1993) 'RIVPACS – a technique for evaluating the biological quality of rivers in the UK', *European Water Pollution Control*, vol 3, pp15–25.

Chapter 12

Managing freshwater protected areas in the global landscape

C.M. Finlayson, N.C. Davidson, P.A. Gell, R. Kumar and R.J. McInnes

Key messages

- The increasing demand for freshwater, as well as uncertainties associated with climate change, has placed growing pressure on wetland and natural resource managers to develop sustainable approaches that extend across landscapes and ensure the many benefits we obtain from freshwater wetlands are not further reduced.

- Contracting Parties (countries) to the Ramsar Convention on Wetlands are required to designate at least one wetland for inclusion in the List of Wetlands of International Importance, known as the Ramsar List, which is the keystone of the Convention and is supported by a Strategic Framework.

- There are 1,687 inland freshwater wetlands on the Ramsar List, covering a total area of 195.5 million km² with the largest number, 827 (52 per cent), occurring in Europe, although these comprise 12 per cent only of the total area of these sites.

- A further requirement under the Convention is to prepare and implement an appropriate management plan or other mechanism for protecting listed wetlands. Africa and Asia have management instruments for 58 per cent and 70 per cent of their sites, respectively, rising to more than 80 per cent in other regions.

- Countries are encouraged to establish national wetland inventories; in 2015 only 47 per cent of countries had developed an inventory. The Convention also records reports of adverse change in the ecological character of Ramsar sites, and encourages countries to make wise use of all wetlands.

- Even where good legislative frameworks exist, such as in the case of the European Union, significant challenges remain to fully deliver an appropriate quality and quantity of freshwater to maintain the character of Protected Areas. Challenges remain regarding the degree of fit within institutional structures and practices, while others relate to the efficacy of wider wetland management issues. These issues are particularly relevant when considering the benefits that ecosystem services from wetlands provide for many communities.

- Owing to natural variability as well as directional change the condition of a wetland evolves continually. As the recent condition is merely a subset of the historic conditions experienced by a wetland, management focus should move away from a static view of condition and accommodate variability while also ensuring the wetland retains its listing criteria and the ecological functions that enable it to contribute to the provisions of the Convention. This will require sustainable management approaches that extend across multiple landscape scales.

Introduction to managing at landscape scales

While freshwater wetlands comprise a diverse range of ecosystems and are distributed across many different landscapes, their areal extent and condition globally are not well known given gaps in inventory (Finlayson and van der Valk 1995; Finlayson et al. 2017a; Milton et al. 2017) and assessment (Millennium Ecosystem Assessment 2005; Gordon et al. 2010). Further, there is increasing evidence that many are being degraded and lost (Davidson 2014; Dixon et al. 2015; Gardner et al. 2015) in response to ongoing pressures from agricultural activities, including land clearing and water resource development, pollution and eutrophication, invasive species, industrialisation and urban expansion (Millennium Ecosystem Assessment 2005; Falkenmark et al. 2007). The increasing demand for freshwater, in particular for agriculture and further food production (Gordon et al. 2010), as well as increased pressure from climate change (de Fraiture et al. 2007; Finlayson 2013), has placed further pressure on wetland and natural resource managers to develop sustainable approaches that extend across landscapes and ensure that the many benefits we obtain from freshwater wetlands are not further reduced.

The Ramsar Convention on Wetlands has led efforts to manage wetlands in a sustainable manner including through the use of PAs and the adoption of ecosystem approaches that take into account the role of wetlands within landscapes (Gardner and Davidson 2011; Finlayson et al. 2011). Given that many freshwater wetlands and their biota are affected by pressures from the surrounding landscapes as well as from further afield, we have considered the management of freshwater PAs at a landscape scale. This starts with i) further explanation of the role of the Convention in supporting wetland management at a landscape level in addition to the often reported protection at site level, as well as the importance of networks of sites for migratory waterbirds along flyways of migratory species, ii) the European Union *Water Framework Directive* (or WFD), which was established to provide a framework for the protection of inland surface waters (rivers and lakes), transitional waters (estuaries), coastal waters and groundwater, iii) management planning processes for Pas, iv) issues ensuring that ecosystem services are maintained across landscapes, and v) how baseline conditions are evolving and may herald the development of novel self-sustaining wetlands. These topics are addressed in order to expand on the type of issues that can arise when attempting to manage at a landscape scale.

The Ramsar Convention on Wetlands

The Convention on Wetlands of International Importance arose from concerns of government and non-government organisations to conserve diminishing wetlands (Matthews 1993). It was agreed in the Iranian city of Ramsar in 1971 and was among the first of a set of environmental treaties that were negotiated over several decades in the 20th Century. In addition to directly promoting the conservation of wetlands, the Convention implements the inland waters program of work on behalf of the Convention on Biological Diversity and complements the activities of the Convention on Migratory Species (and related treaties covering the movement of migratory species) for relevant species that use wetlands. While other treaties also cover specific sites or values of freshwater ecosystems, such as the World Heritage Convention, The Ramsar Convention is afforded in-depth discussion here due to its wetlands focus.

Contracting Parties (countries) to the Ramsar Convention are required to designate at least one wetland for inclusion in the List of Wetlands of International Importance, known as the Ramsar List (Figure 12.1). These often constitute PAs under a range of national legislative and management arrangements. Sites are selected for designation as Ramsar sites using nine criteria (Table 12.1). While the Convention has a wide definition of wetlands, and includes coastal and marine, artificial and inland ecosystems, inland freshwater wetlands only are considered here. A description of each designated wetland is provided by means of a Ramsar Information Sheet that includes data on scientific, conservation and management parameters and a map to delimit the boundaries of the site (Ramsar Convention 2012a). Countries are also encouraged to establish national wetland inventories, based on the best scientific information available as a basis for promoting the designation of the largest possible number of appropriate wetland sites, including those types that are currently under represented in the List, including freshwater wetlands in the arid zone, peatlands and streams. In 2012 only 43 per cent of countries had developed a national wetland inventory and in 2015 only 47 per cent.

Figure 12.1 Chilika lake, India, designated a Ramsar wetland of international importance.

Table 12.1 Criteria for listing wetlands of international importance and long-term targets for the Ramsar List.

Specific criterion	Long-term target
1 Contains a representative, rare, or unique example of a natural or near-natural wetland type found within the appropriate biogeographic region.	Include at least one suitable representative of each wetland type, according to the Ramsar classification system, which is found within each biogeographic region.
2 Supports vulnerable, endangered, or critically endangered species or threatened ecological communities.	Include those wetlands which are believed to be important for the survival of vulnerable, endangered or critically endangered species or threatened ecological communities.
3 Supports populations of plant and/or animal species important for maintaining the biological diversity of a particular biogeographic region.	Include those wetlands which are believed to be of importance for maintaining the biological diversity within each biogeographic region.
4 Supports plant and/or animal species at a critical stage in their life cycles, or provides refuge during adverse conditions.	Include those wetlands which are the most important for providing habitat for plant or animal species during critical stages of their life cycle and/or when adverse conditions prevail.

5 Regularly supports 20,000 or more waterbirds.	Include all wetlands which regularly support 20,000 or more waterbirds.
6 Regularly supports 1 per cent of the individuals in a population of one species or subspecies of waterbird.	Include all wetlands which regularly support 1 per cent or more of a biogeographical population of a waterbird species or subspecies.
7 Supports a significant proportion of indigenous fish subspecies, species or families, life-history stages, species interactions and/or populations that are representative of wetland benefits and/or values and thereby contributes to global biological diversity.	Include those wetlands that support a significant proportion of indigenous fish subspecies, species or families and populations.
8 Important source of food for fishes, spawning ground, nursery and/or migration path on which fish stocks, either within the wetland or elsewhere, depend.	Include those wetlands which provide important food sources for fishes, or are spawning grounds, nursery areas and/or on their migration path.
9 Regularly supports 1 per cent of the individuals in a population of one species or subspecies of wetland-dependent non-avian animal species.	Include all wetlands which regularly support 1 per cent or more of a biogeographical population of one non-avian animal species or subspecies.

The Ramsar List is a keystone of the Convention (Gardner and Davidson 2011) and is supported by a Strategic Framework (Ramsar Convention 2012b) that provides a vision for the List, namely to:

> develop and maintain an international network of wetlands which are important for the conservation of global biological diversity and for sustaining human life through the maintenance of their ecosystem components, processes and benefits/services.

The Strategic Framework also has the following objectives:

1 To establish national networks of Ramsar sites under each Contracting Party which fully represent the diversity of wetlands and their key ecological and hydrological functions;
2 To contribute to maintaining global biological diversity through the designation and management of appropriate wetland sites;
3 To foster cooperation among Contracting Parties, the Convention's International Organization Partners, and local stakeholders in the selection, designation, and management of Ramsar sites;
4 To use the Ramsar site network as a tool to promote national, supranational/regional, and international cooperation in relation to complementary environment treaties.

The List has been developed progressively since the first site, the Cobourg Peninsula, Australia, which contains coastal and freshwater wetlands, was designated in May 1974. It currently (October 2016) contains 2,242 sites covering 2.15 million km², which represents 17 per cent of the estimated 12.1 million km² of global wetlands, although the area of global wetlands is itself an underestimate (Fluet-Chouinard et al. 2015).

There are 1,818 inland wetlands, covering a total area of 199.2 million km² on the Ramsar List (as at October 2016). Of these, 1,687 (93 per cent) include inland freshwater wetlands, covering a total area of 195.5 million km² (the other 131 inland Sites are wholly brackish or saline). The largest number of freshwater Ramsar Sites, 827 (49 per cent), is in Europe, although these comprise only 12.4 per cent of the total area globally of these designated sites (Figure 12.2, Table 12.2). The largest area of Sites is in Africa with 295 sites (17 per cent) covering 90.4 million km² (46.2 per cent) of the total area listed. The number and area of sites reflects the geopolitical differences between regions, with many smaller wetland sites remaining in most European countries.

Landscape protection along waterbird flyways

Many waterbirds – species and populations ecologically-dependent on wetlands – are highly migratory, many moving broadly north-south around the globe between their breeding and non-breeding grounds and using intermediate staging areas for refuelling to complete their migrations. Others are nomadic and largely move in response to rainfall patterns and the availability of freshwater and suitable habitat, such as across much of Australia (Bellio et al., 2016). Migratory and nomadic

Figure 12.2 Distribution of inland freshwater Ramsar sites. Source: Ramsar Sites Information Service (https://rsis.ramsar.org) [Accessed 14 October 2016].

species pose particular conservation challenges since their survival depends on inter-connected landscape-scale networks of wetlands. Whilst many such species are, outside their breeding season, highly aggregatory in relatively small numbers of sites, others are more widely dispersed and depend on landscape-scale conservation measures.

The term 'flyway' is a nature conservation construct used in several ways to describe the inter-annual movements of migratory waterbirds (Davidson and Stroud, 2016). These include: single species migration systems; multi-species flyways, grouping species that follow broadly geographically similar north-south migration routes; and at a larger scale 'geo-political' global regions for waterbird conservation management (Figure 12.3).

There has been a long history (since the early 20th century) of the establishment of bilateral and multilateral agreements aimed specifically at conserving migratory waterbirds. These vary between being facilitative, formal and legally-binding. As well as expecting broad landscape-scale conservation measures to be undertaken, most flyway initiatives include the establishment of wetland site networks for migratory waterbirds. Non-governmental organisations (NGOs) have played key roles in the establishment of these initiatives and continue to significantly support their implementation.

Globally, since 1971, the Ramsar Convention on Wetlands has supported the designation of "coherent and comprehensive" networks of Wetlands of International Importance (Ramsar Sites), including specific attention to waterbirds through the designation of all wetlands which qualify under two of its nine designation criteria (Table 12.1): Criterion 5 (sites regularly supporting >20,000 waterbirds) and Criterion 6 (sites regularly supporting >1 per cent

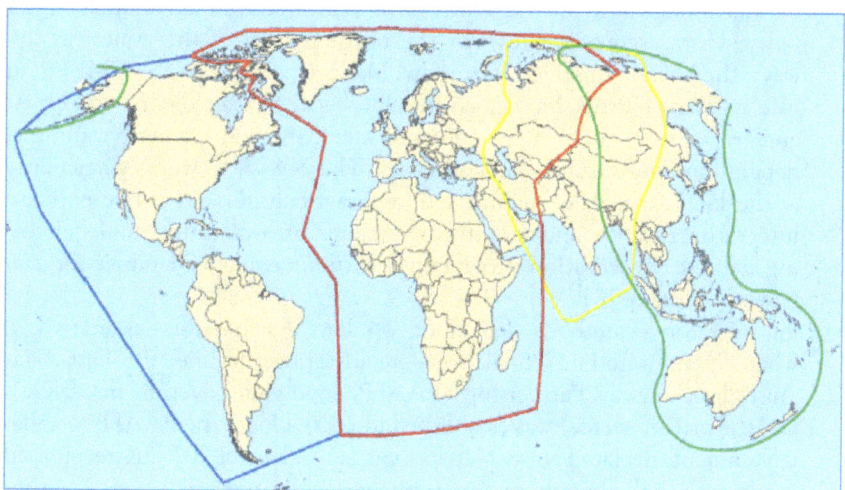

Figure 12.3 The main 'geo-political' flyways for migratory waterbird conservation. From Boere and Stroud (2006).

of a waterbird biogeographic population). To date (as at 17 October 2016), Ramsar Contracting Parties (member governments) have designated 526 freshwater Ramsar Sites under Criteria 5 and 593 freshwater Sites under Criterion 6 (a number of which have been designated under both Criteria). Whilst this appears an impressive site network, major gaps are still recognised: over one half of all these Ramsar Sites are in Europe, and the freshwater Ramsar Site network for waterbirds is patchy in other regions. Analyses in the early 2000s of BirdLife International's Important Bird and Biodiversity Areas (IBAs) network (www.birdlife.org/worldwide/programmes/sites-habitats-ibas), suggested that, at that time, of qualifying IBAs only 24 per cent in Europe, 14 per cent in Africa and 11 per cent in Asia had been Ramsar designated (BirdLife International 2001; 2002; 2005).

At the geopolitical flyway scale a number of agreements exist (Davidson and Stroud, 2016; Finlayson, 2016a):

- for flyways covering Eurasia and Africa the African-Eurasian Migratory Waterbird Agreement (AEWA) was established in 1995 through the Convention on Migratory Species (CMS). Whilst the designation of Ramsar Sites is the main tool for AEWA for site network conservation, to support identification of the full network of key migratory waterbird sites on this flyway, a GIS-based "Critical Site Network Tool" (CSN) has been developed (see www.wingsoverwetlands.org/).

- for the Americas, the Western Hemisphere Shorebird Reserve Network (WHSRN) was established in 1986 as a voluntary mechanism supporting local communities to recognise and safeguard their key wetlands for migratory shorebirds. To date, 96 sites in 15 countries, from Alaska in the north to Tierra del Fuego in the south have been declared as WHSRN Sites (www.whsrn.org/whsrn-sites). In the northern part of the Americas flyways, the North American Waterfowl Management Plan (NAWMP) is an international partnership signed in 1986 between Canada and the USA, and by Mexico in 1994, to conserve abundant and resilient waterfowl populations and sustainable landscapes. The NAWMP works particularly at the landscape scale, shaping land-use, agricultural and public policies, integrating science and monitoring systems into planning, and delivering habitat conservation programmes (http://nawmp.wetlandnetwork.ca/nawmp-revision-2012/).

- for migratory waterbirds depending on East Asian–Australasian flyway, where many waterbird populations are in serious decline, the East-Asia-Australasia Flyway Partnership (EAAFP, involving governments, NGOs and the private sector) was established in 2006. Under the EAAFP member governments declare Flyway Network Sites. To date, of 900 sites recognised as internationally important along the flyway, this network covers 124 sites (covering 20,213,845 ha) in 17 countries (www.eaaflyway.net/about/the-flyway/flyway-site-network/).

While these flyway agreements have engendered cooperation and activities across national borders they generally do not have a strong legal basis under national legislation (Pittock et al., 2010) and are not as binding as other agreements, such as those on trade and defence that may be supported by specific legislation to ensure their enactment.

European Union Water Framework Directive

Freshwater ecosystems are vital natural resources which provide drinking water for humans, provide habitats for many different species of wildlife, and are an important resource for *inter alia* agriculture, industry and recreation. The European Union (EU) has recognised the importance of the future protection and improvement of the water environment as being essential for the sustainable development and the long term health, well-being and prosperity of European citizens.

The 'Directive 2000/60/EC of the European Parliament and of the Council of 23 October 2000 establishing a framework for Community action in the field of water policy' (European Commission, 2000) (or in short the *Water Framework Directive* or WFD) was adopted by EU member states and came into force in December 2000. The Directive aims to establish a framework for the protection of inland surface waters (rivers and lakes), transitional waters (estuaries), coastal waters and groundwater. Ultimately it seeks to ensure that all aquatic ecosystems and, with regard to their water needs, terrestrial ecosystems and wetlands achieve 'good status'.

The WFD has adopted the river basin as a basis for a single system of water management, even where these cross administrative or political boundaries. The Directive requires a cyclical process to be established for each river basin district where management plans are prepared, implemented and reviewed every six years. Four distinct elements are identified within the management process: characterisation and assessment of impacts on river basin districts; environmental monitoring; the setting of environmental objectives; and the design and implementation of the programme of measures needed to achieve them.

The Directive sets out the following environmental objectives for freshwater bodies:

- preventing deterioration in status;
- achieving good surface water status or, for artificial or heavily modified surface water bodies, good ecological potential and good surface water chemical status;
- good groundwater status; or
- any less stringent objective applicable under Article 4.5 of the Directive.

The focus of the WFD on water bodies, and particularly their relationships within a river basin, helps to emphasise the functional role the conservation and management of protected freshwater sites, such as wetlands or river reaches, has within the wider hydrological cycle.

Considerable progress has been made on the implementation of the WFD (European Union, 2010) especially with regard to supporting the objectives of the protected network of Natura 2000 sites established under the 'Habitats Directive' (European Commission, 1992) and the 'Birds Directive' (European Commission, 1979; Evans, 2012). However, significant challenges remain to fully achieve the ambitious objectives of the Directive. Some of these challenges revolve around the degree of fit within institutional structures and practices (Moss, 2004), while others relate, both directly and indirectly, to wetland management issues such as the concept of environmental flows as a key measure for restoring and managing freshwater ecosystems (Acreman and Ferguson, 2010).

Management planning – effectiveness of management and monitoring

Management of freshwater ecosystems in PAs across landscapes is strongly connected with management processes elsewhere within the catchment, such as when dealing with water flows, and also beyond, such as when dealing with migratory species that transcend catchment boundaries. The former is exemplified by the efforts in many countries or regions to establish catchment management authorities that can operate across jurisdictional boundaries, such as the Murray-Darling Basin Authority (MDBA) in Australia that operates across five sub-national jurisdictions (www.mdba.gov.au/ accessed 6 November 2016), the Mekong River Commission (www.mrcmekong.org/ accessed 6 November 2016), and the Permanent Okavango River Basin Water Commission (OKACOM) (www.okacom.org/okacom-commission accessed 6 November 2016). The Mekong River Commission coordinates activities between four riparian countries along the Mekong River in South-East Asia, but not including China which contains much of the headwaters of the river, and OKACOM operates between the three riparian countries along the Okavango River in Southern Africa. The success of such processes for managing freshwater protected areas has been investigated on a number of occasions with mixed outcomes, as well as different views on their organisational structure and effectiveness within the context of national sovereignty and prerogatives (Hooper, 2005; Schmeier, 2010, 2012).

The MDBA has agreed a water sharing plan, incorporating environmental and consumptive needs, across jurisdictions that had previously argued and continued to argue about equitable access to water for consumptive purposes (Connell, 2011), although the merits of the plan have been questioned and challenged on the basis of environmental and social-economic outcomes (Pittock et al. 2010; Bond et al., 2014). The operational constraints that face basin authorities that cross national borders have been documented (Schmeier, 2010) and include the institutional design, that is, the way its organizational bodies are designed and interact with one another, the institutions' link to its member states and the distribution of tasks between the different governance levels, and the financing of the institution.

The ability to manage the water flows to a freshwater PA is critical for success-ful management and yet many PA management plans do not allow for activities outside of the PA itself – this places the PA at the mercy of the actions of others who may have different priorities, such as water resource authorities or energy utilities (Arthington et al., Chapter 8, this volume). This is the situation for the freshwater PAs in Australia's Murray-Darling Basin and has necessitated the establishment of complex institutional arrangements to ensure adequate environ-mental flows are maintained (Connell, 2011). There are also differences within PAs depending on the priority given to managing the freshwater component. Kakadu National Park in northern Australia covers almost 2 million hectares including most of the catchment of the South Alligator River and substantial portions of other river catchments. The wetlands, including coastal saline as well as freshwater sites, comprise about 15 per cent of the Park and have attracted considerable management attention including the successful control of the inva-sive shrub *Mimosa pigra*, and to some extent the floating weed *Salvinia molesta*, but have not been able to contain the spread of invasive pasture species such as *Brachiaria mutica*, nor the cane toad *Bufo marina* (Finlayson, 2005; Finlayson et al., 2006; Bayliss et al., 2012). The successful approaches have entailed dedi-cated and targeted efforts over many years. Water pollution is a further problem for managers of freshwater PAs, especially those where the pollution comes from upstream sources that cannot be readily influenced by the PA managers.

In order to assist wetland managers the Ramsar Convention has developed guidelines and a framework for managing Wetlands of International Importance and other wetland sites. Despite the existence of such guidance, as well as many national guidelines or regulations in support of management planning, some 49 per cent of 861 freshwater Ramsar sites did not have management plans in 2016 (Table 12.2). The guidelines are relevant to conservation of all wetlands not just those included in the List of Wetlands of International Importance. They further emphasise that Ramsar site management plans should be integrated into the public development planning system at local, regional or national level. The integration of site management plans into spatial and economic planning at the appropriate level, including at the integrated river basin and coastal zone scales, will ensure implementation, public participation and local ownership. The man-agement plan itself should be a technical document, though it may be appropriate for it to be supported by legislation and in some circumstances to be adopted as a legal document.

A management plan should suit the local requirements and only be as large or complex as the site requires. The size and complexity of a plan, and the resources made available for its production and implementation, must be in proportion to the needs of the site, and in proportion to the resources available. Thus, for small sites, a brief and concise plan may suffice. Alexander (2016) points out that

Table 12.2 Number and area of inland freshwater wetlands included in the Ramsar List (October 2016).*

Region	Number of wetlands	Area of wetlands (million km²)	Number of wetlands with management plans
Africa	295 (19%)	90.4 (46%)	84 (28%)
Asia	207 (13%)	13.8 (7%)	102 (49%)
Europe	827 (52%)	24.2 (12%)	471 (57%)
Neotropics	143 (7%)	39.4 (20%)	65 (45%)
North America	150 (6%)	21.0 (11%)	86 (57%)
Oceania	65 (3%)	6.8 (3%)	53 (82%)
Total	**1,687**	**195.5**	**861 (51%)**

*These numbers may rise due to recent additions to the management plan database.

management planning should be adaptive and developed in an iterative and ongoing way and can be applied at any site regardless of size. The key elements include being adaptable to the changing biophysical, socio-economic and political factors that affect the site, such as those outlined in the plans for the Okavango Delta in Botswana (Ramberg, 2016; Finlayson, 2016b) and Chilika lagoon in India (Pattnaik and Kumar, 2016; Finlayson, 2016a).

Managing for ecosystem services, trade-offs

The Ramsar Convention's wise use approach calls for ensuring compatibility of human use of wetlands with the goal of maintaining their ecological character (Finlayson et al., 2011). Ecosystem services have been included in the definition of ecological character as a means of bridging the gap that had developed between efforts to maintain the ecological functions of wetlands and those that sought to promote their usefulness for supporting human well-being. It is increasingly appreciated that in a human-dominated world, ecosystem services are not generated by ecosystems alone, but by socio-ecological systems of which humans form an integral part (Reyers et al., 2013; Levin et al., 2013). Wetland ecosystem uses and management choices have led to changes in the type, magnitude or relative mix of ecosystem services. As ecosystem services are not independent of each other and comprise complex non-linear interactions, attempts at optimizing a single service may, desirably or even unknowingly, lead to the reduction or loss of other services, thus creating trade-offs (Raudsepp-Hearne et al., 2010). Trade-offs involving agricultural production and water quality, land use and biodiversity, water use and aquatic biodiversity, and current water use for irrigation and future agricultural production have been identified as being significant when considering the future of water and wetlands (Millennium Ecosystem Assessment, 2005).

Wise use of wetlands entails stakeholder engagement and transparency in negotiating trade-offs between ecosystem services in order to determine equitable conservation outcomes (Finlayson et al., 2011). Addressing these requires an understanding of the underlying biophysical constraints, as well as societal preferences for bundles of ecosystem services. The values humans hold for ecosystem services are important institutions in themselves to engender changes in the ways societies structure and manage trade-offs (Jacobs, 1997) by acting as an essential feedback mechanism on the consequences of consumption choices and behaviour (Zavestoski, 2004). Wetlands can be valued in a number of ways given the diverse ontologies and epistemologies which influence the constitution as well as the conceptualization of value (Farber et al., 2002).

The predominant economic model underlying public decision-making fails to take into account the value of freshwater ecosystems, as markets in their conventional sense do not exist for these largely public goods and services that flow from ecosystems and biodiversity (TEEB, 2010). This generates incentives which are perverse to wetlands wise use, instead favouring their conversion for alternate usages often having well-defined cost and benefit flows. Economic valuation, one of the several ways in which the value of nature can be expressed, enables articulation of impact of human use on ecosystem services values in units that allow for their incorporation in such societal decisions (Mooney et al., 2005), thus demonstrating the interrelationship between wetland 'values' and societal objectives of water, food and climate security and to propagate the idea of wetlands as 'natural capital' (Russi et al., 2013).

Different stakeholders often attach different values to ecosystem services, largely depending on the impact the service has on their well-being. While developing response strategies for managing trade-offs, specific attention needs to be given to the social distribution of costs and benefits related to ecosystem service provision, and the role of institutional arrangements in mediating these linkages (Lakerveld et al., 2015; Paavola, 2007; Robards et al., 2011).

Changing baselines and novel ecosystems

As outlined above, ecosystem change is continuous and occurs across different spatial and temporal scales. Over long timeframes ecosystems are subject to many non-biological drivers of change, such as climate and sea-level change which impact on the moisture balance. These can occur in cycles of varying frequency, but also involve stochastic events that can bring on abrupt, long-lasting changes. Also there are directional changes as ecosystems undergo serial change driven by both internal and external dynamics. This is clearly evident in freshwater systems which are impacted by cycles of drying and wetting, large events such as floods and droughts, as well as progressive infilling, but also renovation through channel evulsion or similar large scale geomorphic change. Our understanding of the full history of change requires more than recollection, or even modern monitoring programmes, as many cycles have return times that extend beyond both (Power et al., 1999). In fact, the deeper in time the change the poorer is our appreciation, and so the more likely our surprise when low frequency drivers return to drive the condition of wetlands, as has been revealed by palaeoecological analysis of an increasing number of freshwater systems (Barr et al., 2014).

Permanent and semi-permanent inland lentic waterways are sediment basins as much as they are water basins. Inlets ferry sediments from the surrounding catchment and the still waters allow those to settle within the wetland. Deep lakes beyond the limit of the glacial ice fields have potentially accumulated those sediments for many tens of thousands of years while others have done so over shorter periods. The sediments also bury as fossils much of the biological and chemical components of the waterway and so provide an archive of wetland limnology continuously through time. So, given that all Ramsar sites were listed and described after 1974, their perceived condition reflects the boundary conditions that were at play over a relatively short period of their history. However, each wetland holds secure a record of its ancestry beyond 1974 and through palaeoecological approaches it is possible to reveal a much more dynamic state in many of the world's most significant wetlands.

The identification of a static ecological character for each Ramsar site then denies this record of past, present and future change in wetland state in response to a broad array of drivers. Over long time frames there are records of giant salinas of the present day, such as Lake Eyre, being permanent lakes (Magee et al., 1995) for millennia and many examples of the late-20th-century state of a wetland significant for birds and fish being unrepresentative of that leading up to the major changes that came after the industrial revolution. In fact, given the clear evidence for the Great Acceleration and reporting of the decline in the world's freshwaters (Millennium Ecosystem Assessment, 2005), it is reasonable to expect that few of the world's wetlands remain as they were 200 years ago. Palaeolimnological evidence shows this clearly for at least the Murray-Darling Basin (Gell and Reid, 2014, 2016) and the Yangtze River Basin (Dong et al., 2016), and Jenny et al. (2016) attest to widespread eutrophication and hypoxia, particularly in northern hemisphere wetlands from the 19th century.

The condition of wetlands whose character has been described since the 1970s may be relatively static, but is much more likely to have changed recently (Reeves et al., 2015; Bhattacharya et al., 2016), or be in transition (Figure 12.4; Boon et al., 2016), or be accommodating pressures until a threshold is reached (Wang et al., 2012) whereupon it may change abruptly to a contrasting state. Further, while the direct impact of catchment change on wetlands is clear we need to remind ourselves that they will continue to be subject to low frequency climate cycles as well as the hydroclimatic shifts that will come as the global climate warms.

Figure 12.4 The ecological character of the Gippsland Lakes Ramsar site in south-eastern Australia is considered to be in transition, including changes in the health of the vegetation, such as the dieback of wetland trees.

However, 'stationarity is dead' (Milly et al., 2008) is a new truism as changes in global circulation will shift the water balance in many regions driving sites through states and across wetland classes. The combination of cyclical change, with directional (infilling, species invasions, ontogeny), mean that future wetlands will not have past analogues. To manage for a state identified in the late 20th century not only neglects anthropogenic change, and ongoing cycles of hydroclimate, but also the evolution of novel ecosystems (Acreman et al., 2014; Gell et al., 2016). For the management of sites that have evolved into novel systems it needs to be recognised that they still retain capacity to support natural ecological assets and important components of aquatic biodiversity. Here, as Kopf et al. (2015) suggest, their management should not be directed at returning them to an historical baseline, but to an anthropogenic baseline that attends to both the ecological and socio-economic forces that drove their shift in condition, which in turn needs to take into account their position in the landscape and trade-offs between uses. Under Ramsar this could be seen as relieving signatory nations of the pressure to maintain their ecological character and to identify limits of acceptable change, especially when the principal drivers are outside their jurisdiction or are regional or global in nature. The landscape mechanisms mentioned in the above text need to be considered with care. If they are interpreted and used as a means of maintaining the status quo in a wetland, or returning it to a past condition that does not reflect current landscape change, they could fail.

Protected Areas need to take into account the dynamics and trends of the wider global landscape, including social values, but also biophysical conditions that are subject to change, and at times change that cannot be readily reversed, and could result in the formation of novel or emerging wetland types (Kopf et al., 2015). Such changes present new challenges for conservation thinking and planning and while the scientific knowledge about changes in the landscape is increasing rapidly (Capon et al., 2015; Gell et al., 2016) the institutional processes, such as those derived from the extensive guidance provided by the Ramsar Convention that largely guide our views on managing freshwater PAs, may not be keeping pace. This is evident in the challenge the Ramsar Convention has faced when considering global climate change and how it impinges on the conservation and planning concepts developed in recent decades (Finlayson et al., 2017b), including how baselines are set and used as part of the complex processes of responding to ecological change (Gell et al., 2016). This extends to understanding how landscape processes, social and biophysical, affect the values of a PA and how these are managed across and within landscapes at different scales.

References

Acreman, M.C. and Ferguson, A.J.D. (2010). 'Environmental flows and the European Water Framework Directive'. *Freshwater Biology*, 55(1), pp32–48.

Acreman, M., Arthington, A.H., Colloff, M.J., Couch, C., Crossman, N., Dyer, F., Overton, I., Pollino, C.A., Stewardson, M. and Young, W. (2014). 'Environmental flows for natural, hybrid and novel riverine ecosystems in a changing world'. *Frontiers in Ecology and Environment*, vol 12, pp466–473.

Alexander, M. (2016). 'Adaptive Management Planning'. In Finlayson, C.M., Milton, G.R., Prentice, C. and Davidson, N.C. (eds). *The Wetland Book: Distribution, Description and Conservation*, Springer Publishers, Dordrecht.

Barr, C., Tibby, J., Gell, P., Tyler, J., Zawadzki, A. and Jacobsen, G. (2014). 'Climatic variability in southeastern Australia over the last 1500 years inferred from the fossil diatom records of two crater lakes'. *Quaternary Science Reviews*, 95: pp115–131.

Bayliss, P., Van Dam, R. and Bartolo, R.E. (2012). 'Quantitative ecological risk assessment of the Magela Creek Floodplain in Kakadu National Park, Australia: Comparing point source risks from the Ranger Uranium Mine to diffuse landscape-scale risk'. *Human and Ecological Risk Assessment*, 18: pp115–151.

Bellio, M., Minton, C. and Veltheim, I. (2016). 'Challenges faced by shorebird species using the inland wetlands of the East Asian-Australasian flyway: the little curlew example'. *Marine* and *Freshwater Research*, dx.doi.org/10.1071/MF15240.

Bhattacharya, R., Hausmann, S., Hubeny, J.B., Gell, P. and Black, J.L. (2016). 'Ecological response to hydrological variability and catchment development: insights from a shallow oxbow lake in Lower Mississippi Valley, Arkansas'. *Science of the Total Environment*, 569–570: pp1087–1097.

BirdLife International (2001). *Important Bird Areas and potential Ramsar Sites in Europe*. Wageningen, The Netherlands: BirdLife International.

BirdLife International (2002). *Important Bird Areas and potential Ramsar Sites in Africa*. Cambridge, UK: BirdLife International.

BirdLife International (2005). *Important Bird Areas and potential Ramsar Sites in Asia*. Cambridge, UK: BirdLife International.

Boere, G.C. and Stroud, D.A. (2006). 'The flyway concept: what it is and what it isn't'. In Boere, G.C., Galbraith, C.A. & Stroud, D.A. (eds). *Waterbirds around the World*, pp40–47. The Stationery Office, Edinburgh, UK.

Bond, N.R., Costello, J., King, A., Warfe, D., Reich, P. and Balcombe, S. (2014). 'Risks and opportunities from artificial watering as a means of achieving environmental flow objectives'. *Frontiers in Ecology and Environment*, 12, pp386–394.

Boon, P.I., Cook, P. and Wood, R. (2016). 'The Gippsland Lakes: management challenges posed by long-term environmental change'. *Marine & Freshwater Research*, 67: pp721–737.

Capon, S.J., Lynch, J.J., Bond, N., Bruce, C., Chessman, B.C., Davis, J., Davison, N, Finlayson, C.M., Gell, P.A., Hohnberg, D., Humphrey, C., Kingsford, R.T., Nielsen, D., Thomson, J.R., Ward, K. and MacNally, R. (2015). 'Regime shifts, thresholds and multiple stable states in freshwater ecosystems; a critical appraisal of the evidence'. *Science of the Total Environment*, 534, pp122–130.

Connell, D. (2011). 'Water reform and the federal system in the Murray-Darling Basin', *Water Resources Management*, vol. 25, no. 15, pp3993–4003.

Davidson, N.C. (2014). 'How much wetland has the world lost? Long-term and recent trends in global wetland area', *Marine and Freshwater Research* 65(10): pp934–941, http://dx.doi.org/10.1071/MF14173.

Davidson, N.C. and Stroud, D.A. (2016). 'Waterbird flyways – and the history of international co-operation for waterbird conservation'. In Finlayson, C.M., Middleton, B., McInnes, R.J., Everard, M., van Dam, A., Irvine, K. and N.C. Davidson, N.C. (eds) *The Wetlands Book. Structure & function, management and methods*. Springer, Dordrecht.

de Fraiture, C., Smakhtin, V., Bossio, D., McCornick, P., Hoanh, C., Noble, A., Molden, D., Gichuki, F., Giordano, M., Finlayson, M. and Turral, H. (2007). 'Facing climate change by securing water for food, livelihoods and ecosystems', *SAT eJournal*, 4, p12.

Dixon, M.J.R., Loh, J., Davidson, N.C., Beltrame, C., Freeman, R. and Walpole, M. (2015). 'Tracking global change in ecosystem area: The Wetland Extent Trends index'. *Biological Conservation*, 193: pp27–35, doi.org/10.1016/j.biocon.2015.10.023.

Dong, X., Yang, X., Chen, X., Liu, Q., Yao, M., Wang, R. and Xu, M. (2016). 'Using sedimentary diatoms to identify reference conditions and historical variability in shallow lake ecosystems in the Yangtze floodplain', *Marine and Freshwater Research*, 67: pp803–815.

European Commission (1979). 'Directive 79/409/EEC of the European Parliament and of the Council of 2 April 1979 on the Conservation of Wild Birds', *Official Journal of the European Communities*, 1–18. Brussels: European Commission.

European Commission (1992). 'Directive 92/43/EEC of the European Parliament and of the Council of 21May 1992 on the Conservation of Natural Habitats and of Wild Fauna and Flora', *Official Journal of the European Communities*, 35, pp7–50. Brussels: European Commission.

European Commission (2000). 'Directive 2000/60/EC of the European Parliament and of the Council of 23 October 2000 Establishing a Framework for Community Action in the Field of Water Policy', *Official Journal of the European Communities*, 22 December, L 327/1. Brussels: European Commission.

European Union (2010). *Water is Life: How the Water Framework Directive helps safeguard Europe's resources*. Luxembourg: Publications Office of the European Union, p28.

Evans, D. (2012). 'Building the European Union's Natura 2000 network'. *Nature Conservation*, 1, p11.

Falkenmark, M., Finlayson, C.M. and Gordon, L. (coordinating lead authors) (2007). 'Agriculture, water, and ecosystems: avoiding the costs of going too far'. In Molden, D. (ed.), *Water for Food, Water for Life: A Comprehensive Assessment of Water Management in Agriculture*. Earthscan, London, UK, pp234–277.

Farber, S.C., Costanza, R. and Wilson, M.A. (2002). 'Economics and ecological concepts for valuing ecosystem services', *Ecological Economics*, 41(3), pp375–392.

Finlayson, C.M. (2005). 'Plant ecology of Australia's tropical floodplain wetlands: a review', *Annals of Botany*, 96, pp541–555, doi:10.1093/aob/mci209.

Finlayson, C.M. (2013). 'Climate change and the wise use of wetlands – information from Australian wetlands', *Hydrobiologia*, 708, pp145–152.

Finlayson, C.M. (2016a). 'North American Waterfowl Management Plan (NAWMP)'. In Finlayson, C.M., Everard, M., Irvine, K., McInnes, R.J., Middleton, B.A., van Dam, A.A. and Davidson, N.C. (eds), *The Wetland Book I: Structure and Function, Management and Methods*. Springer Publishers, Dordrecht, doi:10.1007/978-94-007-6172-8_134-1.

Finlayson, C.M. (2016b). 'Wetland management planning: Okavango Delta (Botswana)'. In Finlayson, C.M., Everard, M., Irvine, K., McInnes, R.J., Middleton, B.A., van Dam, A.A. and Davidson, N.C. (eds), *The Wetland Book I: Structure and Function, Management and Methods*. Springer Publishers, Dordrecht.

Finlayson, C.M. (2016c). 'Wetland management planning: Lake Chilika (India)'. In Finlayson, C.M., Everard, M., Irvine, K., McInnes, R.J., Middleton, B.A., van Dam, A.A. and Davidson, N.C. (eds), *The Wetland Book I: Structure and Function, Management and Methods*. Springer Publishers, Dordrecht.

Finlayson, C.M. & van der Valk, A.G. (1995). 'Wetlands classification and inventory: A summary', *Vegetatio*, 118, pp185–192.

Finlayson, C.M., Lowry, J., Bellio, M.G., Walden, D., Nou, S., Fox, G., Humphrey, C.L. and Pidgeon, R. (2006). 'Comparative biology of large wetlands: Kakadu National Park, Australia'. *Aquatic Sciences*, 68, pp374–399.

Finlayson, C.M., Davidson, N., Pritchard, D., Milton, G.R. and MacKay, H. (2011). 'The Ramsar Convention and ecosystem-based approaches to the wise use and sustainable development of wetlands'. *Journal of International Wildlife Law and Policy*, 14, pp176–198, http://dx.doi.org/10.1080/13880292.2011.626704.

Finlayson, C.M., Milton, G.R. & Prentice, C. (2017a). 'Wetland types and distribution'. In Finlayson, C.M., Milton, G.R., Prentice, C. and Davidson, N.C. (eds). *The Wetland Book: Distribution, Description and Conservation*, Springer Publishers, Dordrecht.

Finlayson, C.M., Capon, S.J., Rissik, D., Pittock, J., Fisk, G., Davidson, N.C., Bodmin, K.A., Papas, P., Robertson, H.A., Schallenberg, M., Saintilan, N., Edyvane, K. and Bino, G. (2017b). 'Policy considerations for managing wetlands under a changing climate'. *Marine and Freshwater Research* (in press), published early online at: https://doi.org/10.1071/MF16244.

Fluet-Chouinard, E., Lehner, B., Rebelo, L.-M., Papa, F. and Hamilton, S.K. (2015). 'Development of a global inundation map at high spatial resolution from topographic downscaling of coarse-scale remote sensing data', *Remote Sensing of Environment*, 158: pp348–361.

Gardner, R.C. and Davidson, N.C. (2011). 'The Ramsar Convention'. In Lepage, B. (ed.) *Wetlands – Integrating Multidisciplinary Concepts*, pp189–203. Springer, Dordrecht.

Gardner, R.C., Barchiesi, S., Beltrame, C., Finlayson, C.M., Galewski, T., Harrison, I., Paganini, M., Perennou, C., Pritchard, D.E., Rosenqvist, A. and Walpole, M. (2015). *State of the World's Wetlands and their Services to People: A compilation of recent analyses.* Ramsar Convention Secretariat, Ramsar Scientific and Technical Briefing Note No. 7, Gland, Switzerland.

Gell, P. and Reid, M. (2014). 'Assessing change in floodplain wetland condition in the Murray Darling Basin'. *The Anthropocene*, 8: pp39–45.

Gell, P. and Reid, M. (2016). 'Muddied waters: the case for mitigating sediment and nutrient flux to optimise restoration response'. *Frontiers in Ecology and Evolution*, doi: 10.3389/fevo.2016.00016.

Gell, P.A., Finlayson, C.M. and Davidson, N.C. (2016). 'Understanding change in the ecological character of Ramsar wetlands: perspectives from a deeper time – synthesis'. *Marine and Freshwater Research*, 67 pp869–879.

Gordon, L., Finlayson, C.M. and Falkenmark, M. (2010). 'Managing water in agriculture to deal with trade-offs and find synergies among food production and other ecosystem services'. *Agricultural Water Management*, 97, pp512–519, doi:10.1016/j. agwat.2009.03.017.

Hooper, B.P. (2005). *Integrated River Basin Governance: Learning From International Experiences.* London, IWA Publishing.

Jacobs, M. (1997). 'Environmental valuation, deliberative democracy and public decision-making'. In J. Foster (ed.), *Valuing Nature? Economics, Ethics and Environment*, pp211–231. Routledge, London.

Jenny, J-P., Francus, P., Normadeau, A., Lapointe, F., Perga, M-E., Ojala, A., Schemmelmann, A. and Zolitschka, B. (2016). 'Global spread of hypoxia in freshwater ecosystems during the last three centuries caused by rising local human pressure', *Global Change Biology*, doi: 10.1111/gcb.13193.

Kopf, R.K., Finlayson, C.M., Humphries, P., Sims, N.C. and Hladyz, S. (2015). 'Anthropocene baselines: assessing change and managing biodiversity in human-dominated aquatic systems', *Bioscience*, 65: pp798–811.

Lakerveld, R.P., Lele, S., Crane, T.A., Fortuin, K.P.J. and Springate-Baginski, O. (2015). 'The social distribution of provisioning forest ecosystem services: Evidences and insights from Odisha, India', *Ecosystem Services*, 14, pp56–66.

Levin, S.T., Xepapadeas, A.S., Crépin, J., Norberg, A., de Zeeuw, C., Folke, T., Hughes, K., Arrow, S., Barrett, G., Daily, P., Ehrlich, N., Kautsky, K., Mäler, G., Polasky, S., Troell, M., Vincent, J.R. and Walker, B. (2013). 'Social-ecological systems as complex adaptive systems: modeling and policy implications', *Environment and Development Economics*, 18: pp111–132.

Magee, J.W., Bowler, J.M., Miller, G.H. and Williams, D.L.G. (1995). 'Stratigraphy, sedimentology, chronology and palaeohydrology of Quaternary lacustrine deposits at Madigan Gulf, Lake Eyre, South Australia'. *Palaeogeography, Palaeoclimatology, Palaeoecology*, 113: pp3–42.

Matthews, G.V.T. (1993). *The Ramsar Convention on Wetlands: Its History and Development.* Bureau of the Ramsar Convention: Gland, Switzerland.

Milly, P.C.D., Betancourt, J., Falkenmark, M., Hirsh, R.M., Kundewicz, Z.W., Letenmaier, D.P. and Stouffer, R.J. (2008). 'Stationarity is dead: whither water management'. *Science*, 319: pp573–574.

Millennium Ecosystem Assessment (2005). *Ecosystems and Human Well-being: Wetlands and Water Synthesis*. World Resources Institute, Washington, DC.

Milton, G.R., Prentice, C. and Finlayson, C.M. (2017). 'Wetlands of the world'. In Finlayson, C.M., Milton, G.R., Prentice, C. and Davidson, N.C. (eds), *The Wetland Book: Distribution, Description and Conservation*, Springer Publishers, Dordrecht.

Mooney, H., Cropper, A. and Reid, W. (2005). 'Confronting the human dilemma: how can ecosystems provide sustainable services to benefit society?'*Nature*,434: pp561–562.

Moss, T. (2004). 'The governance of land use in river basins: prospects for overcoming problems of institutional interplay with the EU Water Framework Directive'. *Land Use Policy*, 21(1), pp85–94.

Pattnaik, A.K. and Kumar, R. (2016). 'Lake Chilika (India): Ecological restoration and adaptive management for conservation and wise use'. In Finlayson, C.M., Milton, G.R., Prentice, C. and Davidson, N.C. (eds), *The Wetland Book: Distribution, Description and Conservation*, Springer Publishers, Dordrecht.

Paavola, J. (2007). 'Institutions and environmental governance: a reconceptualization'. *Ecological Economics*, 63: pp93–103.

Pittock, J., Finlayson, C.M., Gardner, A. and McKay, C. (2010). 'Changing character: the Ramsar Convention on Wetlands and climate change in the Murray-Darling Basin, Australia'. *Environmental and Planning Law Journal*, 27, pp401–42.

Power, S., Casey, T., Folland, C., Colman, A. and Mehta, V. (1999). 'Inter-decadal modulation of the impact of ENSO on Australia'. *Climate Dynamics*, 15, pp319–324.

Ramberg, L. (2016). 'Okavango Delta, Botswana (Southern Africa)'. In Finlayson, C.M., Milton, G.R., Prentice, C. and Davidson, N.C. (eds), *The Wetland Book: Distribution, Description and Conservation*, Springer Publishers, Dordrecht.

Ramsar Convention (2012a). 'Information Sheet on Ramsar Wetlands – 2012 revision'. Ramsar COP11 Resolution XI.8 Annex 1.

Ramsar Convention (2012b). 'Strategic Framework and Vision for the List of Wetlands of International Importance – 2012 revision'.

Raudsepp-Hearne, C., Peterson, G.D. and Bennett, E.M. (2010). 'Ecosystem service bundles for analyzing tradeoffs in diverse landscapes'. *Proceedings of the National Academy of Sciences*, 107 (11): pp5242–5247.

Reeves, J.M., Gell, P.A., Reichman, S.M., Trewarn, A.J. and A. Zawadzki, A. (2015). 'Industrial past, urban future: using palaeo-studies to determine the industrial legacy of the Barwon Estuary, Victoria, Australia'. *Marine and Freshwater Research*, 67, pp837–849.

Reyers, B., Biggs, R., Cumming, G.S., Elmqvist, T., Hejnowicz, A.P. and Polasky, S. (2013). 'Getting the measure of ecosystem services: a social-ecological approach'. *Frontiers in Ecology and the Environment* 11(5): pp268–273.

Robards, M.D., Schoon, M.L., Meek, C.L. and Engle, N.L. (2011). 'The importance of social drivers in the resilient provision of ecosystem services'. *Global Environmental Change*, 21: pp522–529.

Russi, D., ten Brink, P., Farmer, A., Badura, T., Coates, D., Förster, J., Kumar, R. and Davidson, N. (2013). *The Economics of Ecosystems and Biodiversity for Water and Wetlands*. IEEP, London and Brussels; Ramsar Secretariat, Gland.

Schmeier, S. (2010). 'The organizational structure of river basin organizations: lessons learned and recommendations for the Mekong River Commission (MRC)'. Available online at: www.mrcmekong.org/assets/Publications/governance/MRC-Technical-Paper-Org-Structure-of-RBOs.pdf.

Schmeier, S. (2012). 'Navigating cooperation beyond the absence of conflict: mapping determinants for the effectiveness of river basin organisations', *International Journal of Sustainable Society*, 4, pp11–27.

TEEB (2010). *The Economics of Ecosystems and Biodiversity: Mainstreaming the Economics of Nature: A Synthesis of the Approach, Conclusions and Recommendations of TEEB*.

Wang, R., Dearing, J. A., Langdon, P. G., Zhang, E., Yang, X., Dakos, V. and Scheffer, M. (2012). 'Flickering gives early warning signals of a critical transition to a eutrophic lake state'. *Nature*, 492, pp419–422.

Zavestoski, S. (2004) 'Constructing and maintaining ecological identities: the strategies of deep ecologists'. In Clayton, S. and Opotow, S. (eds), *Identity and the Natural Environment: The Psychological Significance of Nature*, The MIT Press, Cambridge, MA, pp297–316.

Chapter 13

Climate change and the management of freshwater protected areas

C.M. Finlayson and J. Pittock

Key messages

- While Protected Areas (PAs) have played an important role in species and ecosystem conservation, managing them with the same rulebook for coming decades without reflection or flexibility may result in adverse consequences that increase the pressures on critical biodiversity targets. Strategies for promoting more climate-resilient approaches are needed rather than focussing on maintaining past reference states. Climate change and flow regulation are leading to the development of novel ecosystems that may require new thinking and a range of novel approaches to water management to cope with increasingly uncertain futures.
- A range of climate change adaptation interventions has been proposed to better conserve freshwater biodiversity. These involve identifying parts of the freshwater landscape that can provide refugia and sustain ecological complexity, and guidance to manage environmental flows to counter climate change impacts. Free-flowing rivers that do not require day-to-day management to provide the flow variability and connectivity needed to conserve aquatic biodiversity may become conservation priorities.
- Many climate change adaptation measures can be seen as 'no regrets' measures that offer benefits for the environment and people regardless of changes in the climate. As all adaptation options have risks and costs as well as benefits the adoption of a suite of different but complementary interventions is likely to result in better practice by spreading the risk while seeking to maximize the benefits for PA ecosystems and people.
- The development of an international framework and guidance for managing freshwater protected areas in changing climates is overdue and is currently a major gap. The Ramsar Convention in particular is well placed to fill this gap by building on its wise use and conservation handbooks.

Promoting climate resilient management

By their nature, freshwater PAs tend to be focussed on a specific locality with clearly demarcated boundaries, while successful management of PAs is normally expressed in terms of outcomes, compared with a reference state, and generally outlined in a management plan. Ideally, PA managers have a record of quantitative observations that can serve as a reference state to guide their management, but more often this is likely to consist of qualitative and historical records and depends on operational priorities. In many instances there is increasing uncertainty about the usefulness of baselines for wetland management given past changes in land use and more so with changes in climate (Acreman et al., 2014; Finlayson et al., 2016; Gell et al., 2016).

Whether a PA is intended to protect a specific 'place' (defined both geographically and in terms of a set of ecological or physico-chemical properties), a set of species located in a particular location, or both, or even specifically focussed on iconic or vulnerable species, climate change presents a significant, or even fundamental challenge to how a PA can be managed. From the species and population perspective, for instance, climate has both a primary, direct, and indirect set of influences on the location, phenology, phenotypic expression, and the interactions within populations and between species (Parmesan, 2006). Against this background steps to promote climate resilient management and adaptation measures are discussed in the text below and placed in an international policy context with a number of guiding principles.

Promoting climate-resilient management

The boundaries of a PA that were intended to protect organisms and places from outside influences cannot "fence out" climate shifts in precipitation, storm intensity and frequency, droughts and floods, or temperature extremes. As a result, PAs may in effect be restricting some of the autonomous climate adjustments that could be made by the species within. This is evident for migratory and nomadic waterbird species that are known to respond to weather conditions with large fluctuations in populations and movements between breeding and feeding areas in both cold-temperate (Ridgill and Fox, 1990) and arid zones (Bellio et al., 2016). Communities can be remarkably plastic, with the repeated appearance, evolution, and disappearance of so-called no-analogue communities that bear little resemblance to extant community composition, often in response (at least in part) to changes in climate. The extension of the range of a "new" native species in a PA, for instance, may signal effective autonomous adaptation rather than a species invasion that should be resisted and discouraged (Finlayson, 2009). Likewise, declines in abundance may be evidence for a range shift. Managing for a fixed community definition, particularly with narrowly defined relative abundance, may in fact be counterproductive to effective climate-adaptive management (Catford et al., 2012; Acreman et al., 2014).

Ecological processes may also be sensitive to shifts in climate. Flow regime and hydroperiod typically reflect strong climate signals such as precipitation and snowpack melt (Null et al., 2013; Poff, 2009; Poff and Matthews, 2013). Fire regimes can reflect drought frequency and severity as well as seasonal variations in rainfall and the growth patterns of critical components of the vegetation. Both types of disturbance regime are important for regulating a broad range of ecological processes across multiple levels and spatial scales. In addition, climate-driven alteration of PA physical properties, such as warmer water, may promote shifts in dissolved oxygen levels or the growth of pathogenic organisms.

While PAs have arguably played the single most important role in species conservation globally for the past century, managing PAs with the same rulebook for coming decades without reflection or flexibility may result in adverse consequences that increase the pressures on critical biodiversity targets. This calls for 1) the development and use of management actions that take into account the consequences of climate changes, or 2) the adaption of existing management actions to also address the consequences of climate changes (Lukasiewicz et al., 2013a, 2016). Strategies for promoting a more climate-resilient approach to PA management include:

1   Managing and monitoring climatic variables and target species within and outside of PA boundaries. A regional approach to management will be critical for anticipating and adjusting to factors that may prompt shifts in the management regime (Poff et al., 2010).
2   Viewing range shifts as a (potential) adaptation by some species to climatic variables—or at least a question worthy of investigation. Corridors between PAs or other high-quality habitat may be critical for the long-term survival of some species, while more sessile or isolated species may require assistance to disperse to, and establish in, new habitats (Hannah, 2010).
3   Assessing our philosophical approach to PAs whereby managers may need to consider transitory ecological conditions that enable particular species to respond to rapidly changing habitats, rather than continuing to focus on maintaining past reference states (Matthews et al., 2011; Acreman et al., 2014; Gell et al., 2016).

In making these recommendations it is recognized that many of the proposed actions are already being used by wetland managers or have been proposed. In this respect the usefulness of existing management approaches for dealing with the consequences of climate change should continue to be investigated, as outlined by Arthington (2012), Lukasiewicz et al., (2013a, 2016), and Finlayson (2013) when considering the status of PAs in the Murray-Darling Basin, Australia (Box 13.1).

Climate change adaptation

A range of climate change adaptation interventions has been proposed to better conserve freshwater biodiversity in wetland PAs and river systems, including a set of options detailed in Australia (Arthington, 2012; Lukasiewicz et al., 2013a, 2016) and referred to below. These involve identifying parts of the freshwater landscape that may be more resilient to climate change and which can provide refugia and ecological (and microclimatic) complexity, such as river reaches shaded by riparian vegetation or even by mountains, or those that form corridors that may enable species to move to more favourable habitats. Another option is to manage environmental flows to counter climate change impacts by maintaining or providing appropriate water quality and volumes for targeted biodiversity (Olden and Naiman, 2010; Poff and Matthews, 2013). Generally these measures are only possible on rivers with storage dams that have outlets suitable for releasing the right volume of water at the right temperatures (Figure 13.1), or other infrastructure to enable water of a suitable temperature to be directed to specific locations (Pittock and Hartmann, 2011). These approaches are also dependent on management institutions maintaining the infrastructure and making decisions consistently and in a timely manner, for instance, to release water from dams, or even to reoperationalize dam infrastructure (Watts et al., 2011). By contrast, free-flowing rivers do not require day-to-day management to provide the flow variability and connectivity needed to conserve aquatic species. Free-flowing rivers may then become conservation priorities, but could still be at risk from climate-induced changes that cannot be addressed without the presence of infrastructure needed to manipulate flows (Pittock and Finlayson, 2011) or constraints on water abstraction. In an increasingly complex world with increasing development of agriculture and water infrastructure, PA managers may in fact find themselves hemmed between and dependent on that same infrastructure to deliver ecological outcomes, and yet have little influence over operational matters.

Many climate change adaptation measures can be seen as 'no regrets' measures that offer benefits for the environment and people regardless of changes in the climate. The restoration of riparian forests (Figure 13.2) to shade adjoining freshwater ecosystems and provide other conservation benefits is one example (Davies, 2010). The control of invasive species (which are not adapting autonomously to climate shifts) is another example. The co-benefits for different groups of people associated with these no regrets adaptation measures offer opportunities to build greater stakeholder support for conservation.

We may have entered a period of 'regretful' adaptation, however, when choices are more difficult and their consequences more lasting. Engineering interventions are often promoted as adaptation options in freshwater systems (Pittock and Finlayson, 2013; Pittock et al., 2012; Bond et al., 2014). As climate change may require the safety standards of existing infrastructure in freshwater ecosystems to be upgraded (Figure 13.1), this could provide an opportunity for protected area managers to secure further changes to reduce environmental

Figure 13.1 In-stream structures can obstruct the movement of aquatic species as well as be used to direct water to assist species complete their life cycles, although much existing infrastructure may need to be modified to achieve these outcomes (photographs © C.M. Finlayson).

Figure 13.2 The restoration of riparian vegetation brings many benefits to the many species that depend on the shade provided or the provision of nutrients or food items (photographs © C.M. Finlayson).

impacts and aid climate change adaptation, such as by installing fish passages on dams and controlling thermal pollution (Pittock and Hartmann, 2011; Lukasiewicz et al., 2016). There are also many proposals for "environmental water demand management" or "environmental works and measures" based on engineering interventions (e.g., pipes and pumps) that deliver and use less water to conserve aquatic biodiversity (Pittock and Lankford, 2010). While these options may be politically appealing they also have high risks of unforeseen environmental impacts and management failure and should be considered with caution (Pittock et al., 2012; Bond et al., 2014). In Australia's Murray-Darling Basin, for example, identified risks include: barriers and constraints to fauna movement, reduced water quality, high opportunity costs, institutional failure, and path dependency (Pittock et al., 2012). The Murray-Darling provides an exemplar of both the opportunities and the pitfalls that can occur when relying on engineering works and measures to manage and manipulate flows in a river that has been inexorably changed (Box 13.1).

The definition and management of infrastructure and the decision-making systems that manage such infrastructure are clearly in a period of rapid evolution. However, to deliver effective outcomes for freshwater species and ecosystems the management and construction of infrastructure must evolve to be viewed broadly as both built and "natural" eco-hydrological components of the landscape, while our decision-making processes must shift to more flexible, humble, and incremental approaches (Matthews and Wickel, 2009; Matthews et al., 2011; Parmesan et al., 2011). While raising such issues and opportunities it is recognized that in many river systems the extent of change wrought by infrastructure that is unsympathetic to maintaining freshwater species is already profound (Nilsson et al., 2005), possibly irreversible in many cases (Kopf et al., 2015; Gell et al., 2016), and escalating (Poff and Matthews, 2013). The choices to install such infrastructure also invoke opportunity costs and path dependencies that may serve the present but are likely to further limit future adjustments (Pittock and Finlayson, 2011; Lukasiewicz et al., 2013a, 2016).

Too often decision-makers fix their attention on one intervention or adaptation measure as the solution to climate change impacts. In practice, each adaptation option has risks and costs as well as benefits – freshwater protected area managers should identify these and evaluate the most suitable options. The adoption of a suite of different but complementary interventions is likely to result in better management practice by spreading the risk while seeking to maximize the benefits and avoiding perverse outcomes. The use of environmental flows on regulated rivers and the protection of free-flowing rivers is an example. With this in mind, Lukasiewicz et al. (2013a) developed a catchment-scale framework for assessing options for climate change adaptation. The framework can be used to assess each option by considering: relevance and benefits; effectiveness under a range of climate change scenarios; the risks of maladaptation; the complementarity of different options; the ecosystem service and socio-economic outcomes; and constraints to implementation. This approach emphasizes consultation with stakeholders to identify co-benefits that encourage and help build support for climate change adaptation interventions.

Many institutions, such as the United Nations Environment Programme, are promoting greater conservation of the environment to increase resilience to climate change impacts and aid adaptation. Various jargon with similar or overlapping meaning is used to describe this approach, including 'green infrastructure', 'natural capital', 'ecosystem management', 'ecosystem based adaptation' and 'ecosystem services' (IEMP, 2011). Conservation of freshwater ecosystems is among the interventions favoured by these approaches.

Responding to climate change

As climate change will impact to some degree on all PAs the measures outlined above can help PA managers achieve the best possible outcomes. While adaptation measures are available, and other management actions may also be adapted to specifically address the consequences of climate change, there is an omnipresent need for international and national guidance that can assist PA mangers to make the choices they need to face. While Lukasiewicz et al., (2013a), for example, provide a framework for assessing the relative merits of adaptation measures, international guidance on how to address climate change in freshwater PAs is lacking. As pointed out by Finlayson et al. (2017) the Ramsar Convention has not addressed these important management issues given ongoing dissension about the role of the Convention that has limited its ability to provide such guidance.

The adoption in mid-2012 by the Convention of a decision to specifically develop guidance for wetland management in the face of climate change, including *inter alia* the determination of appropriate reference conditions and specified limits of change as well as the provision of advice on ecosystem-based adaptation (Finlayson, 2013), has not yet been enacted (Finlayson et al., 2017). As pointed out by Gell et al. (2016), the Convention has provided specific guidance for the wise use of wetlands and, while these guidelines address many of the technical issues that PA managers need to consider, they have not hitherto provided guidance on establishing baselines for determining change in ecological character, nor how to respond to variations and change due to climate change, despite requests to do so (see Finlayson, 1996, 2013).

Finlayson et al., (2017) responded to the inertia in the Ramsar Convention and provided a set of key principles to inform wetland conservation and management policy within the context of climate change.

1   Objectives and targets for wetland management should look mainly (but not exclusively) to accommodate and compensate for climate change, rather than accept or avoid impacts.
2   Objectives for wetland management under climate change should include ecological, social and economic targets across multiple scales and consider issues such as representativeness, connectivity, refugial values, etc.

3   Flexible governance and adaptive co-management frameworks across multiple scales and sectors are essential to managing wetlands under climate change.

4   Easily reversed, no-regret or low-regret adaptation options with multiple, cross-sectoral benefits should be implemented in the initial phases of adapting wetland management.

5   Long-term management strategies should identify triggers for new actions including novel/high risk adaptation options (e.g., species translocations) and plan for such eventualities.

6   Scientific monitoring and evaluation of management strategies are needed.

In contrast to the limited response of the Convention, a number of scientific analyses have provided information that can be used to assist PA managers to adapt to climate change. As an example, Arthington (2012) provides a number of strategies and processes based on an analysis of environmental flow requirements that equally could apply to the management of freshwater PAs. A shortened version, with direct application to the management of freshwater PAs, is provided below.

1   **Manage freshwater PAs**: Identify and establish freshwater protected areas as a key focus for management throughout entire catchments, using systematic conservation planning approaches to identify priority habitats and waterbodies, including planning for mitigation of threats, or to establish priorities for environmental flow delivery.

2   **Protect and restore flow regimes**: Ensure sound environmental flow and socio-economic assessments of new and old dams, and of plans for increasing diversions from rivers and groundwater pumping. Restore flow regimes or provide more suitable environmental flows to meet conservation obligations for endangered species, key habitats and processes. Improve the science and modeling of hydro-ecological scenarios and incorporate the potential effects of climate change in scenario modeling.

3   **Alter dam operations and floodplain management**. Introduce time-limited licensing for all dams, allowing for 5–10 year reviews of safety and risk of failure, and socio-economic and environmental impacts. Remove or retrofit dam structures to improve environmental flow outcomes and downstream water quality, and review the effectiveness of fishways. Release water to mimic natural flow variability wherever possible, and recognize the ecological benefits of low-flow periods and flow intermittency in dryland rivers. Maintain or enhance hydrological and ecological connectivity in fragmented landscapes.

4   **Legislate and establish policy and governance mechanisms for adaptive management**: Identify responsibilities at the catchment scale to ensure a coherent approach to legislation, policy, governance and environmental flow management linked to monitoring. Review and implement

conservation legislation that supports biodiversity and resilience of rivers and wetlands.

5 **Engage with local communities, and promote science and education**: Increase efforts to engage local communities, including indigenous peoples, and to incorporate value systems and beliefs into existing or novel management systems. Support the synthesis of scientific and indigenous knowledge generated from place-based studies into a wider body of knowledge, principles and management 'rules' for rivers of contrasting hydrologic and geomorphic character, and socio-economic setting. Encourage and support river ecosystem science, education and knowledge dissemination.

The development of an international policy framework for managing freshwater PAs in the face of climate change is seen as a necessary, but not the only step for ensuring such PAs are effective. Adaptation measures, such as those described above, are also seen as necessary and will require a flexible and at times imaginative approaches to ensure they are effective.

Box 13.1   Murray-Darling Basin Ramsar wetlands, Australia.

The Murray-Darling Basin contains the longest rivers in Australia and covers about a million square kilometres (around a seventh of the continent's land mass). Large floodplain forests and other wetlands cover more than 5.7 million ha (5.6 per cent of the Basin), with 16 sites, covering 636,300 ha, designated as Ramsar wetlands of international importance (Pittock et al., 2010). The tenure of these wetlands includes nature reserves (IUCN category II) managed by state governments and non-government organisations, forestry and hunting reserves (IUCN category VI) managed by state governments, and small areas of privately managed pastoral lands (IUCN category VI).

The waters of the Basin are extensively exploited. Median annual end of river flows have fallen to 29 per cent of levels that occurred before extensive water extraction started, and the mouth of the River Murray was closed from 2002 to 2010 due to over extraction and climatic variability. Vast areas of wetlands have suffered from reduced and/or seasonal water flows, desiccation, salinity and acid sulphate generation with consequential loss of floodplain forest and other biodiversity (Pittock and Finlayson, 2011). Climate change threatens to further reduce water flows in the Basin particularly in the south (Pittock et al., 2010). Additional threats include invasive species, cold water pollution from dam releases, and barriers to movement of aquatic wildlife.

(continued)

(continued)

The ecological health of the wetlands depends on the delivery of water that mimics the volume, timing and quality of natural flows. Contested use of water resources has dominated Basin management. In 2007–08 a national Water Act was adopted and established Federal Government control over water allocations in the Basin. The Act is based on Australia's obligations to implement the Convention on Biological Diversity and Ramsar Convention on Wetlands and requires conservation of key environmental assets, ecosystem functions and services (Pittock, 2013; Pittock et al., 2010). In 2012 a basin plan was adopted that may see up to 3,200 GL/yr (29 per cent of the water diverted for consumption) returned to the environment by 2024. However, increased groundwater diversions, carbon sequestration plantings and "works and measures" may result in less water for the environment (Pittock, 2013).

The decision not to return more water to the environment through the basin plan was justified in part on the notion that "environmental works and measures" comprising engineering interventions using pumps, levees and weirs will help to conserve wetland biodiversity using less water. This option has risks, for instance, failing to provide or disrupting habitat connectivity, concentrating salt in wetlands and relying heavily on timely state government operations and maintenance (Pittock et al., 2012). Further, reallocations to reduce the impact of climate change on water availability have not been proposed despite models projecting climate-induced reductions of up to 37 per cent by 2030 (Pittock, 2013). While restoring adequate flows is a primary measure for conserving the basin's wetlands, other important actions have been overlooked, including restoring riparian forests, protecting the few remaining free-flowing rivers, and re-engineering the thousands of dams to eliminate cold-water pollution and restore fish passage (Pittock and Finlayson, 2011).

The buy-back of water entitlements from irrigators is significant for the conservation of wetland PAs. These entitlements are now owned and independently managed for conservation by the Federal Government's Commonwealth Environmental Water Holder, reducing the loss of environmental water through administrative reallocation in times of scarcity (Connell, 2011). Importantly, as the Basin Plan is to be revised at least every ten years, there is increased potential for further adaptive management of water allocations and other measures.

While progress is being made in restoring flows needed for the ecological health of the main wetlands, further decisions are required for a range of different but complementary management interventions to better adapt to climate change. A framework for assessing catchment-scale climate change adaptation has been developed to help environmental managers choose adaptation options that reduce risk, maximize benefits and minimize perverse outcomes under climate change (Lukasiewicz et al., 2013b).

References

Acreman, M., Arthington, A.H., Colloff, M.J., Couch, C., Crossman, N., Dyer, F., Overton, I., Pollino, C.A., Stewardson, M. and Young, W. (2014). Environmental flows for natural, hybrid and novel riverine ecosystems in a changing world. *Frontiers in Ecology and Environment*, 12, 466–473.

Arthington, A.H. (2012). *Environmental Flows. Saving Rivers in the Third Millennium.* University of California Press, Berkeley.

Bellio, M., Minton, C. and Veltheim, I. (2016). *Challenges Faced by Shorebird Species Using the Inland Wetlands of the East Asian-Australasian Flyway: The little Curlew Example.* Marine and Freshwater Research (in press), http://dx.doi.org/10.1071/MF15240.

Bond, N.R., Costello, J., King, A., Warfe, D., Reich, P. and Balcombe, S. (2014). Risks and opportunities from artificial watering as a means of achieving environmental flow objectives. *Frontiers in Ecology and Environment*, 12, 386–394.

Catford, J., Naiman, R., Chambers, L., Roberts, J., Douglas, M. and Davies, P. (2012). Predicting novel riparian ecosystems in a changing climate. *Ecosystems*, 16, 382–400.

Connell, D. (2011). Water reform and the federal system in the Murray-Darling Basin. *Water Resources Management*, 25, 3993–4003.

Davies, P.M. (2010). Climate change implications for river restoration in global biodiversity hotspots. *Restoration Ecology*, 18, 261–268.

Finlayson, C.M. (1996). The Montreux Record: a mechanism for supporting the wise use of wetlands. Proceedings of the 6th Meeting of the Conference of the Contracting Parties of the Convention on Wetlands (Ramsar Convention Bureau, Gland, Switzerland). Technical Sessions: Reports and presentations, Brisbane, Australia, Vol. 10/12 B, pp 32–37.

Finlayson, C.M. (2009). Biotic pressures and their effect on wetland functioning. In Maltby, E. and Barker, T. (eds), *The Wetlands Handbook.* Wiley-Blackwells, Oxford, UK, pp 667–688.

Finlayson, C.M. (2013). Climate change and the wise use of wetlands – information from Australian wetlands. *Hydrobiologia*, 708, 145–152.

Finlayson, C.M., Clarke, S.J., Davidson, N.C. and Gell, P.A. (2016). Role of palaeoecology in describing the ecological character of wetlands. *Marine and Freshwater Research*, 67, 687–694.

Finlayson, C.M., Capon, S.J., Rissik, D., Pittock, J., Fisk, G., Davidson, N.C., Bodmin, K.A., Papas, P., Robertson, H.A., Schallenberg, M., Saintilan, N., Edyvane, K. and Bino, G. (2017). Adapting policy and management for the conservation of important wetlands under a changing climate. *Marine and Freshwater Research* (in press), published online early at: www.publish.csiro.au/MF/MF16244, https://doi.org/10.1071/MF16244.

Gell, P.A., Finlayson, C.M. and Davidson, N.C. (2016). Understanding change in the ecological character of Ramsar wetlands: perspectives from a deeper time – synthesis. *Marine and Freshwater Research*, 67, 869–879.

Hannah, L. (2010). Climate, ecology, and conservation: an editorial essay. *WIREs Climate Change*, 1, 624–626.

IEMP (2011). Restoring the natural foundation to sustain a green economy. UNEP Policy Series on Ecosystem Management, Issue No. 6, International Ecosystem Management Partnership, United Nations Environment Programme, Nairobi.

Kopf, R.K., Finlayson, C.M., Humphries, P., Sims, N.C. and Hladyz, S. (2015). Anthropocene baselines: assessing change and managing biodiversity in human-dominated aquatic ecosystems. *BioScience*, 65, 798–811.

Lukasiewicz, A., Finlayson, C.M. and Pittock, J. (2013a). Identifying low risk climate change adaptation in catchment management while avoiding unintended consequences. National Climate Change Adaptation Research Facility, Gold Coast, Australia.

Lukasiewicz, A., Finlayson, C.M. and Pittock, J. (2013b). *Incorporating Climate Change Adaptation into Catchment Managment: A User Guide*. Charles Sturt University, Albury, Australia.

Lukasiewicz, A., Pittock, J. and Finlayson, C.M. (2016). Are we adapting to climate change? An adaptation assessment framework for managing freshwater ecosystems. *Climatic Change*, DOI 10.1007/s10584-016-1755-5.

Matthews, J.H. and Wickel, B. (2009). Embracing uncertainty in freshwater climate change adaptation: a natural history approach. *Climate and Development*, 1, 269–279.

Matthews, J.H., Wickel, B.A. and Freeman, S. (2011). Converging currents in climate-relevant conservation: water, infrastructure, and institutions. *PLoS Biology*, 9, 1001159.

Nilsson, C., Reidy, C.A., Dynesius, M. and Revenga, C. (2005). Fragmentation and flow regulation of the World's large river systems. *Science*, 308, 405–408.

Null, S.E., Viers, J.H., Deas, M.L., Tanaka, S.K. and Mount, J.F. (2013). Stream temperature sensitivity to climate warming in California's Sierra Nevada: impacts to coldwater habitat. *Climatic Change*, 116, 149–170.

Olden, J.D. and Naiman, R.J. (2009). Incorporating thermal regimes into environmental flows assessments: Modifying dam operations to restore freshwater ecosystem integrity. *Freshwater Biology*, 55, 86–107.

Parmesan, C. (2006). Ecological and evolutionary responses to recent climate change. *Annual Review of Ecology Evolution and Systematics*, 37, 637–669.

Parmesan, C., Duarte, C., Poloczanska, E., Richardson, A.J. and Singer, M.C. (2011). Overstretching attribution. *Nature Climate Change*, 1, 2–4.

Pittock, J. (2013). Lessons from adaptation to sustain freshwater environments in the Murray–Darling Basin, Australia. *Wiley Interdisciplinary Reviews: Climate Change*, 4, 429–438.

Pittock, J. and Finlayson, C.M. (2011). Australia's Murray-Darling Basin: freshwater ecosystem conservation options in an era of climate change. *Marine and Freshwater Research*, 62, 232–243.

Pittock, J. and Finlayson, C.M. (2013). Climate change adaptation in the Murray-Darling Basin: reducing resilience of wetlands with engineering. *Australian Journal of Water Resources*, 17, 161–168.

Pittock, J., Finlayson, C.M. and Howitt, J.A. (2012). Beguiling and risky: "Environmental works and measures" for wetlands conservation under a changing climate. *Hydrobiologia*, 708, 111–131.

Pittock, J., Finlayson, M., Gardner, A. and McKay, C. (2010). Changing character: The Ramsar Convention on Wetlands and climate change in the Murray-Basin. *Environmental and Planning Law Journal*, 27, 401–425.

Pittock, J. and Hartmann, J. (2011). Taking a second look: climate change, periodic re-licensing and better management of old dams. *Marine and Freshwater Research*, 62, 312–320.

Pittock, J. and Lankford, B.A. (2010). Environmental water requirements: demand management in an era of water scarcity. *Journal of Integrative Environmental Sciences*, 7, 75–93.

Poff, N.L. (2009). Managing for variation to sustain freshwater ecosystems. *Journal of Water Resources Planning and Management*, 135, 1–4.

Poff, N.L. and Matthews, J.H. (2013). Environmental flows in the Anthropocene: past progress and future prospects. *Current Opinion in Environmental Sustainability*, 5, 667–675.

Poff, N.L., Richter, B.D., Arthington, A.H., Bunn, S.E., Naiman, R.J., Kendy, E., Acreman, M., Apse, C., Bledsoe, B.P., Freeman, M.C., Henriksen, J., Jacobson, R.B., Kennen, J.G., Merritt, D.M., O'Keeffe, J.H., Olden, J.D., Rogers, K., Tharme, R.E. and Warner, A. (2010). The ecological limits of hydrologic alteration (ELOHA): a new framework for developing regional environmental flow standards. *Freshwater Biology*, 55, 147–170.

Ridgill S.C. and Fox A.D. (1990). *Cold Weather Movements of Waterfowl in Western Europe*. IWRB Special Publication No. 13, Slimbridge, UK.

Watts, R.J., Richter, B.D., Opperman, J.J. and Bowmer, K.H. (2011). Dam reoperation in an era of climate change. *Marine and Freshwater Research*, 62, 321–327.

Chapter 14

Freshwater ecosystems in protected areas

A synthesis

C.M. Finlayson, A.H. Arthington and J. Pittock

Key messages

- In 2014 the IUCN issued a call for a specific focus on the coverage and management of freshwater ecosystems in their own right rather than as components of Protected Areas (PAs) established for the protection of terrestrial ecosystems. The argument that freshwater PA managers need to apply freshwater-specific conservation principles and tools to ensure that PAs are effective in protecting aquatic biodiversity and ecosystem services has been extended by others.
- Chapters of this book have explored five high-level ecological principles and numerous practices for conserving the biodiversity and ecosystem services of freshwater PAs, including rivers and other freshwater ecosystems, brackish waters and estuaries.
- Decisions about the management of freshwater PAs in relation to national policy settings as well as the settings provided by the Ramsar Convention and the Aichi Biodiversity Targets will be particularly relevant to achieving the Sustainable Development Goals for 2030. These goals are likely to drive the conservation agenda in the coming decades. Given the mixed outcomes from more than 40 years of international wetland conservation under the framework provided by the Ramsar Convention, and expected further changes in freshwater ecosystems and for their species in response to global change, we end with a plea for civic authorities, scientists and wider society to develop and embrace effective approaches to ensure that freshwater ecosystems are conserved and restored.
- We place particular stress on the importance of the five high-level principles developed herein as a framework for management of freshwater PAs,

and also, possibly even more importantly, as a basis for further communication, and the development of more sophisticated messages, motivations and methods for the conservation of freshwater biodiversity, ecological goods and ecosystem services now and in the future.

Introduction

Recent publications have confirmed the ongoing loss and degradation of freshwater wetlands and continued decline in the species that depend on them, despite an increase in attention to wetland conservation internationally, nationally and locally (Gardner et al., 2015; Pittock et al., 2015). While the effort through the Ramsar Convention on Wetlands (www.ramsar.org, accessed 21 May 2017) to have 2,266 wetland sites covering 2.19 10^6 km^2 listed as internationally important (known as Ramsar sites) has been widely recognised, it represents a paradox for wetland and biodiversity managers. Although this outcome provides a sound basis for meeting the Aichi Biodiversity Target 11, whereby at least 17 per cent of inland water areas are conserved through effectively and equitably managed, ecologically representative and well-connected systems of protected areas (Milton and Finlayson, Chapter 2, this volume), it is at odds with the wetland losses and declines that are being recorded. In the introduction to this book *Freshwater Ecosystems in Protected Areas: Conservation and Management* Finlayson et al. (Chapter 1, this volume) point to the paradox "whereby the effort to develop and agree on international policy, along with a substantial and expanding information base, has not halted and reversed the global degradation and loss of wetlands." The Ramsar Convention has also made many far-reaching and forward-looking decisions for ensuring the conservation and wise use of wetlands over the past four decades (Finlayson et al. 2011; Finlayson 2012; Davidson 2016); however, these have not reduced the loss and decline of wetlands and their species globally (Davidson, 2014; Dixon et al., 2016).

Given the importance of Protected Areas (PA) in the global efforts to conserve freshwater wetlands (Figure 14.1) there have been many questions asked about their effectiveness for protecting freshwater ecosystems and species (Pittock et al., 2015). Amongst these has been concern that many PAs have not adequately considered the particular ecological processes and conservation priorities for freshwater ecosystems. In response to such questions IUCN (2014) issued a call for a specific focus on the coverage and management of freshwater ecosystems in their own right rather than as components of PAs established for the protection of terrestrial ecosystems. Against this background the chapters of this book have explored the principles and practices for conserving freshwater biodiversity and ecosystems (including rivers and other freshwater ecosystems, brackish waters and estuaries) in PAs, and

Figure 14.1 Protected Areas containing wetlands have been established in many parts of the world with many of them listed as Ramsar sites (wetlands of international importance) such as the Bundala National Park, Sri Lanka and Kakadu National Park, Australia (photographs © C.M. Finlayson).

extended the argument made in Pittock et al. (2015) that PA managers need to apply freshwater-specific conservation principles and tools to ensure that PAs are effective in protecting aquatic biodiversity and ecosystem services. These comments have been made within the context that it is unlikely that PAs alone could effectively conserve freshwater ecosystems (Finlayson 2006).

As a step in the complex process of identifying and managing PAs established specifically for freshwater ecosystems Arthington et al. (Chapter 3, this volume) provide five ecological principles that could assist with their better management. These principles relate to the:

1   intimate links that exist between freshwater ecosystems and the surrounding catchment (basin or watershed);
2   importance of the water regime (standing, flowing or sub-surface) for managing freshwater ecosystems;
3   hydrological, biogeochemical and ecological connectivity that occurs within and between freshwater (and estuarine) ecosystems;
4   protection of evolving patterns and 'hot spots' of native freshwater biodiversity, endemism and endangered species; and
5   maintenance of ecological resilience in the freshwater ecosystems of the present and the future.

These principles offer a framework to synthesise the key messages provided by the other chapters included in this book. Each principle is considered in turn in the text that follows; the background for each is presented in Arthington et al. (Chapter 3, this volume) and is not repeated here. The treatment of each principle is not even, reflecting the perspectives of other authors when considering specific aspects of managing freshwater PAs, and possibly also the state of our combined knowledge about the specific issues raised by each of the principles. The latter point raises the need for further effort to explore and extend our understanding of the core elements and processes captured by each principle.

Synthesis of key messages

Milton and Finlayson (Chapter 2, this volume) point out that freshwater ecosystems comprise a wide variety of types, such as rivers, lakes and swamps, and can be permanently, seasonally or intermittently flooded. They also support a high diversity of species (Figure 14.2) and provide important ecosystem services. However, despite their global importance for biodiversity and ecosystem services an accurate and comprehensive inventory and global map of freshwater ecosystems does not exist, undermining efforts to determine how well represented they are in PAs, and whether freshwater PAs are an effective global conservation measure. These issues are explored further in the synthesis that follows.

Figure 14.2 Freshwater wetlands support a high diversity of species (photographs © C.M. Finlayson).

Principle 1   Catchment characteristics and processes

A universal guiding principle for PA management is that the entire catchment with its land, water, biogeochemical resources and processes is the ideal unit to be protected and managed. Where full protection is not possible, the catchment needs to be managed in a sustainable way that minimises threats and impacts from non-reserved lands. Protected area management that fails to recognise and address the threats and pressures arising in the catchment risks loss of environmental quality, species diversity and ecological resilience.

Flitcroft et al. (Chapter 10, this volume) consider the benefits of integrating the management of rivers and other wetlands with the management of the wider landscape. In particular, integrated management approaches are seen as having the potential to address both scientific questions and applied challenges with the former including the management of multi-scale ecosystem functions, processes, and interactions, and the latter the management of land uses in diverse multi-user landscapes. Further, rather than viewing rivers as disconnected reaches where organisms could be analysed in isolation of a catchment or network context, they recommend the use of more holistic approaches and broader conceptual frameworks that link ecological communities to underlying geophysical systems. These frameworks include the river continuum concept, hierarchical organization of instream habitat, the natural flow regime, process domains, the network dynamics hypothesis, and the riverscape concept. Catchment-management plans were seen as a means of integrating the diverse land uses and owners of freshwater ecosystems, who, in combination, may directly or indirectly influence the quality of a shared river system. They also pointed out that there is no clear template for catchment management that works well everywhere, but rather, there are a variety of successful and innovative approaches and tools that work under particular circumstances.

Nel and Roux (Chapter 11, this volume) consider the inherent connectivity of most freshwater ecosystems that drives the need for PA authorities to engage in water planning processes both within their PA boundary and beyond in associated catchments. They emphasized that for planning to be effective and achieve its desired outcomes, it needed to be embedded in a management approach that acknowledged the inherent complexity and uncertainty not only of ecosystems, but of linked social–ecological systems. With this in mind, planning within a PA should ensure that freshwater ecosystems were managed explicitly as part of the PA landscape. PA strategies and management plans should therefore include freshwater conservation objectives, such as freshwater-friendly design and management of PA boundaries, visitor facilities, road-river crossings, and weirs and dams inside PAs. Management plans should also allow for monitoring of fresh-water indicators to inform the management of freshwater ecosystems. They also considered the necessity of planning beyond the boundary of the PA to be embed-ded within a regional water management framework that provided a platform for the cross-sectoral, cooperative governance of water resources at multiple scales of governance. This may be seen as an ideal situation but when in place it will enable the PA authority to engage with the social–ecological linkages that may impact on freshwater ecosystems within their PAs. It also offers the opportunity to identify new societal roles that PAs may contribute to in the future, such as the provision of ecosystem services to downstream communities.

Finlayson et al. (Chapter 12, this volume) consider the need to manage fresh-water PAs in a broader context – the global landscape – where the increasing demand for freshwater, as well as uncertainties associated with climate change, has placed growing pressure on wetland and natural resource managers to develop sustainable approaches that extend across landscapes and ensure the many ben-efits obtained from freshwater ecosystems are not further reduced. They point out that even where good legislative frameworks existed, such as those of the European Union's Water Framework Directive, significant challenges remained to fully deliver an appropriate quality and quantity of freshwater to maintain the character of PAs. Challenges include the degree to which the management of freshwater PAs fits within institutional structures and practices, while others relate to the efficacy of wider wetland management, such as managing invasive species and water quality. Issues such as these are particularly relevant when con-sidering the benefits that ecosystem services from freshwater ecosystems provide for many local people and communities. They further consider how well managers addressed the natural variability as well as successional change that occurs within these ecosystems; natural variability and successional change are part of the eco-logical character of these ecosystems and vary in space and time. As the recent condition of any particular freshwater ecosystem is a subset of the historic eco-logical conditions that it has experienced, the management focus needs to move away from a static view of condition and instead accommodate variability while also ensuring the wetland retains its important ecological features. Much of the reporting required by the Ramsar Convention and the listing of Ramsar sites does not seem to accommodate these important concepts. Accommodating natural variability and successional change will require the adoption of sustainable man-agement approaches that extend across multiple temporal and landscapes scales.

Arthington et al. (Chapter 8, this volume) consider the management of specific freshwater ecosystems and comment that a number of relatively simple changes to the way PAs are designed and managed could help to further improve the conser-vation benefits for surface and groundwater-dependent freshwater ecosystems and estuaries. These include: avoid using a river as the boundary of a PA; incorporate natural large-scale catchment processes into PAs; ensure that the water regimes of rivers, lakes, peatlands and groundwater-dependent ecosystems, as well as their linkages and interactions, are recognized and well managed within PAs and their catchments; avoid the development of visitor infrastructure on priority freshwater ecosystems in PAs; encourage the expansion of existing PAs to incorporate biodiversity hotspots, functional processes and connectivity; and promote new PAs for the last remaining free-flowing rivers and other high priority freshwater ecosystems. They further recognise that PAs on their own are unlikely to fully protect freshwater ecosystems, especially where large-scale catchment threats impinge on the PA and its ecosystems. Therefore, conservation efforts should not stop at the boundaries of PAs, but extend into developing cooperative relationships and activities among various entities with overlapping water management mandates. Integrated and adaptive management approaches that facilitate engagement and empowerment of all stakeholders, inclusive and iterative learning, and purposeful action amidst inherent complexities, are recommended.

Dudley et al. (Chapter 5, this volume) consider the management of freshwater ecosystems within the context of catchment processes. As freshwater PAs are particularly influenced beyond their boundaries by watershed-scale, regional or international influences, the adoption of landscape-scale approaches is an essential component of their management. This means that managing with other stakeholders is particularly important in freshwaters; fortunately, experience in using these approaches is developing fast. The most effective unit for managing freshwater PAs is the river basin or catchment in which the PA is situated and from which it received its surface water (Athington et al. Chapter 3, this volume; Finlayson et al. Chapter 12, this volume). To be effective for the conservation of freshwater biodiversity the management within a freshwater PA must take into account the particularities of these systems, such as the key role of spatial–temporal connectivity for maintaining ecological processes and the effective propagation of threats along these systems (Hermoso et al. Chapter 4, this volume). New mechanisms that could encourage the adoption of a mix of public and private funding to achieve improved outcomes for biodiversity conservation in freshwater systems (e.g., payments for ecosystem services, water reserves, biodiversity offsets and system-wide planning) are promoted by Hermoso et al. in Chapter 4.

Principle 2   Water regimes and eco-hydrological principles

The flow of water is one of five dynamic environmental regimes that regulate much of the structure and functioning of every running water ecosystem and many aspects of lentic and groundwater systems. The naturally dynamic flow regime plays a critical role in sustaining native biodiversity and ecosystem integrity in streams and rivers. Likewise the characteristics and variability of lentic (lakes and estuaries) and subsurface water regimes are critical to their dynamics, management and conservation.

Arthington et al. (Chapter 8, this volume) point out the importance of establishing environmental flows and maintaining the water regimes of rivers, lakes and groundwater dependent ecosystems as critical for sustaining the biodiversity and ecosystem services within freshwater and estuarine PAs. Conservation managers should aim to protect the natural water regimes of specific freshwater ecosystems, or restore them where changes in flow threaten species and processes. Setting a limit on hydrologic alteration remains the most challenging aspect of environmental water management, however numerous methods and practical guidelines are available to assist in this task. Pittock et al. (Chapter 9, this volume) emphasize the opportunities provided by the re-operation of dams to improve safety with respect to changing hydrology with climate change. These engineering works can enable provision of environmental flows and restoration of the water regimes of rivers. The effective management of groundwater-dependent ecosystems requires integration of associated surface and groundwater resources and an understanding of the origins, pathways and storages of water. Some groundwater ecosystems are entirely dependent on continuous groundwater discharges, whilst others are maintained by minor but critical groundwater inflows restricted to particular seasons or inter-annual episodes.

Nel and Roux (Chapter 11, this volume) emphasize that when PA authorities engaged with regional water planning they should insist that scientifically-credible outputs from freshwater ecosystem assessments, such as environmental flow assessment, systematic conservation planning and ecosystem service assessment, are used to assess the needs of freshwater ecosystems when negotiating trade-offs between different water development options. PA authorities could act as powerful stakeholders in the negotiation for freshwater ecosystem protection within regional water planning processes. They further explored the requirements for successful Integrated Water Resources Management (IWRM) and how these could be used to ensure the provision of sufficient water flows for freshwater PAs, as well as the value of ecosystem or nature-based adaptation and adaptive management to respond to multiple pressures. These approaches, and their underlying principles, provide useful guidance for water planning, whether protected area authorities are planning inside or outside protected areas. Arthington et al. (Chapter 8, this volume) also describe the value of Integrated Lake Basin Management (ILBM) that similarly supports the provision of sufficient water to sustain the biodiversity and ecosystem services provided by lakes.

Principle 3   Connectivity of freshwater ecosystems

The spatial and temporal connectivity patterns and processes of aquatic ecosystems in their natural state are important elements for consideration in PA design and management. Connectivity in rivers is defined in three spatial dimensions: longitudinal (upstream–downstream), lateral

(continued)

(continued)

(interactions between channel and riparian/floodplain systems), and vertical (connections between the surface and groundwater systems), with temporal dynamics influencing all spatial dimensions of connectivity. Thus minimizing the impacts of dams and levee banks as barriers, and changes to water flows, is crucial for freshwater conservation. The hydrological connectivity between lakes, streams, estuaries and subsurface environments also requires special attention in PA design and management, and forms a central pillar of Integrated Lake Basin Management (ILBM).

Hermoso et al. (Chapter 4, this volume) emphasize that to be effective for conserving freshwater biodiversity, conservation efforts must consider the particularities of these systems, including the key role of spatial and temporal connectivity for sustaining ecological processes and the problem of the effective propagation of threats along these systems. Pittock et al. (Chapter 9, this volume) more specifically consider the importance of riparian and floodplain corridors that are particularly biodiverse and often form key habitat for animals in the terrestrial landscape. Maintaining or restoring connectivity pathways and processes is a conservation priority for both freshwater and terrestrial ecosystems with considerable benefits to be gained from their protection and restoration. The importance of riparian forests derives from the key roles they play by providing organic matter that drives major elements of aquatic food chains, forming physical habitat, filtering pollutants, and providing shade, appropriate light environments and water temperatures.

Pittock et al. in Chapter 9 (this volume) further consider the important question of "how wide is wide enough" for riparian corridors. At the very least, a riparian corridor should be wide enough to enable full development of the vegetation canopy to maximize shade across the relevant water body and form an adequate mesic (moist, humid) micro-climate. They then explore a more complex answer whereby the full width of the regularly inundated riparian and floodplain land should be restored. Systems of river corridor protected areas have been established in many jurisdictions around the world based on criteria such as biological importance, maintaining free-flowing ecological processes and supporting cultural values. Increasingly, dams and levee banks are being removed from rivers and their floodplains to restore ecosystem functions and services. Restoration of functional riparian and floodplain systems may aid flood management and enhance other climate change adaptation measures.

Finlayson et al. (Chapter 12, this volume) consider the management of freshwater ecosystems in global landscapes and in particular highlight the importance of continental-scale connectivity in support of migratory species, in particular waterbirds that depend on networks of sites across continents. The Ramsar Convention and various migratory species agreements provide the opportunity to ensure these networks are maintained. They point out that the loss or degradation of individual sites could have ramifications across the network, and that whilst the importance of such networks is recognised there is an ongoing need to ensure that they are effectively managed. The latter is a key issue for countries with Ramsar sites that do not yet have effective management processes in place. The importance of managing PAs using integrated frameworks, such as ILBM and IWRM, is essential given the complexities that arise when considering the hydrological connectivity between disparate freshwater systems within catchments, and the need to ensure the provision of environmental flows and water regimes (Arthington et al. Chapter 8, this volume; Nel and Roux Chapter 11, this volume). Dudley et al. (Chapter 5, this volume) outline how the IUCN categories for PAs enable different kinds of 'non-traditional' freshwater PAs, such as heritage rivers, to be recognised in national and international PA systems.

Principle 4   Aquatic biodiversity, endemism and conservation

A primary goal of biodiversity conservation is to delineate PAs that conserve species-rich habitats and vital resources, important species radiations and the greatest number of threatened endemic species. Significant inter-basin differences in biodiversity and levels of endemism may mean a lack of 'substitutability' among freshwater habitat units, adding to the complexity of freshwater biodiversity conservation. The tools of systematic conservation planning lend themselves to identification of the most beneficial options for biodiversity protection.

Milton and Finlayson (Chapter 2, this volume) point out that wetlands are found on all continents, including Antarctica, and global mapping initiatives have been used to display the general distribution of many wetland types, but not all have been adequately mapped, in particular the smaller wetlands. Combined areal estimates for rivers, lakes and reservoirs (≥ 1 ha) and rice fields indicate that freshwater ecosystems globally cover 12.54–14.44 million km^2, and while this figure is larger than many previous estimates it is considered an underestimate given gaps in coverage (Milton and Finlayson, Chapter 2, this volume). Mapping initiatives have shown the complexity and variability that occurs within and among freshwater wetlands even when they have been similarly defined and classified. Wetland heterogeneity is a consequence of many factors, including the hydrological regime and the ecological processes that themselves vary temporally and spatially. While there are gaps there is also an increasing amount of information about individual wetlands, including comparative studies of large sites and inventories of wetland complexes.

Efforts to conserve freshwater biodiversity through the Ramsar Convention on Wetlands are examined by Finlayson et al. (Chapter 12, this volume). The flagship of the Convention is the List of Wetlands of International Importance, known as the Ramsar List, which contains 1,687 inland freshwater wetlands covering a total area of 1.96 million km². While the List is seen as a successful expression of the global effort to conserve wetlands, the distribution of Ramsar sites is uneven with 827 (52 per cent), covering only 12 per cent of the total area, occurring in Europe. However, there are concerns that not all sites are effectively managed, for example, only 58 per cent and 70 per cent of sites in Africa and Asia respectively, have management instruments. There is also a lack of evidence that current Ramsar sites and freshwater PAs more generally have been established on a systematic basis using spatial planning tools and biodiversity data to ensure they effectively represent the range of freshwater ecosystems and species within any particular biogeographical region. Efforts to ensure effective representativeness of freshwater ecosystems in PA networks could be enhanced by making greater use of spatial data sets and frameworks, such as the Freshwater Ecoregions of the World (Abell et al., 2007) and the increasing information provided for water-bird flyways (Davidson and Stroud, 2016). Bush et al. (2014) also demonstrate an approach for including predicted species distributions in systematic reserve designs for rivers under climate change, and explore the impact of varying con-nectivity requirements for different species.

Turak and Pittock (Chapter 7, this volume) consider species diversity in freshwater ecosystems and point out that while they support a huge diversity of life the information base behind this is not well documented nor communicated adequately. Better information is needed to understand which freshwater species are present in freshwater PAs, and which are not, as well as those that may be re-introduced as part of ecosystem restoration and species recovery programs. Dudley et al. (Chapter 5, this volume) add that while freshwater ecosystems contain a disproportionately high level of biodiversity they have experienced very high lev-els of loss particularly over the last few decades and, as a consequence, PAs play a critical role in reducing and reversing this decline. Turak and Pittock (Chapter 7, this volume) also note that despite inadequacies it is accepted that PAs have a critical role in protecting freshwater species. It is also widely accepted that con-servation actions within PAs are essential but unlikely to be sufficient to protect populations of many freshwater species, including nomadic and migratory species that move between PAs, whether locally or globally (Finlayson et al., Chapter 12, this volume). Threats and pressures that originate outside PAs must be recog-nized and managed at catchment and wider scales (Arthington et al., Chapter 8, this volume; Pittock et al., Chapter 6, this volume). In response to the complex of threats to freshwater ecosystems and species from within an individual PA and from beyond its boundaries, Turak and Pittock in Chapter 7 conclude that there is a need for system-level understanding of the requirements of freshwater species and drivers of variability and change as well as the inter-dependencies that exist between terrestrial, freshwater, estuarine and marine ecosystems.

Principle 5   Ecological resilience

Freshwater species have long histories of exposure and adaptation to variable environmental conditions and extremes (e.g., drought and flood cycles), conferring resistance and resilience at the individual, community and ecosystems levels. Maintaining catchment integrity, natural flow and standing water regimes, the spatial and temporal dimensions of connectivity, and native biodiversity hotspots will help to maintain the ecological resilience of aquatic systems in protected areas, and support societal adaptations to shifting environmental and climatic regimes.

Pittock et al. (Chapter 6, this volume) highlight that freshwater ecosystems are among the most threatened in the world, are under-represented in PA policy and have the highest portion of species threatened with extinction. The biodiversity of these ecosystems is particularly threatened because its conservation depends on maintaining hydrological connectivity (longitudinally along rivers, laterally between a river or water body and its floodplain, as well as the groundwater–surface water interactions), managing exogenous threats (that are propagated across catchments), and integrating governance by multiple management authorities. They emphasize the importance of managing the impacts of agriculture, aquaculture and fishing and minimizing the impacts of water infrastructure, invasive species incursion, pollution prevention, and reducing impacts of visitor facilities and activities.

Finlayson and Pittock (Chapter 13, this volume) point out that while PAs have played an important role in species and ecosystem conservation it is necessary to manage them in a reflective and flexible way to avoid adverse consequences that could increase the pressures on critical biodiversity targets. Rather than focussing on maintaining past reference states it will be necessary to develop strategies for promoting more climate-resilient approaches. This is particularly important as climate change, flow regulation and invasive species are leading to the development of novel ecosystems that may require a range of new approaches for water (and land) management to cope with increasingly uncertain futures. These include the identification of those parts of the freshwater landscape that can provide refuges and sustain ecological complexity, and the provision of guidance to manage environmental flows to counter the impacts of new water infrastructure and climate change. Conservation priorities include free-flowing rivers that do not require day-to-day management to provide the flow variability and connectivity needed to conserve aquatic biodiversity. As all adaptation options involve risks and costs as well as benefits, the adoption of a suite of different but complementary interventions is likely to result in better practice by spreading the risk while seeking to maximize the benefits for PA ecosystems and people. Other measures can be seen as 'no regrets' measures that offer benefits for the environment and people regardless of changes in the climate.

Hermoso et al. (Chapter 4, this volume) also encourage the development of new mechanisms to support public and private funding to achieve improved outcomes for biodiversity conservation in freshwater systems, including making payments for ecosystem services, establishing water reserves and biodiversity offsets, and implementing system-wide planning to limit the impacts of water infrastructure on aquatic ecosystems. They also point out that despite increasing and innovative efforts in recent decades to ensure more effective conservation of freshwater ecosystems, there remains an urgent need to ensure that freshwater PAs are effectively managed and monitored to ensure their effectiveness. The widely anticipated impacts of climate change on freshwater ecosystems add to the need for ensuring that PAs are effectively managed, and in an adaptive way. In this respect the development of an international framework for managing freshwater PAs in changing climates is overdue and is currently a major gap in freshwater conservation (Finlayson and Pittock, Chapter 13, this volume). The Ramsar Convention in particular is well placed to fill this gap, although Finlayson et al. (Chapter 12, this volume) point out inadequacies in the existing approaches to assessing change in relation to the reliance on reference conditions that may not be sustainable in the future, or attainable if a site is in need of restoration.

Concluding comments

The synthesis presented in the above text supports the contentions of Finlayson et al. (Chapter 1, this volume) that rather than treating freshwater conservation as a marginal part of PA management, it is important in its own right and central to sustaining the biodiversity contained within the world's PAs. In this respect the establishment and management of PAs specifically for freshwater biodiversity protection needs far more attention, in particular to ensure that PAs contain a representative sample of the diversity of freshwater species and are effectively managed within a landscape context and in accord with established ecological principles (Arthington et al., Chapter 3, this volume). Further, the conservation effort in freshwater PAs will require specific conservation and management tools, including catchment management plans, appropriate water management practices such as the use of environmental flows, and natural resource governance processes to ensure effective co-ordinated management at a landscape scale, taking into account on-site and off-site influences and the importance of connectivity within and between sites.

Freshwater and estuarine ecosystems are among the most threatened in the world, are under-represented in PA policy and have the highest portion of species threatened with extinction. The conservation of freshwater biodiversity depends on a complex suite of interacting processes, including hydrological and ecological processes that operate in longitudinal, lateral and vertical dimensions, and the management of threats that are propagated across catchments. Effective management is dependent on integrated governance arrangements between multiple management authorities including those operating across and between catchments. This is particularly important given the type and complexity of threats to freshwater ecosystems, many of which may be exacerbated by climate change. Decisions about the management of freshwater PAs in relation to national policy settings will also be influenced by settings provided by international frameworks, including the Ramsar Convention on Wetlands, the Aichi Biodiversity Targets and the UN Sustainable Development Goals for 2030.

In addition to the outcomes that may be achieved through international frameworks, there are examples of new approaches being developed at a national level. An outstanding example is the conferral of legal rights on rivers, such as has occurred for the Whanganui River in New Zealand and the Ganga and Yamuna Rivers in India (www.thenewsminute.com/article/ganga-and-yamuna-are-now-legal-entities-what-does-mean-and-it-good-move-58999, accessed 18 May 2017). The foundations of these moves are steeped in the cultural and social settings of the countries concerned, but could equally represent the vanguard of further innovative and ethical efforts to ensure the conservation of the biodiversity and multiple values of freshwater ecosystems. The imperative to consider new or novel approaches to ensuring the future of freshwater PAs is illustrated by recent changes in drainage patterns as glaciers retreat, such as recorded for the Kaskawulsh Glacier in Canada (Shugar et al., 2017) as well as by changes in the range of keystone species such as the beaver (*Castor canadensis*) in North America (Jung et al., 2017).

Given the mixed outcomes from more than 40 years of international wetland conservation under the framework provided by the Ramsar Convention, and expected further changes in freshwater ecosystems and for their species in response to global change, we conclude with a plea for civic authorities and wider society to develop and embrace effective approaches to ensure that freshwater ecosystems are conserved and restored. Opportunities abound through more effective management of traditional and established PAs, the establishment of further PAs, including those established in response to changing species distributions, and those that support multiple uses, as well as newer approaches that enable people and the diverse biodiversity represented in freshwater ecosystems to thrive as the ecological conditions change in response to increasing global pressures on aquatic diversity and ecosystem services. This could encompass more innovative or imaginative land tenure and institutional arrangements for land ownership and stewardship in addition to governmental processes in support of PAs in a changing landscape. In support of this plea we stress the importance of the five high-level principles developed herein as a framework for management of freshwater PAs, and also, possibly even more importantly, as a basis for further communication, and the development of more sophisticated messages, motivations and methods for the conservation of freshwater biodiversity, ecological goods and ecosystem services now and in the future.

References

Abell, R., Allan, J.D. and Lehner, B. (2007) 'Unlocking the potential of protected areas for freshwaters', *Biological Conservation*, vol 134, pp48–63.

Bush, A., Hermoso, V., Linke, S., Nipperess, D., Turak, E. and Hughes, L. (2014) 'Freshwater conservation planning under climate change: demonstrating proactive approaches for Australian Odonata'. *Journal of Applied Ecology*, vol 51, pp1273–1281.

Davidson, N.C. (2014) 'How much wetland has the world lost? Long-term and recent trends in global wetland area', *Marine and Freshwater Research*, vol 65, pp934–41.

Davidson, N.C. (2016) 'The Ramsar Convention on Wetlands'. In Finlayson, C.M., Everard, M., Irvine, K., McInnes, R.J., Middleton, B.A., van Dam, A.A. and Davidson, N.C. (eds), *The Wetland Book I: Structure and Function, Management and Methods*. Springer Publishers, Dordrecht, doi 10.1007/978-94-007-6172-8_113-1.

Davidson, N.C. and Stroud, D.A. (2016) 'Waterbird flyways – and the history of international co-operation for waterbird conservation'. In Finlayson, C.M., Everard, M., Irvine, K., McInnes, R.J., Middleton, B.A., van Dam, A.A. and Davidson, N.C. (eds), *The Wetland Book I: Structure and Function, Management and Methods*. Springer Publishers, Dordrecht.

Dixon, M.J.R., Loh, J., Davidson, N.C., Beltrame, C., Freeman, R. and Walpole, M. (2016) 'Tracking global change in ecosystem area: The Wetland Extent Trends index'. *Biological Conservation*, vol 193, pp27–35.

Finlayson, C.M. (2006) 'Freshwater protected areas: can we expand our options to include private wetlands?' *Ecological Management and Restoration*, vol 7, pp77–78.

Finlayson, C.M. (2012) 'Forty years of wetland conservation and wise use'. *Aquatic Conservation: Marine and Freshwater Ecosystems*, vol 22, pp139–143.

Finlayson, C.M., Davidson, N., Pritchard, D., Milton, G.R. and MacKay, H. (2011) 'The Ramsar Convention and ecosystem-based approaches to the wise use and sustainable development of wetlands'. *Journal of International Wildlife Law and Policy*, vol 14, pp176–198.

Gardner, R.C., Barchiesi, S., Beltrame, C., Finlayson, C.M., Galewski, T., Harrison, I., Paganini, M., Perennou, C., Pritchard, D.E., Rosenqvist, A. and Walpole, M. (2015) 'State of the World's wetlands and their services to people: A compilation of recent analyses'. Ramsar Convention Secretariat, Ramsar Scientific and Technical Briefing Note No. 7, Gland, Switzerland.

IUCN (2014) 'The promise of Sydney: innovative approaches for change. A strategy of innovative approaches and recommendation to reach conservation goals in the next decade'. Available at: worldparkscongress.org/downloads/approaches/Stream1.pdf.

Jung, T.S., Frandsen, J., Gordon, D.C. and Mossop, D.H. (2017) 'Colonization of the Beaufort coastal plain by Beaver (*Castor canadensis*): a response to shrubification of the Tundra?' *The Canadian Field-Naturalist*, vol 130, doi: http://dx.doi.org/10.22621/cfn.v130i4.1927.

Pittock, J., Finlayson, M., Arthington, A.H., Roux, D., Matthews, J.H., Biggs, H., Harrison, I., Blom, E., Flitcroft, R., Froend, R., Hermoso, V., Junk, W., Kumar, R., Linke, S., Nel, J., Nunes da Cunha, C., Pattnaik, A., Pollard, S., Rast, W., Thieme, M., Turak, E., Turpie, J., van Niekerk, L., Willems, D. and Viers, J. (2015) 'Managing freshwater, river, wetland and estuarine protected areas'. In Worboys, G.L., Lockwood, M., Kothari, A., Feary, S. and Pulsford, I. (eds) *Protected Area Governance and Management*, ANU Press, Canberra.

Shugar, D.H., Clague, J.J., Best, J.L., Schoof, C., Willis, M.J., Copland, L. and Roe, G.H. (2017) 'River piracy and drainage basin reorganization led by climate-driven glacier retreat'. *Nature Geoscience*, vol 10, pp370–375.

Index

For Product Safety Concerns and Information please contact our EU
representative GPSR@taylorandfrancis.com
Taylor & Francis Verlag GmbH, Kaufingerstraße 24, 80331 München, Germany